Modelling Complex Ecological Dynamics

Fred Jopp • Hauke Reuter • Broder Breckling
Editors

Modelling Complex Ecological Dynamics

An Introduction into Ecological Modelling for Students, Teachers & Scientists

Title Drawings by Melanie Trexler
Foreword by Sven Erik Jørgensen
& Donald L. DeAngelis

Editors

Dr. Fred Jopp
Department of Biology
University of Miami
P.O. Box 249118
Coral Gables, FL 33124
USA
fredjopp@bio.miami.edu

Dr. Hauke Reuter
Department of Ecological Modelling
Leibniz Center for Tropical Marine Ecology
GmbH (ZMT)
Fahrenheitstraße 6
28359 Bremen
Germany
hauke.reuter@zmt-bremen.de

Dr. Broder Breckling
General and Theoretical Ecology
Center for Environmental Research and
Sustainable Technology (UFT)
University of Bremen
Leobener St.
28359 Bremen
Germany
broder@uni-bremen.de

ISBN 978-3-642-05028-2 ISBN 978-3-642-05029-9 (eBook)
DOI 10.1007/978-3-642-05029-9
Springer Heidelberg Dordrecht London New York

Library of Congress Control Number: 2011921703

Cover design: F. Jopp and WMXDesign GmbH, Heidelberg, Germany

Printed on acid-free paper

Springer is part of Springer Science+Business Media (www.springer.com)

Foreword

Natural systems are complex, heterogeneous and diverse. If we look in detail, in fact, we see that each system is unique, differing from all others in various characteristics. Scientific investigation is largely a process of simplifying and selecting from such systems a small set of key components, governing factors, and relationships that are sufficient to describe how the system works. From these, ecologists try to develop generalizations across many systems. By this process, they improve their understanding of nature. This knowledge may also help where guidance in management is necessary.

There is no one "right" way to perform the simplifications used in the study of natural systems. This text is about the quantitative modelling of natural systems, but makes the point that a number of different approaches to such modelling have evolved and may be valid. Thus, ecologists have at their disposal alternative ways of specifying the aspects that are needed for describing of how an ecological system works and which aspects can be left out of consideration.

Ecological modelling today plays an increasingly important part in facilitating insights into how organisms interact with their environment and each other, and how this creates the properties of ecological systems. The general use of quantitative models in studying nature developed historically as a specific part of the advancement of science. Where one locates the starting point of modelling depends on one's particular perspective. Some of the methods we use today – differential equations – were developed during the seventeenth century. One prediction of such equations, exponential growth, representing an important component today in many ecological models, became well known through the famous work of Malthus in the late eighteenth century in an economical context. Verhulst's formula of logistic growth was formulated in the nineteenth century. With the equations for a predator–prey interaction by Lotka and Volterra in the early twentieth century, quantitative ecology started to use models of successively increasing complexity.

In the early stages of development, ecological modelling was largely based on differential equations, which were fundamental primarily for the development of classical mechanics. This may have contributed to the notion that modelling in ecology was merely an application of differential equations or other mathematical formalisms. However, if this were actually true, it would not be reasonable to consider ecological modelling as a distinct discipline. Ecological modelling would in that case be more properly viewed as a subdiscipline of applied physics or mathematics. The point of view taken in this book is a different one. It presents the modelling of complex ecological dynamics as a part of ecology, thus a subdiscipline of biology. It makes use of a wide variety of techniques imported from various sources, among which there are numerous mathematical methods, but also techniques from computer science and operations research. In addition, systems theory, quantitative methods from geography, and methods from a variety of other fields have helped supply formal methods to solving ecological problems.

It is the understanding of the organizers of this text that modelling should start with the specific ecological questions at hand and then the most appropriate ways of representation and formalisation should be selected. That is, ecological modelling should not be primarily steered by knowledge of applied mathematics, but should start from the foundations of ecological and biological knowledge and insight. Then the quantitative methods that are most suitable can be chosen and applied. Using and adapting methods from outside biology for ecological purposes requires a broad overview of the methodological repertoire that is applicable for representing and understanding patterns and processes arising from the interaction of organisms with their environment.

The spectrum of what can be applied in this field of science has indeed grown considerably – quantitatively as well as structurally. Ecological modelling has grown very rapidly during the last 35 years. When the journal Ecological Modelling was launched in 1975, only 300 pages were published per year – around 20 papers. Today, the journal publishes 4,000 pages of a larger format and about 400 papers per year. Ecological models are used much more widely today and are indispensable tools in ecological research and environmental management. Models are also able today to solve a wider range of problems, because we have a wider range of different model types available. Thirty-five years ago most of the published models were either biogeochemical models or population dynamic models. We have today many more types of models that can account for the spatial distribution, the shifts in structure, adaptation, individuality and the quality of the available data sets. This textbook presents both the quantitative and qualitative progress in ecological modelling and draws a clear up-to-date image of the field of ecological modelling.

The progress of the last few decades results not only in ecological modellers now being able to make use of an enormous range of mathematical tools and specialised software. It also results in a great expansion in the different ways of looking at ecological systems. Therefore, an overview of ecological modelling cannot be given today simply by introducing a single approach or technique. An overview of how to model complex ecological dynamics is a task that by its nature can only be addressed by bringing together experts on different methods. This is similar to the approach used in large-scale modelling efforts, where scientists specialised in different fields work together with other collaborators who have statistical, geographical, or computer science expertise.

Accordingly, the editors of this textbook did not attempt to summarise second-hand information from literature, but instead asked leading specialists in the most relevant domains of modelling to contribute chapters, starting from the level of elementary introductions and leading up to summaries of more advanced topics and studies. This required structuring criteria on how to guide the reader through the field. The editors decided on a historical and conceptual introduction, followed by an ordering of topics, based on increasing complexity, introducing a range of different modelling techniques in 11 subsequent chapters. The main part of the book presents the most relevant modern approaches, starting with equilibrium methods of ecosystem mass transfer balances and ending with object-oriented systems approaches allowing for time variation, as well as structurally varying self-organizing networks. To illustrate applications of these methods, the final section of the book describes a number of selected prominent case studies, which also emphasise the necessity of cooperation in the application of different techniques to solve complex tasks.

Producing a coherent text through the efforts of many collaborators was only possible through productive interaction. In this respect, this collaboration reflects in miniature the way ecological modelling is usually done in the world. It is not a field for "lone wolves", but requires considerable team spirit. This book creates a proper ambience for such spirit, by going beyond just the compilation of facts and how-to's, to demonstrate concrete examples of cooperation in this field. Reading the lines and in between the lines the reader experiences practical application of modelling – and how to bring complex things together.

Copenhagen, Denmark — Sven Erik Jørgensen
Coral Gables, USA — Donald L. DeAngelis

Preface

The idea for this book emerged in 2007 at the annual meeting of the Ecological Society of Germany, Austria and Switzerland. When the European Ecological Federation met in Leipzig, in the following year, we had drafted a concept that was well received by various colleagues, and which also caught the interest of the publisher.

For many years we have been working in different fields of ecological modelling with the purpose of solving ecological questions, broadening existing approaches and exploring new advances in the modelling of plants, animals, communities, landscapes, terrestrial and aquatic environments and the application of simulation models. In the light of this background and encouraged by discussions with our colleagues, we agreed that a textbook was needed to provide a broad overview in the field of ecological modelling to our students. Thus, we wanted to compile a broad scope of different perspectives and practices in ecological modelling: an introduction to the diversity of model approaches, model development and model evaluation.

Such a compendium and orientation is vital, in particular for young scientists who are less experienced with the various levels of complexity in ecological research and who are looking for the right model type to help solve a specific scientific question. We believe that the era of one-trick ponies in ecological modelling will soon be phased out. No user should be limited to an inappropriate tool, spending endless time and energy "adapting" and working around inherent limitations before being able to apply a model – there are many alternative pathways. Therefore, the focus of this book is to highlight the diversity of different views, methods and approaches. Being able to choose from a multitude of approaches allows a much better understanding of diversity and variability within natural systems. Studying ecology means an attempt at understanding the complex dynamics of natural processes and we must be aimed at capturing these with a maximum of clarity and conceptual ease. This can only be achieved by considering the available options. International experts and competent colleagues have been invited to communicate the key areas of their expertise in a clear and straightforward way, emphasizing the merits and limitations of individual methods along case studies.

Favouring a theory-guided, application-oriented perspective, we reduced the extent of mathematical formalisms. This does not mean that the book is free of mathematical expressions, but it is written in a comprehensive and encompassing style providing easy access to the central ideas and concepts. The entire textbook can serve as a curriculum for studying ecological modelling, but it is equally suitable for reading only single chapters that cover your focal interest. Since the rule "everyone to his/her own taste!" is also true for modellers, we do not advocate any specific model or software programme: please feel free to develop your own applications and codes, and provide them to your colleagues when possible, as this will aid in advancing the repertoire of options.

Finally, we very much enjoyed the inspiring teamwork that made this book possible. Many individuals have helped us during the last years by providing feedback, ideas, and in particular, revisions of parts of this book. Especially, we would like to thank Kathryn L. Berry for her invaluable help with all aspects of text handling. Sincere thanks are expressed to Stefanie Wolf, editor at Springer, who pleasantly and professionally guided us through the project ... and to the entire Springer team.

Coral Gables, FL, USA — Fred Jopp
Bremen, Germany — Hauke Reuter
Bremen, Germany — Broder Breckling

Supplementary material for the book is available at
www.mced-ecology.org

Contents

Part I Introduction

1 **Backgrounds and Scope of Ecological Modelling: Between Intellectual Adventure and Scientific Routine** 3
Broder Breckling, Fred Jopp, and Hauke Reuter

2 **What Are the General Conditions Under Which Ecological Models Can be Applied?** 13
Felix Müller, Broder Breckling, Fred Jopp, and Hauke Reuter

3 **Historical Background of Ecological Modelling and Its Importance for Modern Ecology** 29
Broder Breckling, Fred Jopp, and Hauke Reuter

Part II Modelling Techniques and Approaches

4 **System Analysis and Context Assessment** 43
Broder Breckling, Fred Jopp, and Hauke Reuter

5 **Steady State Models of Ecological Systems: EcoPath Approach to Mass-Balanced System Descriptions** 55
Matthias Wolff and Marc Taylor

6 **Ordinary Differential Equations** 67
Broder Breckling, Fred Jopp, and Hauke Reuter

7 **Partial Differential Equations** 93
Michael Sieber and Horst Malchow

8 **Cellular Automata in Ecological Modelling** 105
Broder Breckling, Guy Pe'er, and Yiannis G. Matsinos

9 **Leslie Matrices** 119
Dagmar Söndgerath

10 **Modelling Ecological Processes with Fuzzy Logic Approaches** 133
Agnese Marchini

11 **Grammar-Based Models and Fractals** 147
Winfried Kurth and Dirk Lanwert

12 **Individual-Based Models** 163
Hauke Reuter, Broder Breckling, and Fred Jopp

13 **Modelling Species' Distributions** 179
Carsten F. Dormann

14 **Decision Trees in Ecological Modelling** 197
Marko Debeljak and Sašo Džeroski

Part III Application Fields, Case Studies and Examples

15 **Neutral Models and the Analysis of Landscape Structure** 215
Robert H. Gardner

16 **Stage-Structured Integro-Differential Models: Application to Invasion Ecology** 231
Aurélie Garnier and Jane Lecomte

17 **Modelling Resilience and Phase Shifts in Coral Reefs: Application of Different Modelling Approaches** 241
Andreas Kubicek and Esther Borell

18 **Trophic Cascades and Food Web Stability in Fish Communities of the Everglades** 257
Fred Jopp, Donald L. DeAngelis, and Joel C. Trexler

19 **Lake Glumsø: Case Study on Modelling a Small Danish Lake** 269
Søren Nors Nielsen and Sven Erik Jørgensen

20 **Biophysical Models: An Evolving Tool in Marine Ecological Research** 279
Alejandro Gallego

21 **Modelling the Everglades Ecosystem** 291
Fred Jopp and Donald L. DeAngelis

22 Model Integration: Application in Ecology and for Management 301
Dietmar Kraft

Part IV Integrative Approaches in Ecological Modeling

23 How Valid Are Model Results? Assumptions, Validity Range and Documentation 323
Hauke Reuter, Fred Jopp, Broder Breckling, Christoph Lange, and Gerd Weigmann

24 Perspectives in Ecological Modelling 341
Fred Jopp, Broder Breckling, Hauke Reuter, and Donald L. DeAngelis

Glossary 349

References 355

Index 389

Contributors

Esther M. Borell The Interuniversity Institute (IUI) for Marine Sciences, P.O.B. 469 Eilat, 88103 Israel, estherborell@yahoo.co.uk

Broder Breckling General and Theoretical Ecology, University of Bremen, Leobener Str., 28359 Bremen, Germany, broder@uni-bremen.de

Donald L. DeAngelis Department of Biology, University of Miami, P.O. Box 249118, Coral Gables, FL 33124, USA, ddeangelis@bio.miami.edu

Marko Debeljak Department of Knowledge Technologies, Jozef Stefan Institute, Jamova 39, 1000 Ljubljana, Slovenia, marko.debeljak@ijs.si

Carsten Dormann Department of Computational Landscape Ecology, Helmholtz Centre for Environmental Research – UFZ, Permoserstr.15, 04318 Leipzig, Germany, dormann@ufz.de

Sašo Džeroski Department of Knowledge Technologies, Jozef Stefan Institute, Jamova 39, 1000 Ljubljana, Slovenia, saso.dzeroski@ijs.si

Alejandro Gallego Marine Scotland - Science, Marine Laboratory, P.O. Box 101, 375 Victoria Road, Aberdeen AB11 9BD, UK, A.Gallego@marlab.ac.uk, A. Gallego@MARLAB.AC.UK

Robert H Gardner Appalachian Laboratory, 301 Braddock Rd, Frostburg, MD 21532, USA, gardner@al.umces.edu

Aurélie Garnier Office National de la Chasse et de la Faune Sauvage, Direction des Études et de la Recherche, CNERA Petite Faune Sédentaire de Plaine, 39 Boulevard Einstein, CS 42355, 44323 Nantes Cedex3, France, au.garnier@gmail.com

Fred Jopp Department of Biology, University of Miami, P.O. Box 249118, Coral Gables, FL 33124, USA, fredjopp@bio.miami.edu

Sven Erik Jørgensen University of Copenhagen, Faculty of Pharmaceutical Sciences, Universitetsparken 2, 2100 Copenhagen, Denmark, sej@farma.ku.dk

Dietmar Kraft Institute for Chemistry and Biology of the Marine Environment (ICBM), Carl-von-Ossietzky-University of Oldenburg, 26111 Oldenburg, Germany, dkraft@icbm.de

Andreas Kubicek Department of Ecological Modelling, Leibniz Center for Tropical Marine Ecology GmbH (ZMT), Fahrenheitstrasse 6, 28359 Bremen, Germany, andreas.kubicek@zmt-bremen.de

Winfried Kurth Department for Computer Science, Chair for Computer Graphics and Ecological Informatics, Georg-August University of Göttingen, Buesgenweg 4, 37077 Göttingen, Germany, wk@informatik.uni-goettingen.de

Christoph Lange Fraunhofer MEVIS, Universitätsallee 29, 28359 Bremen, Germany, christoph.lange@mevis.fraunhofer.de

Dirk Lanwert Coordinator for e-learning, Georg-August University of Göttingen, Stabsstelle Lehrentwicklung und Lehrqualität, MZG, room 5.142, Platz der Göttinger Sieben 5, 37073 Göttingen, Germany, dirk.lanwert@uni-goettingen.de

Jane Lecomte Univ. Paris-Sud 11, Unité Ecologie Systématique et Evolution, UMR8079, Orsay F-91405, France; AgroParis Tech, Paris F-75231, France, jane.lecomte@u-psud.fr

Horst Malchow Institut für Umweltsystemforschung, Fachbereich Mathematik/Informatik, Universität Osnabrück, 49069 Osnabrück, Germany, malchow@uos.de

Agnese Marchini DET – Dipartimento di Ecologia del Territorio, University of Pavia, Via S. Epifanio 14, 27100 Pavia, Italy, agnese.marchini@unipv.it

Yiannis G. Matsinos Department of Environment, University of the Aegean, Mytilene 81100, Greece; Sector of Ecosystems Management, University Hill Xenia Building, Mytilene 81100, Greece, matsinos@aegean.gr

Felix Müller Ecology Center, Department of Ecosystem Research, Christian-Albrechts-University Kiel, Olshausenstraße 75, 24118 Kiel, Germany, fmueller@ecology.uni-kiel.de

Søren Nors Nielsen ECO-Soft, Kålagervej 16, DK-2300 Copenhagen, Denmark, soerennorsnielsen@gmail.com

Guy Pe'er Department of Conservation Biology, Helmholtz Centre for Environmental Research, Permoserstr. 15, 04318 Leipzig, Germany, guy.peer@ufz.de

Hauke Reuter Department of Ecological Modelling, Leibniz Center for Tropical Marine Ecology GmbH (ZMT), Fahrenheitstraße 6, 28359 Bremen, Germany, hauke.reuter@zmt-bremen.de

Michael Sieber Department of Mathematics and Computer Science, Institute of Environmental Systems Research, University of Osnabrück, 49076 Osnabrück, Germany, msieber@uni-osnabrueck.de

Dagmar Söndgerath Abt. Umweltsystemanalyse, Institut für Geoökologie, Zimmer 115, Germany; Technische Universität Braunschweig, Langer Kamp 19c, 38106 Braunschweig, Germany, d.soendgerath@tu-bs.de

Joel C. Trexler Department of Biological Sciences, Florida International University, Miami, FL 33199, USA, trexlerj@fiu.edu

Gerd Weigmann Institute for Zoology, Raum 307, Königin-Luise-Str. 1-3, 14195 Berlin, Germany; Freie University, Berlin, Germany, weigmann@zedat.fu-berlin.de

Matthias Wolf Department of Ecological Modelling, Leibniz Center for Tropical Marine Ecology GmbH (ZMT), Fahrenheitstr. 6, 28359 Bremen, Germany, matthias.wolf@zmt-bremen.de

Part I
Introduction

Chapter 1
Backgrounds and Scope of Ecological Modelling: Between Intellectual Adventure and Scientific Routine

Broder Breckling, Fred Jopp, and Hauke Reuter

Abstract The biological environment is full of diversity, changing situations, and dynamic alterations. This makes the field exciting and demanding, and sometimes confusing to understand. Models are one of the means to gain and maintain an overview of the various phenomena emerging in the different biotopes and relating to a wide range of scientific questions.

1.1 Getting Started: Motivations for Ecological Modelling

Nature is in a state of continuous dynamic change. Through the millenia it has accumulated the diversity and complexity as it is found today. Anthropogenic influences have altered many habitats, often with the consequence of a dramatic loss of biological diversity. Many species vanish while others colonize new environments. We need to understand the implications of natural dynamics and human interventions.

How far can ecology as an empirical science and modelling as a set of analysis and synthesis tools help us to enable an adequate understanding of the ongoing dynamics (Fig. 1.1a–g)? Different modelling approaches have been applied to assess a wide range of questions relating to most ecological systems. The book contains several examples to illustrate questions asked and systems studied with different modelling approaches. For instance, how important is the impact of agriculture for local biodiversity loss and what role does dispersal of organisms play in this context (Fig. 1.1a; Chap. 16); how important are seasonal variations of wetlands for the structure and the dynamics of small fish communities in a disturbed marshland (DeAngelis et al. 2010; Fig. 1.1b; Chap. 19); how are food webs structured and what are the driving forces for oscillating dynamics (Fig. 1.1c; Chaps. 5 and 12); what are the environmental factors determining phase shifts in coral reef systems and at which points should management intervene (Fig. 1.1c;

B. Breckling (✉)
General and Theoretical Ecology, Center for Environmental Research and Sustainable Technology (UFT), University of Bremen, Leobener St., 28359 Bremen, Germany
e-mail: broder@uni-bremen.de

F. Jopp et al. (eds.), *Modelling Complex Ecological Dynamics*,
DOI 10.1007/978-3-642-05029-9_1,

Fig. 1.1 Environments and examples (see text) where ecological modelling contributed important aspects to the understanding of ecological dynamics. (**a**) North European pastures, (**b**) Everglades marshland (Florida, USA), (**c**) coral reef system, Indo-West Pacific (Foto courtesy of E. Borell), (**d**) school of Mouth Mackerels feeding on zooplankton, Red Sea, Egypt, (**e**) tropical mangrove forest, Malaysia, (**f**) conventional and genetically modified maize fields in Northern Germany (gene flow), (**g**) distribution of landcover types in Northern Germany (remote sensing image)

Chap. 18); what mechanisms rule swarms and schools of individuals and how do these swarms interact with the environment (Fig. 1.1d; Chaps. 2 and 12); Can we reliably forecast stocks in fisheries to support a sustainable resource use and what impact does Global Change have here (Fréon et al. 2005; Chap. 21); what mechanisms control competition in Mangrove forests (Fig. 1.1e; Chap. 2); how can we assess and compare the gene flow of conventional and genetically-modified crops (Fig. 1.1f; Chap. 16); how can Geographical Information Systems (GIS) be used for model integration and what is the benefit of this procedure for ecology (Fig. 1.1g; Chap. 22)?

In all the above fields ecological models were set up to help in gaining insights, to understand the implications of change and to identify knowledge gaps. In some cases plausible predictions can be made while in others we can only study scenarios and learn more about the type and range of uncertainties. Ecological systems are structurally diverse and complex systems. These systems are known to express complex interaction networks with a high number of interrelationships and context-specific feedback processes. The outcome finally evolving from this interaction network is highly dynamic. Only parts of the entire system are accessible for empirical measurements. Any kind of modelling that tries to fill these empirical gaps will have to take the complexity and the dynamics of ecological systems into account. This was also eponymous for the title of this book: *Modelling Complex Ecological Dynamics.* The conceptual approaches, techniques and applications compiled in the following chapters will attempt to give an overview of what is feasible and what can be achieved. *You, the reader, are invited to share the findings in this field, and eventually pick up the thread as a researcher and expand the knowledge in an interesting discipline of environmental science.*

For a long time ecological work had focused on making an inventory of biota, collecting, gathering and classifying organisms in the diverse range of habitats (e.g., Linné 1748; Lamarck 1815–1822; Darwin 1859; Haeckel 1866; all of which also included the development of criteria and hypotheses on processes leading to the diversification of biota). This was then followed by studying how the biological entities respond to the environment and relate with each other. In physical systems, the experimental setup to decide on hypotheses of relations of material objects can be intentionally constructed ("framework constellation", McCarthy 1963). However, the precision in assessing ecological relations is usually much lower. In an early phase of scientific development, a general formalization and mathematization of natural systems was seriously considered. The French mathematician and astronomer Pierre Simon de Laplace (1749–1842) developed the idea, that in an ideal situation, when the precise location and impetus of all components moving in the universe would be known, the physical laws of dynamic interaction should allow to forecast any state in the future, and would equally enable to recalculate (backwards) all states in the past. Achieving this goal, of course would require infinite calculation capacities with infinite precision. Such an infinitely fast and capable calculator, called Laplace's demon, certainly was meant only as a theoretical consideration (de Laplace 1814), but outlined what was considered as the field of scientific intelligence.

Today in ecology, we are satisfied if we can make some cautious steps towards better understanding. To globally summarize all aspects of how organisms relate to each other and their environment in strict and quantitative cause-effect relationships does not seem to be reasonable. In this context, understanding ecological dynamics will remain incomplete and approximate, as suggested by theoretical consideration as well as practical experience. Nevertheless, it remains a challenge to find out where, to what extent and why forecasts are possible. Or to set up rules that lead to satisfying explanations and rationales when this is not the case.

Understanding complex environmental relations also requires inclusion of an abstract representation of the phenomena in focus. This abstract representation can contain theoretical or empirical elements or assumptions, which are brought together for making final statements that deal with cause and effect. In the following, we will refer to this abstract representation by using the term *model.* A model can be a conceptual setting, a verbal description, a simplified physical representation, and of course also a description in quantitative mathematical relations. While mathematical representations and verbal descriptions sometimes are seen as something entirely different, we would like to emphasize that there are usually transitions and translations between both. This is meant as an invitation to all ecologists not only to observe organisms and patterns in the field but equally to think how observations can be captured in formal settings and thereby used to investigate the underlying processes and mechanisms. When we arrive at such a turning point between empirical investigations and their transformation in mathematical forms, we are able to look at our observations from another angle. Taking a modelling perspective, we will discover new implications which were not entirely obvious at first glance and lead to new insights. The ecological complexity we observe is not always and necessarily complicated: There are surprisingly simple avenues to complexity, which are starting directly in the heart of simplicity.

1.2 Simplicity and Formal Representations: Appetisers to Model Complex Ecological Dynamics

Models need to be simpler than the original. A model representation as complex as the relations it represents would not really be enlightening and helpful. A model should aid in the understanding by allowing an overview and focusing on certain important aspects. The challenge is to make a simplification to capture the essence of a specific focus of interest. In fact, this is one of the most important challenges in modelling and a source of lasting controversies (DeAngelis and Gross 1992).

Now, let us see how complex the relations will be that can result from very simple approaches. The black box approach is perhaps the most radical simplification that made many phenomena accessible for modelling. In this approach, theoretically, in the considered context, everything is put into a box. It is assumed that the internal dynamics in the box are irrelevant as long as they do not greatly change the amount of the relevant content. Only an accounting for the input and output of the desired variables is done. In principle, this is what Alfred Lotka and Vito Volterra (Lotka 1925; Volterra 1926) did in their famous model approach on the relation of a predator and a prey population that they tackled with a simple differential equation model (see Chap. 6).

But how can such an approach help to understand complexity? Some empirical ecologists used to critisize (specifically mathematical) models as being too theoretical because the simplifications, manifested in the black-box approach, seemed to ignore what is biologically interesting to them: the diversity, variability and heterogeneity of ecological entities.

Is the expectation of capturing and expressing the essence of ecological entities in forms of simplicity a contradiction? Some aspects of life and technology can be precisely forecast and operated. For instance, a computer would not work in the case its states of operation were not fully determined. In other fields projections are impossible, e.g., the efforts to forecast earthquakes precisely have only very limited success. Comparatively, weather forecasts work sometimes but only within very narrow time frames. Biological relationships are notoriously difficult to predict in models. However, there are surprising applications of these relations. We can use simple models to demonstrate the causes of complex dynamics. This will not tell us which example out of the possible range we will actually meet next time in the field. But it will help us to link different aspects of causal structures and consider joint contexts of the ecological systems.

To create complex dynamics, we only need simple relations, slight modifications, and iterative repetition. Then we can show: Not all complex phenomena are based on equally complex relations. Simple settings can generate complexity, and hence complexity can be based on simple mechanisms. Here we start with a few of the most simplistic examples. In the subsequent chapters we will then successively turn to more sophisticated approaches and solutions, which will relate to advanced ecological theory and application.

One Formal Step into the Kingdom of Chaos

To show how close simplicity and complexity are, we leave out all biological realism for a moment. We only link numbers with each other and define a specific predecessor and successor for each. Since numeric operations are frequently used in ecological modelling, we are not too far away from our subject.

We define the successor (y) of each number (x) as the value we obtain if we subtract the inverse. This is a simple mathematical operation: $y = x - (1/x)$

Since the result is again a real number, we can apply the same operation again and thus create chains of operations. These chains have interesting properties. They tend to approach 0; however, when becoming smaller than $+1$ (and larger than -1) the algebraic sign changes and they then alternate between positive and negative values. Since we can use any starting point, the procedure successively intertwines the chains like infinitely long and thin spaghetti. This becomes apparent if we follow the fate of an interval that contains any possible starting points between two close numbers that are far outside the interval between $[-1; +1]$ (see arrows in Fig 1.2). It is apparent, that the numbers will become subsequently smaller (when being positive – or larger when being negative). The interval itself will increase in extent after each step.

This kind of movement, increasing the differences between originally close values through continuous operations, can be observed in this simple equation. It exhibits what is called deterministic chaos. In this case the chaos expands any small interval by iteration successively until they become independent from each other. A graphic representation of the process is shown in Fig. 1.2.

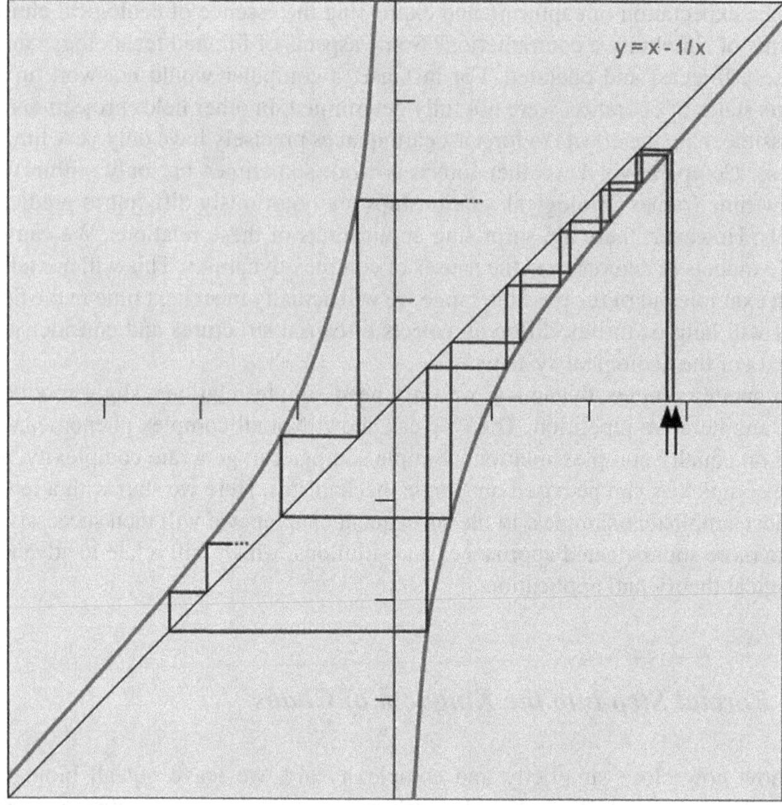

Fig. 1.2 Graphical iteration of the function $y = x_{suc} = x - (1/x)$. The *arrows* show two initially close starting values. For iteration, the starting value is drawn towards the function line. To obtain the next iteration, the determined value is connected to the diagonal (where y and the according x-value equate) and again move to the function. It can be seen, that initially small intervals become successively expanded. Initial correlations tend to get weaker and get successively lost. This is a characteristic feature of what is called deterministic chaos

Complexity in a Simplistic Ecosystem Model

In ecological modelling it is often discussed to what extent an ecosystem model can be simplified. An extreme simplification is putting all biomass into one big box (Y) and distinguishing only growth (biomass production) and decay (decomposition, re-mineralization). Growth could be simplified as a constant increase of an amount C_1. The decrease will be simplified as the vanishing quantum of an amount proportional to the current amount of Y with C_2 as a proportionality factor. We decide for a step-wise operation with either biomass growth or decay occurring in each step. If the biomass Y exceeds a threshold value (+1 for simplicity), decomposition is operated, otherwise increase occurs. This yields the form:

IF Yn < 1 THEN

$Y_{n+1} = Y_n + C_1$

ELSE

$Y_{n+1} = Y_n - C_2{*}Y_n$

To learn about the properties of this system we investigate its behaviour for $C_2 = 0.5$ and start each iteration process with y = 0.5. The first 1,000 iterations are discarded to eliminate transient behaviour. We then plot the results with C on the *x*-axis and the corresponding simulation results on the *y*-axis. To survey the model outcome across a parameter range we start with $C_1 = 0.0$ and increase it in small steps up to 0.7. The result is a highly structured pattern with interesting changes in periodicities. In the covered range, periods between +2 and infinity approximation occur (Fig. 1.3).

The example shows that simplicity and complexity can be in quite a close relationship: A radically simplified ecosystem model can exhibit aspects of a fractal dynamic (see also Chaps. 3 and 11). If we find such a situation where we have captured complexity in a simple approach, then we have found an interesting underlying mechanism. This is something that we are looking for in various

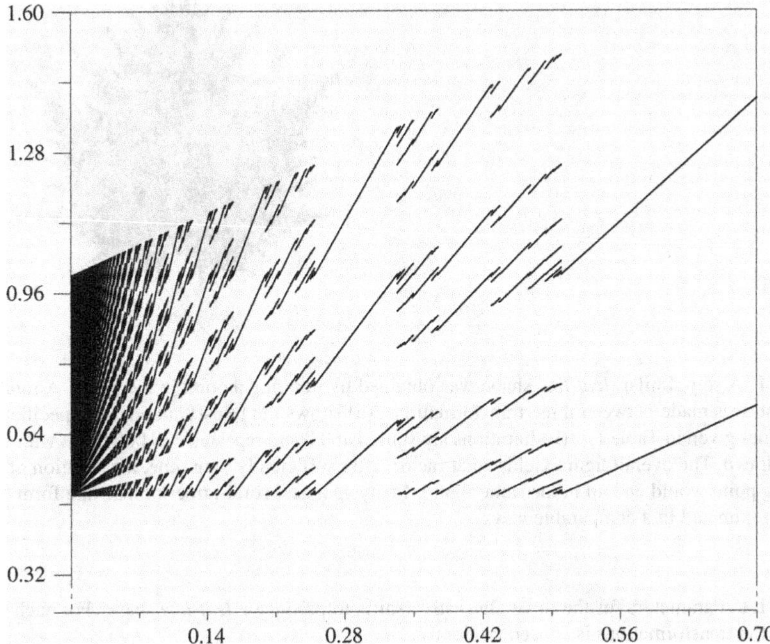

Fig. 1.3 Iterating the model
IF $Y_n < 1$ THEN $Y_{n+1} = Y_n + C_1$
ELSE $Y_{n+1} = Y_n - C_2{*}Y_n$,
$C_2 = 0.5$,
C_1 is drawn on the *x*-axis. Drawn over the according C_1 value, the *y*-axis shows 500 successive calculation results which exhibit different periodicities. The first 1,000 iterations of the simulation for each value of C_1 are not drawn to eliminate transient behaviour

contexts. Here, it shows that a complicated dynamic process does not always need to be based on complicated rules and mathematics.

Regular Random and Organic Forms

Many organic forms have components where certain aspects of its parts repeat the overall shape. When a shape consists of parts of itself, it is called self-similar. Self-similar structures can be highly complex, even though they can be captured in relatively simple models involving random processes. Figure 1.4 was produced by the following instruction: Select any arbitrary point in a plane as an initial value. Then make a random decision between three alternative linear shifts. Keep repeating with the obtained result.

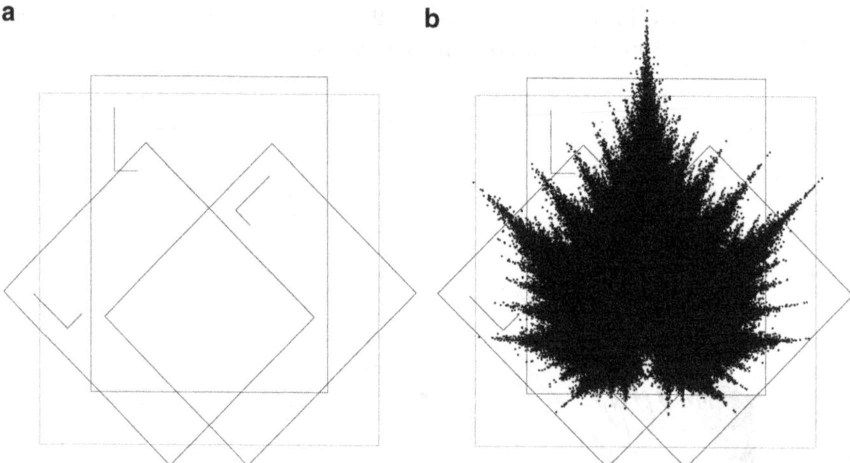

Fig. 1.4 A self-similar, *leaf-like* shape was obtained by iterating a point in the plane. A random decision was made between three transformations. (**a**) shows the transformations as specified by the values given in Table 1.1. (**b**) Iterations are started at 0.0 and repeated 420,000 times with each point drawn. The overall figure yields an attractor: after sufficiently long time, the iteration of *any* starting point would end up in the same figure. Many sponges, coral-, tree- or leaf-like forms can be approximated in a comparable way

Table 1.1 Parameters for the three alternative shifts to produce a *leaf-like* shape. For each step one of the transformations is chosen randomly

Transformation	First	Second	Third
Shrink co-ordinates along the x-axis	0.7	0.6	0.6
Shrink co-ordinates along the y-axis	0.9	0.7	0.7
Shift along the y-axis	0.2	–0.2	–0.2
Shift along the x-axis	–	–0.3	0.3
Turn around the point [0, 0]	–	45°	–45°

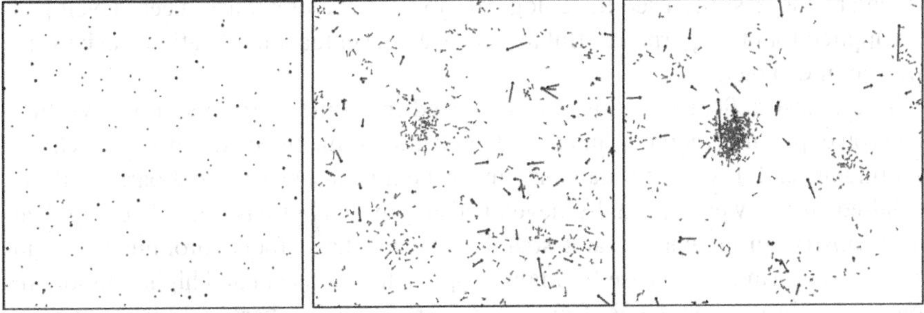

Fig. 1.5 Simulation of an individual-based predator–prey model starts with a random initial configuration (*left*). *Points* indicate the current positions of prey and predators; the line shows the movement from the previous position. *Lighter shades* and *smaller points* represent the prey, *darker* and *larger ones* the predator. The spatial distribution is shown after 50 (*centre*) and 100 steps (*right*). Prey concentrate in dense clusters. For further model specification see Chap. 12

Structure Can Be Created by Blind Random Processes

Let us imagine individual prey, randomly moving in a habitat without any orientation. They regularly reproduce up to a population maximum. Predators move a bit differently and follow a certain direction which is slightly changed stochastically at each step (a detailed model description is given in Chap. 12). Let all individuals be randomly dispersed in the beginning (Fig. 1.5, *left*). Would you imagine finding all prey closely aggregated in one or very few clusters after running the model for some time? Chapter 12 discusses self-organizing phenomena like this in further detail. Apparently, as illustrated with the model, spatial organization of organisms does not need an underlying plan, or a pre-existing gradient. It just can emerge through random interactions.

There are many more examples throughout this book where elementary interactions are specified, which bring up emergent phenomena and surprising results over the long run. The discovery of how these phenomena relate to the underlying interactions can make the construction of ecological models highly fascinating.

Comprehensive modelling and model networking also helps us to assess management and conservation options. You will find some interesting starting points to follow these and comparable model applications for environmental conservation in Chaps. 21 and 22.

1.3 Getting on: Diversity of Approaches in Ecological Modelling

In ecological modelling, not all the approaches are genuine ecological developments. Ecological modelling is in the fortunate situation to be able to do "concept (window-) shopping" in other fields of science. In fact, most of the methodologies which are successfully used in ecological models have previously been developed and applied for other purposes. But the usage and eventual modification can be very specific in ecology.

This is also reflected in how we organized the contents of this book. We first focus on explaining methods and then demonstrate some selected application cases. Continuous and discrete formalisms, linear, nonlinear, ordinary and partial differential equations were originally developed in physics but have also been applied in chemistry, in economics and sociology – and have found prominent use in ecology. The same is the case for matrix approaches. A recent and highly important source for inspiration came from computer science, where cellular automata and in particular the object-orientation paradigm brought important improvements in systems representation and individual-based modelling. Furthermore, network approaches that are linking different model types largely grew with developments in computer application. Still, ecological modelling is not just an application of approaches that were developed elsewhere. Ecological modelling provides room for diverse points of views. It is not a unified, canonical discipline with some procedures legalized and others excluded. It is a space for creativity and experiments.

We are convinced that the central part of ecological modelling must be founded on a basic understanding of the underlying ecology and biology – how organisms grow, how they move, behave, disperse and interact with their environments. A prerequisite for modelling is always a profound biological knowledge. Only then we can go to modelling approaches and select, modify and adapt the most promising techniques. This may at times yield the criticism that the use of the mathematics is not in strict line with all the formal definitions. We think this is not necessarily a problem. As long as the applied techniques adequately represent ecological processes, ecological modelling can be useful, despite this objection. In this regard we want to encourage a creative handling of such situations.

Therefore, the first priority for ecologists should be not only the reproduction of established formal recipes, but, more importantly, the use of them as an inspiration to come up with better, new and improved answers to the large field of remaining open questions in ecology.

Chapter 2
What Are the General Conditions Under Which Ecological Models Can be Applied?

Felix Müller, Broder Breckling, Fred Jopp, and Hauke Reuter

Abstract The purpose of this chapter is to discuss the conditions under which models can be applied. Modelling can help to solve specific problems, but not all questions in ecology require or benefit from the application of a model. It is therefore necessary to have an idea about the criteria under which the development of a model can provide useful information or help to solve questions in ecological analysis and which conceptual and technical approaches are the most appropriate ones. Technical knowledge about the particular modelling techniques is presented in the subsequent chapter of this book. Here, we intend to give an overview of the basic criteria of model application.

2.1 Models as Instruments of System Analysis

Models are abstractions of reality and instruments for the survey and analysis of complex systems (Wainwright and Mulligan 2004; Dale 2003). They are used to reduce the complexity of systems with reference to the specific problem that the observer wants to solve. Ecological models can depict the interactions and changes of environmental elements and simulate the dynamics of spatial and temporal patterns in ecosystems. Thus, they are instruments of environmental systems analysis (Bossel 1992; Gnauck 2000; Hannon and Ruth 2001).

A fundamental system comprehension should be considered as an initial conceptual condition for successful modelling. Ecological systems are complexes of biotic and abiotic elements, which are interrelated by flows of energy, matter and information (Breckling and Müller 1997). These interactions build up a comprehensive and complicated network of heterogeneous direct and indirect effects (Fath and Patten 2000). This network has an extraordinary high connectivity and its complexity rises drastically with the number of elements, relations and nonlinear interactions (Salthe 1993; Grant and Swannack 2007). This has the implication that

F. Müller (✉)
Ecology Center, Department of Ecosystem Research, Christian-Albrechts-University Kiel, Olshausenstraße 75, 24118 Kiel, Germany
e-mail: fmueller@ecology.uni-kiel.de

F. Jopp et al. (eds.), *Modelling Complex Ecological Dynamics*,
DOI 10.1007/978-3-642-05029-9_2, © Springer-Verlag Berlin Heidelberg 2011

we might never be able to fully understand these ecological systems structures and functions and the resulting dynamics. On the other hand, there are many good reasons why we should attempt to do so, e.g., the need to search for solutions of our urgent environmental problems. Systems analysis (see Chap. 4) and modelling provides steps and theories to cope with complexity in ecological systems.

Ecological modelling provides a large set of different approaches to analyse drivers of systems dynamics and extrapolate developments. However, it also has to be applied critically. The modeller should be conscious of the following:

- Models are observer-defined abstractions that can reflect reality only in the framework of the observer's viewpoint, the amount and quality of input information and the basic assumptions of the modeller
- There is an optimal degree of model complexity. This is not the highest complexity because large and complicated models tend to be difficult to handle and can increase uncertainty (Joergensen and Bendoricchio 2001)
- In any case the model *outputs* comprise specific uncertainties. To optimize the results, modelling needs extensive information about the investigated system and about the modelled object or process, as well as a precise question or hypothesis and data for both model development and model testing

2.2 Model Creation Should Be Carried Out in a Systems-Analytical Procedure

To make the general modelling procedure more illustrative, Fig. 2.1 sketches the single steps of a system analysis leading to an applicable ecological model. More technical details are elaborated in Chaps. 4 and 23) on model development, while the conceptual fundament is discussed here. The steps of model preparation begin with basic questions like:

- What is the focal object of the model?
- What is the specific aim of the model and what is its role in solving the focal problem?
- What are the spatial and temporal extents of the model and in what dimensions should the outputs be provided?
- What are the spatial and temporal resolutions of the model and how detailed should the processes be that are represented in the model (*model complexity*)?
- What are the most important issues to be represented and what are the relations between them?
- What data are necessary to (a) develop and (b) test the model?
- What are the forcing functions of the modelled systems and how do these constraints affect the elements?
- How can the interrelations be depicted in a clear and understandable graphical scheme?
- What are the basic assumptions made in model development?

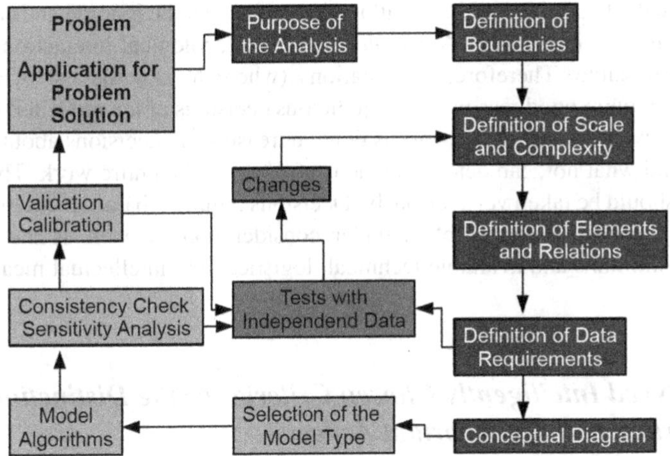

Fig. 2.1 Basic steps of environmental systems analysis and modelling (adapted from Müller 1999)

Although these questions seem to be trivial, they are often not dealt with in a satisfactory manner in practical modelling applications. However, answering these questions and documenting the derived conception needs to be done before technical steps of model development are taken. There are two main reasons for this requirement: On the one hand, this allows one to find the optimal conception for the model without forgetting or neglecting basic preconditions, and on the other hand, the documentation of the respective answers will enable the modeller to return to his original objectives when he has been lost in the complexity of model improvement. Figure 2.1 further illustrates that model development follows a cyclic process: Already with the definition of the data requirements, limitations might become obvious that make it necessary to change the general outline. The results of calibration and validation (see below) usually support this *experience*, and sometimes demand the modeller to go back to the very first steps because some basic requirements could not be met.

Taking into account these working procedures, two focal strategic items should be highlighted to avoid an exaggerated application of the cyclic principle:

Models Require a Clear and Precise Specification of the Focus of Investigation

A clear distinction needs to be made between what is part of the problem and what is left out of the considerations. This may sound fairly straightforward; however, in almost any practical situation this decision poses a serious challenge. The web of ecological interactions is complex. However large the resources for research might be – there will always be additional influences that have to be ignored and cannot be integrated in a given context. A *complete* model of ecological interactions in any field is impossible. Therefore, delimitations (where to end the list of relevant influences) require good reasoning and judicious decisions of the modeller. Regardless of how well the rest of the work is done, unreasonable decisions about what to consider and what not, can determine the usefulness of the entire work. Therefore, this issue should be taken very seriously. Decisions require a balance and linkage of the general nature of the problem under consideration, as well as the specific working conditions and available technical, logistical and intellectual means.

Models Need Intelligently Chosen Criteria for the Distinction of Important and Unimportant Aspects

There has to be a consideration of relevance concerning elements and relations that could make an important part of a model. The decision about which subjects and interactions are considered to be relevant depends on the available background information. During the work, it may turn out that the background information was not sufficient. A careful analysis of what is already known in the field is crucial. The modeller needs a clear view of which influences contribute (sometimes, always or only under specific conditions) in an important way to the given problem. Therefore, a literature survey to attain an overview that goes well beyond the current focus is required. Often the necessary decisions should be prepared in a discussion with co-operators or other experts in the field and experts in related topics.

Furthermore, a consideration of the working conditions for the modelling process is required: What is the available time span, what are the resources, manpower, data bases, etc., and what are the temporal constraints of model development? The answers can be integrated into a synopsis of the requirements to solve a given problem and the compromises that derive from the limitations (and preferences) of the specific situation. This background knowledge will allow the modeller to develop a reasonable work plan. Experience tells us that the duration (time requirement) of model elaboration is difficult to anticipate. Certain steps may be achieved much more easily and faster than expected; however, in most cases unexpected obstacles occur and things usually take longer than expected.

2.3 The Modelling Potential: What Can Models Help to Do?

Models are used for a wide range of purposes. Wainwright and Mulligan (2004) list different application fields for environmental models. They can, for example, be applied as an aid in research, as tools for understanding, tools for simulation and prediction, as virtual laboratories, and integrators between disciplines, and in addition, they are research products and means for communication. In the following, some of these model purposes and potentials are elucidated.

Models Can Help to Analyse the Results of Empirical Investigations or a Theoretical Problem that Is Not Accessible Through Statistical Data Interpretation Alone

Models can generally work within two types of situations – helping to solve empirical problems where a model needs to meet certain requirements resulting from field or laboratory measurements (data), and for theoretical purposes that investigate conditions and possibilities based on assumptions. Models usually go beyond situations and questions that require only data interpretation and statistical analysis. A good example for such a modelling approach is the complex competition situation between hardwood hammocks and mangroves (see Fig. 2.2). In the marshlands of South Florida Everglades (U.S.) hardwood hammocks and mangroves occur with distinct boundaries between their respective areas. Teh et al. (2008) applied a spatially-explicit simulation model to examine the effects of the salinity of the aerated zone of soil overlying a saline body of water, known as the vadose layer, as a function of precipitation, evaporation and plant water uptake (Fig. 2.2 right) on the vegetation. The model predicted that mixtures of saline and freshwater vegetative species represent unstable states, which are highly dependent on initial conditions of the system. The model conceptually explains the mechanism

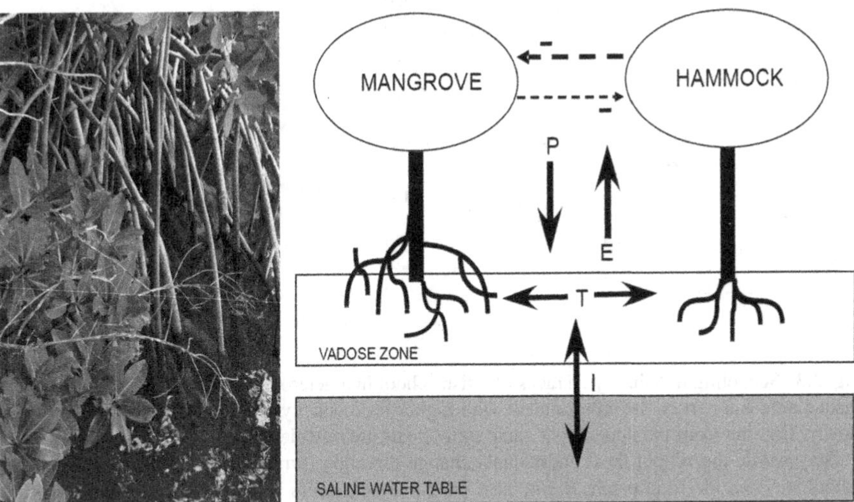

Fig. 2.2 Model on separation of mangroves (*left*) and hardwood hammocks in the marshlands of South Florida Everglades (U.S.). The model focuses on water transport and effects of the salinity on the vegetation (*P* precipitation, *E* evaporation, *T* transpiration, *I* infiltration, Teh et al. 2008)

that allows both vegetation forms to coexist – and why disturbance pattern can have long lasting influences.

Statistics are applied to data, while models are used to interpret systems states and processes, representing the dynamic developed and often applying an iterative procedure. However, models and statistical applications cannot be strictly and consistently delimited, though specific domains of application can be defined – with a minor overlap.

Furthermore, modelling allows to test the coherence and degree of completeness of the understanding of distinct ecological processes. For instance, for a long period it was not clear what kind of behavioural modes would be sufficient to lead to highly aligned fish schools. With an individual-based modelling approach to represent different behavioural patterns of individual fish it was possible to test the existing assumptions (Fig 2.3). Results revealed that, depending on distances between neighbouring fish, attraction, adjustment of direction and swimming speed and repulsion were sufficient to produce schools. Modelling also revealed, that it was only necessary to consider a limited number of nearest neighbours to keep a school together (Huth and Wissel 1994). For aggregation of a school, weighting of neighbours according to distance turned out to be a more efficient model assumption (Reuter and Breckling 1994). Thus models may help to check if knowledge on partial processes is sufficient to represent observed system behaviour.

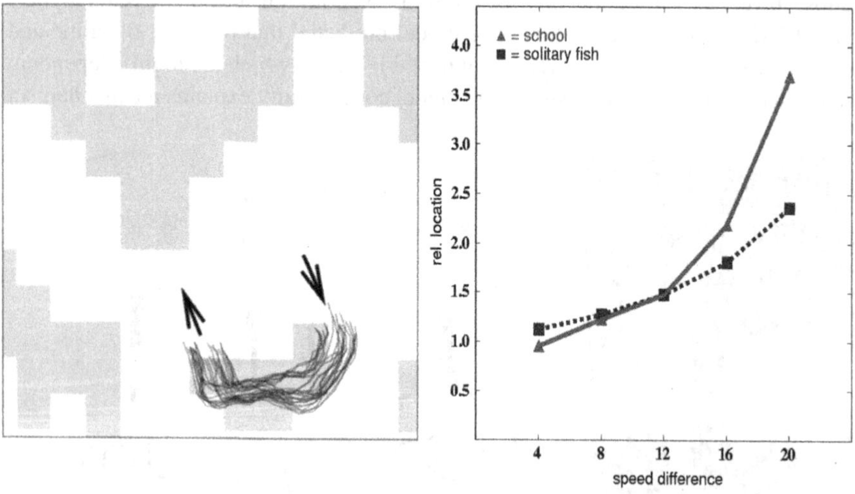

Fig. 2.3 Schooling in fish. *Left*: Traces of a fish school in a heterogeneous environment: the gray shaded area is a part of the environment with higher food density where individuals move more slowly. Coming from the upper right some individuals are outside the food patch and accelerate. To stay within the school these individuals change direction (turn right). As a result the whole school stays on the food patch. *Right*: This phenomenon occurs as an emergent property if the speed difference between preferred and non-preferred parts of the habitat is sufficiently large. Simulated schooling fish stay considerably longer on food patches than do solitary fish. (rel. location indicates how much longer a fish stays on food patches in relation to their coverages; speed difference to normal cruising speed of 25 units/timestep)

Models Can Help to Understand Emergent Properties (Emergent Phenomena)

Usually the modeller works in a situation where he has gathered various information about specific interactions of partial processes but is interested in the overall result to demonstrate how single processes overlay and produce new (emergent) phenomena. Nielsen and Müller (2000) have defined such emergent properties as self-organized features of a system that are not properties of the subsystems. Rather, these properties emerge as a consequence of the interactions within the system. They appear at one organizational level of a system and are not immediately deducible from observation of the single units in isolation, which compose the system. Many examples of emergence can be found in the regulation mechanisms of physiological processes, ranging between the levels of biochemical compounds, organelles, cells, tissues, organisms or populations. On all of these levels certain features emerge, which cannot be provided by the parts: For instance, an isolated chlorophyll molecule cannot use the energy of solar radiation for living processes. Only if it is embedded in a complex system of biochemical compounds, can it help to transform energy to become beneficial for the organism. Or – to consider a larger ecological context – cyclic processes in an ecosystem are only possible due to the interaction of the different subsystems. The storage of nutrients in the system can be comprehended as an important emergent property.

Many interpretations of emergence are based on the information flows between the systems' components. O'Neill et al. (1986) and also Allen and Starr (1992) have founded their ideas on the frequency distribution of ecosystem processes (Fig. 2.4). Following their hypotheses in hierarchy theory, the interactions between a single high frequency process produces the potential for processes of lower frequencies to provide constraints on the higher frequency units due to selected signal filtering procedures.

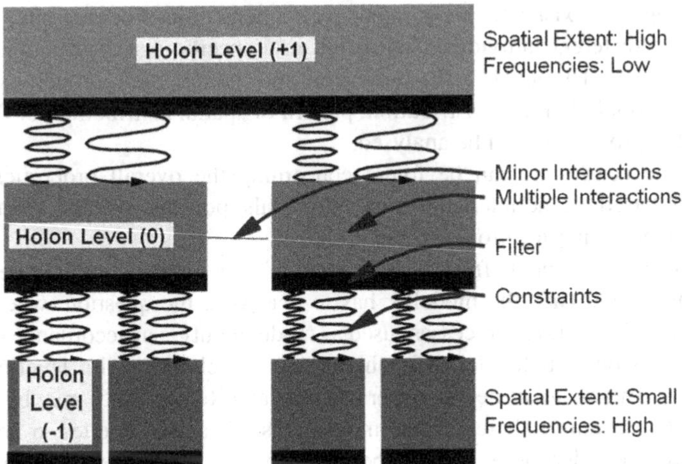

Fig. 2.4 Characteristics of hierarchical structures. Signal transfer between different levels of a hierarchy construct (modified after Müller 1992)

These temporal distinctions are linked with spatial differences: fast processes mostly operate on small spatial units while slow processes tend to have broader spatial extents.

A very practical consequence of hierarchical views on environmental systems is used to make the distinction of working scales for models: The modeller should focus on a certain part of the spatio-temporal continuum of ecological processes: by the selection of the *natural* frequencies and typical spatial extents of the core model variables, the modeller can define the focal level in the hierarchy of ecological relations. To depict the system's organization with minimum information, it is recommended that one works on three scales (with three typical frequencies), that are (a) the focal scale of the main variables (highlighting the interactions between subsystems at the same level of integration) and the two adjacent scales to consider, (b) the fast variables related with the sub-systems as well as (c) the slow variables that act as constraints. The latter frequently can be treated as forcing functions. Often, in model applications, the slow variables are set constant. This represents an approximation that can be reasonable for a limited time horizon but usually limits the long-term applicability of the model: The only constant phenomenon in living systems is change, and the way changes occur.

Models Can Clarify Interaction Implications Between Different Levels of Organization and Can Help to Understand Level-Crossing Phenomena

In the context of hierarchy theory, an organization level is considered as a result of the interactions of a number of elements that bring up specific properties on a broad scale coming into existence due to small scale interactions. For example, interactions among different individuals can result in a specific age distribution on the population level. The level of the individual and the emergent properties on the level of the population (age distribution, pattern of spatial distribution, or distribution of other properties) can be analysed.

The same conditions can be found concerning the overall properties of an ecosystem, based on the interactions of individuals, populations, and abiotic conditions of a particular location (see Fig. 2.5).

Models often use inputs from a lower level of organization and provide results on a higher level. Thus the modeller has to deal with the question, whether the knowledge on lower level processes (used as model input) is in accordance with the overall results on a higher level. In this sense, models deal with level-crossing phenomena. The model of beetle dispersal (Jopp and Reuter 2005) may be used as an illustrative example. It represents movements of carabid beetles in heterogeneous landscapes, depending on the species properties and landscape characteristics. The rules for the step length and angular deviation of single movement steps are derived from empirical data (Fig. 2.6 *left*). It can then be investigated what

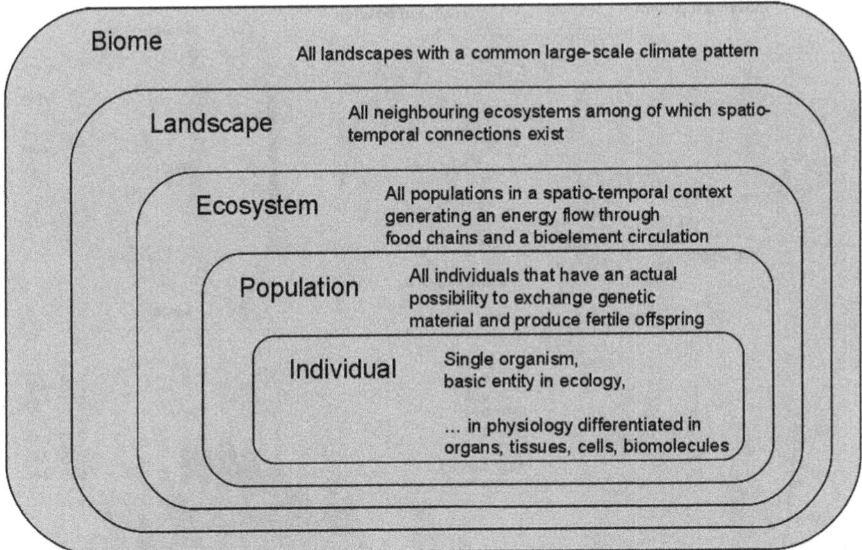

Fig. 2.5 Hierarchichal levels, which are distinguished as organization levels. Ecological modelling deals with individuals, populations, ecosystems, landscapes, and biomes. Still lower levels, from the biomolecule up to the individual, are the domain of physiology; however, sometimes these are also included in models along with ecological levels of interactions

effects these properties have on dispersal in differently structured landscapes (Fig. 2.6 *right*). Depending on the distribution of suitable and less suitable habitats, the model can provide results of how individual behaviour and landscape dispersal patterns relate over the long run.

This field of ecological analysis should not be dealt with by *expert's intuition* alone. In this regard, models allow an extension of conclusions beyond what is accessible in direct empirical investigations. With appropriate model approaches it can be studied how components of a level-crossing interaction network influence each other. For instance, Chap. 18 illustrates how trophic pyramids and trophic cascades of a wetland ecosystem (Everglades) are affected by hydrological changes.

Models Can Illustrate Iterative and Feedback Processes

In ecological models interactions between elements are in most cases specified in a computer programme and executed on a computer. The rationale behind this is that the number of repeated executions of all steps can be several orders of magnitude higher than what can be done by manual calculation. Modelling has special merits when feedback or iterative processes are involved: If we know the elementary interactions – what will be the result if they re-occur 10, 1,000 or 100,000 times? In linear cases, sometimes mathematical calculations can directly lead to a precise

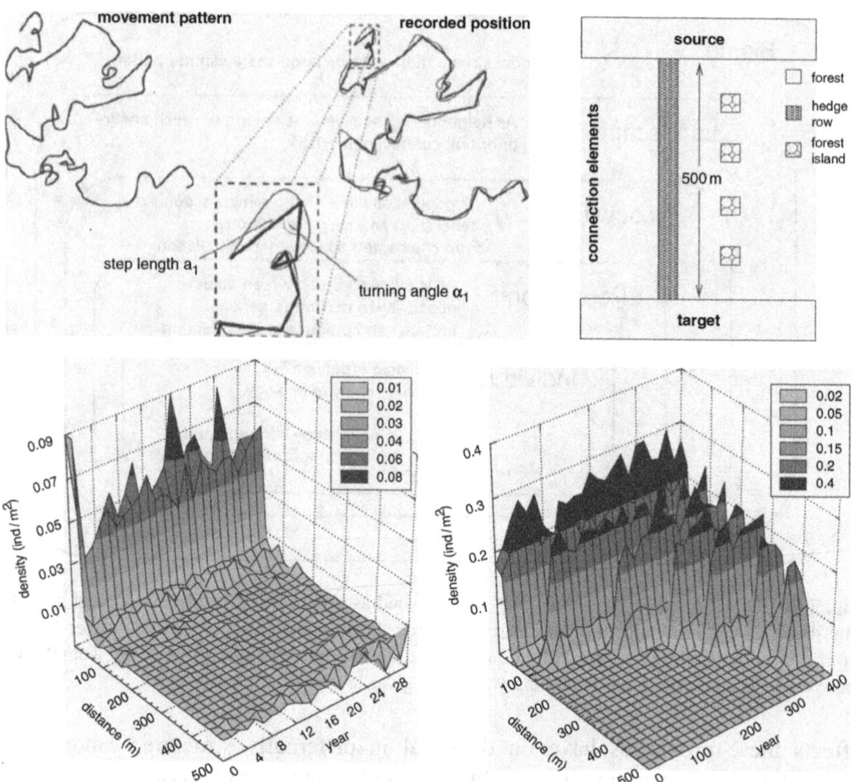

Fig. 2.6 The construction of Move Steps distributions and Model output from a model on beetle dispersal (Jopp and Reuter 2005). *Upper Left and Middle*: Derivation of movement rules from empirical data. *Upper Right*: Simulation set-up for the analysis of connectivity effects of hedgerows and stepping stones. The width of the hedgerows and the number of stepping stones are varied in the scenarios. *Lower Left and Right*: Resulting long term pattern of dispersal for a model population from a source habitat (*top*) to a sink habitat (*bottom*) which are connected by six habitats functioning as stepping stones. Dispersal success and densities result from a combination of movement speed, mortality on hostile lands and probability to cross habitat boundaries. *Abax parallelepipedus*, a slow disperser with high habitat fidelity, has to colonize all stepping stone habitats before reaching the sink habitat (lower right). In contrast, *Carabus hortensis*, which easily crosses borders between habitats, does not colonize the stepping stones, but reaches the sink habitat in a fraction of time (lower left)

result. Nonlinearities frequently are not so easy to extrapolate. Here models are often the only promising way to expand ecological knowledge. For instance, this is frequently the case on grid-based processes (see Chap. 8 on Cellular Automata).

A pattern studied on the basis of grid-based processes are forest fires. The final pattern can be well observed on the regional scale. The overall transition-rules on the small-scale, however, can only be estimated. Model assumptions can be tested to determine whether they lead to patterns that are in line with the observed findings (e.g. Ratz 1995).

New approaches additionally facilitate the potential to work with flexible interaction networks, where the number of elements and the way they are connected can change (see Chap. 11 on L-Systems, and Chap. 12 on Individual-based models). This poses high challenges to the conceptual development of simulation frameworks, especially if not only the involved quantities change in a nonlinear way, but also the structure changes in the course of interactions. This can be the case in modelling the structure and physiological processes in plants (see e.g. Figs. 4.3 and 4.4).

Models Can Facilitate an Understanding of Multi-Scale Problems in Ecology

Phenomena that involve several orders of magnitude in scales are usually difficult to handle. There are examples where model approaches can help to deal with large-scale issues that depend on very small scale interactions.

Nutrient budgets on the landscape level are an example (see Fig 2.7). Here initially the soil physicochemical potentials of different sites are used to calculate local nutrient flows, representing e.g. a patch scale. If the modelling question is

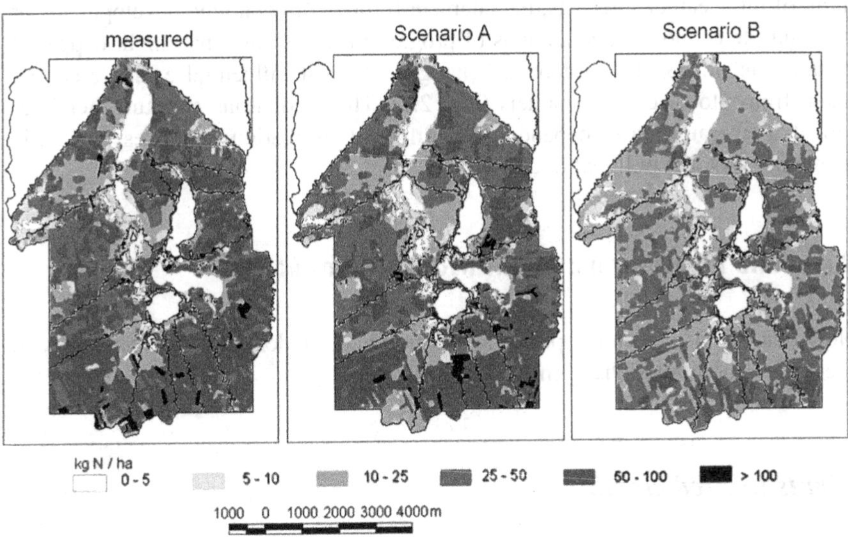

Fig. 2.7 Nitrogen leaching in the Bornhöved Lakes landscape simulated with the WASMOD modelling system (Reiche 1996). *Left*: The measured nutrient retention capacity of different soil types has been taken to calculate a business-as-usual scenario with a dominant small-farm structure. *Right*: Simulation of different scenarios: (**a**) Industrial agriculture – a structure with big farms and efficient land use practice leads to a change in land cover and land-use, providing an overall reduction of nitrogen leaching amounts. (**b**) Green agriculture – due to the reduced use of fertilizers and due to reduced pressures, i.e., on poor soils the nitrogen flows into the groundwater are heavily reduced

related to the budgets of landscapes or catchments, these small items have to be integrated to build up a landscape picture. But as the spatial scale increases, new processes arise, which were not evident at the smaller scale; for instance erosion, groundwater flow or airborne transports connect patches in a horizontal manner. Thus for the respective model analysis, a multi-scale approach has to be used with emergent processes at each level (see Chap. 22 for more examples).

Models Can Support Decision Making Processes

Besides their potential in basic research, ecological models can also be very helpful tools in decision making processes (also see Chap. 22 on Integrated Environmental Models). Often the environmental manager can hardly foresee the effects of certain measures for the states of environmental or social-ecological systems. To reduce this uncertainty, models can illustrate assessment components. In this context scenario modelling is playing a very important role. In that case the model constraints are defined due to the representation of different assumptions on environmental situations, management options or political strategies. From the model application the potential effects can be illustrated and an optimal strategy can be selected. On the other hand, model applications are usual parts of our everyday life; think of the weather forecasts, the characterisations of economic developments or the multiple economic applications of programmes to show what might happen if certain constraints of a system are changed. A most influential example can be taken from global climate models (Fig. 2.8). The predictions of future trends of global temperature rise, depending on different mitigation strategies, are basic elements of global political decisions.

2.4 The Limitations: What Models Cannot Do

Now that we have seen the potential of ecological models, it is also necessary to mention some limitations of models as well.

Limits in Predictability

Ecological models are not a new form of alchemy. You cannot put in cognitive lead, tin, and other low value materials, expect the computer to do magic and hope for intellectual gold as a result. The potential of modelling has limitations, and the "garbage in – garbage out" principle holds. Enthusiasm about modelling sometimes tends to obscure that. Modelling can expand knowledge; however, it cannot replace it. It can derive implications of given knowledge and it can be used to test interactive hypotheses. Because of the complex interactions that take place in ecological

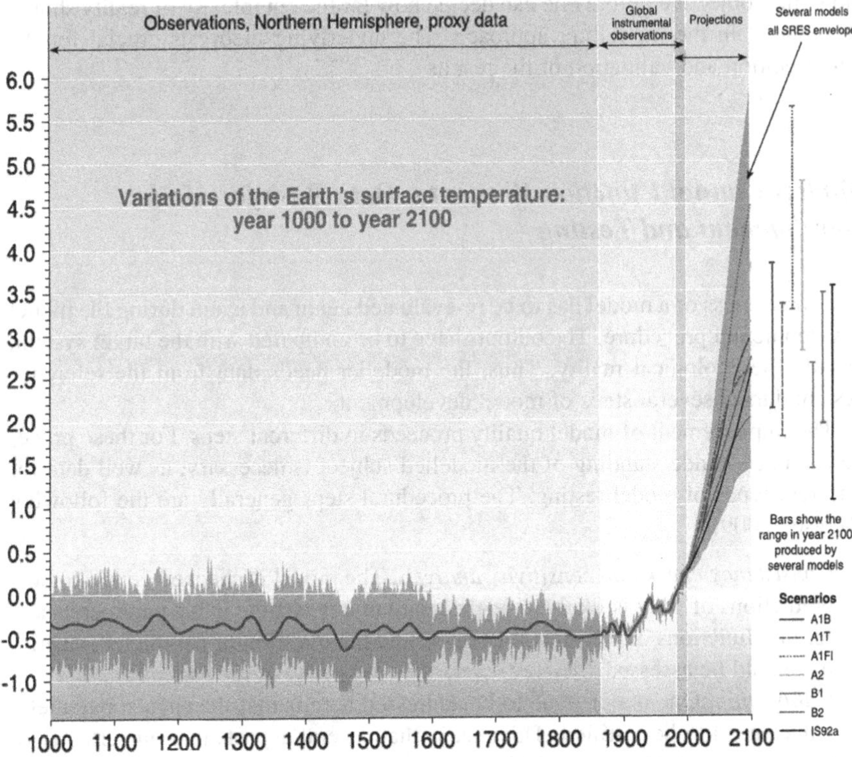

Fig. 2.8 Temperature trends and projections of global average surface temperature
source: UNEP/GRID-Arendal Maps and Graphics Library (2005), Philippe Rekacewicz, http://maps.grida.no/theme/climatechange

systems, the predictive power of models is usually limited. This is especially true for processes with a pronounced singularity, where local specific events largely influence the subsequent dynamics in a way that cannot be precisely forecast. However, probability estimations can often be derived through ecological modelling. To have an idea about this limitation is a precondition to reasonably applying ecological modelling.

Models Cannot Function Without a Precise Question or Hypothesis and an Appropriate Underlying Theoretical Framework

Although this demand seems to be trivial, it has to be stated that a very detailed task specification in connection with explicit knowledge about the purpose is a basic condition for successful modelling. It is essential to know the purpose and the modelling objective before one can decide how far the complexity of reality should be reduced in the modelling approach. The underlying theory is crucial for the interpretation and validation of the results.

Models Cannot Function Without a Data Base for Model Development and Testing

The correctness of a model has to be re-evaluated again and again during the model developmental procedure. The outputs have to be compared with the target system, usually the ecological reality. Thus, the modeller needs data from the reference system during several steps of model development.

The improvement of model quality proceeds in different steps. For these procedures, a deep understanding of the modelled subject is necessary, as well data for different types of model testing. The procedural steps generally are the following (Nielsen 2009):

1. *Consistency check and sensitivity analysis*. The model is checked versus logical predictions of what is likely to be the result of any change in the parameters and forcing functions of the model. Furthermore the question of parameter sensitivity should be assessed.
2. *Calibration*. One major issue to be addressed here is that the chosen parameter values need to be justified. This means that several aspects (e.g. uncertainty of the parameters, their accuracy, their significance for the model) have to be considered. Also, the results of the calibration can be observed by comparing the outputs with data sets.
3. *Validation*. This phase is the highest level of model quality assessment. It is a test of how well, model predictions (prognoses) are matched with actual observations. The higher the potential effect of the modelling results is, the higher should be the emphasis in testing model results against independent data. For validation, one or more datasets are required that describe the modelled situation *independent* of the data used during model development.

Different possibilities to apply data sets for model validation, for quality assessment of models and to secure correctness of model results are described in Chap. 23.

Models Have to Be Treated Skeptically When They Are Applied Outside the Validation Regimes

A consequence of the validation strategy is information on the range of validity of the model. If it is applied within the validation range, the results usually are of

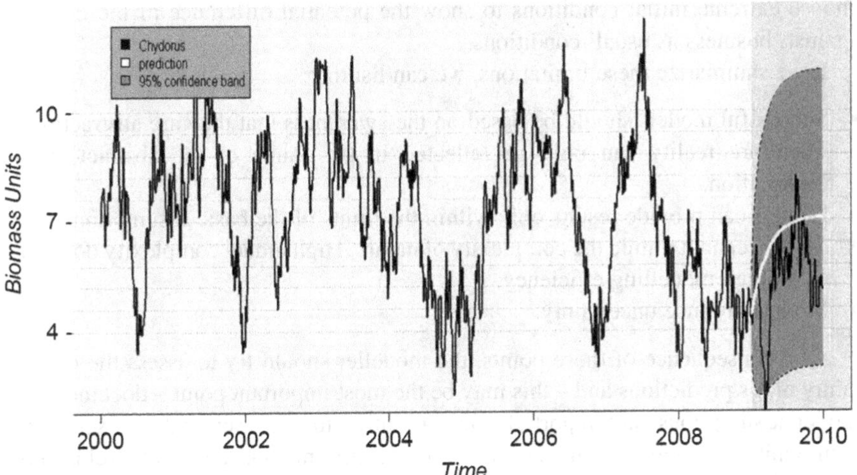

Fig. 2.9 Biomass dynamics of specific Northern-German zooplankton groups from Berlin lakes; *left part* of the figure: real measurements until 2009; *right part*: the projection for a further year on the basis of an autoregressive moving average model (Jopp, unpublished)

a high quality. But regrettably this is usually the most uninteresting case. Most models are developed to show potential future developments, and these dynamics of interest cannot be used for validation, because they are the application cases (However, later, the model quality can be improved on the basis of wrong predictions of the formerly future dynamics). We can never fully know how forcing functions and other input variables for our future model projections will develop through time; we can only know them later, retrospectively . Therefore model results will remain uncertain to a high extent (see e.g. Fig 2.9). If the ranges of the validation data sets are exceeded, the typical nonlinear relations or hysteresis effects can be responsible for extreme modifications of the system's behaviour (see Fig. 6.12). Also, if we apply models to other places than the area or system for which they were developed, there may be new parameter constellations that could not be taken into account during the development phase. Summarizing, a model will never be free of uncertainty and it is essential to respect the range of validity for each part of the model when discussing its results.

Models Rarely Produce Reliable Prognoses, but Can Be Used in Scenarios

Taking this point into account, models should not be used for specific prognoses. But as we still want to benefit from the modelling power, scenarios are a good level for applied modelling. When defining scenario conditions the user has to be aware that his model output may never be realized; i.e.; because it is likely that he will choose extreme initial conditions to show the potential difference of the outputs against 'business as usual' conditions.

If we summarize these limitations, we can list that:

- Successful models should be based on the awareness that they are abstractions. Therefore reality can only be reflected in the frame of the abstract input information.
- Models can provide results only within the limits of the basic assumptions.
- Models cannot mimic the complexity of nature. High model complexity does not mean high modelling efficiency.
- Models produce uncertainty.

As a consequence of these points, the modeller should try to assess the uncertainty of his predictions and – this may be the most important point – document all model assumptions and report the uncertainties to the user and the scientific community. By doing so, modelling as a scientific method does not exclusively follow the golden scientific rules of comprehensibility and reproducibility. It also enables the great opportunities that uncertainties provide: when prior conditions are not fully met and the output allows different interpretations, there is always the chance to follow sidelines of current scientific knowledge. Then, aside from the well-trodden trails, some of the most interesting findings and thrilling discoveries can be made

Chapter 3
Historical Background of Ecological Modelling and Its Importance for Modern Ecology

Broder Breckling, Fred Jopp, and Hauke Reuter

Abstract The chapter outlines major routes of development leading to the current spectrum of concepts and applications in ecological modelling. The field is closely linked to achievements in other sciences, in particular physics, numerics, computer science, and cross-disciplinary adoption of ideas. Ecological modelling emerged initially as a relatively homogeneous field and mainly employed differential equations which originated in classical mechanics. Quantitative ecological dynamics were initially described in a formal analogy to physical processes. In the last few decades, the methodological repertoire in ecological modelling was successively expanded. Nowadays, the whole range of quantitative methods available in numerical mathematics can be seen as a foundation for future model development in ecology. Some pioneers in the field are briefly introduced and their contributions linked to some of the mainstreams and sidelines of the state-of-the-art in ecological sciences. The overview provided here will not be able to provide historical completeness but attempts to facilitate an understanding of the origin of the major approaches presented in this book and how they obtained their role in current ecological modelling.

3.1 A Historical Journey: Mainstream and Sidelines of Model Development in Ecology

Science as a whole has and continues to undergo a long process of advancement. At each period, the current state-of-the-art also represents the background of expectations for how the domains of the unknown might be tackled and structured in the future. The state-of-the-art provides the vocabulary, the grammar, and the paradigms

B. Breckling (✉)
General and Theoretical Ecology, Center for Environmental Research and Sustainable Technology (UFT), University of Bremen, Leobener Street, 28359 Bremen, Germany
e-mail: broder@uni-bremen.de

F. Jopp et al. (eds.), *Modelling Complex Ecological Dynamics*,
DOI 10.1007/978-3-642-05029-9_3,

of world views. When looking back, it becomes apparent that the particular view of each development phase had the tendency to give rise to interesting extrapolations. The existing domains of the unknown get filled with structural analogies of what was known. To this day, existing knowledge creates expectations of what is to come, therefore, truly new knowledge is frequently surprising and controversial. In Italy, Galileo Galilei (1564–1642) established experimental investigations as a targeted approach on scientific questions. In France, René Descartes (1596–1650) provided an elaborated philosophical underpinning of science in a mechanistic world view. In Britain, Newton (1643–1727) formulated a synthesis that showed a validity of the same mechanic laws being applicable on the microscopic scale of the laboratory as well as on the macroscopic scale in astronomy. From 1751 onwards, the French encyclopaedists Denis Diderot (1713–1784), Jean Baptiste D'Alembert (1717–1783), and others attempted to turn a synopsis of science into an emancipatory power during the era of enlightenment. In the following time, science diversified methodologically. Thermodynamics, chemistry, statistical mathematics, and other fields rapidly progressed. In this concert, ecology arrived relatively late. The German chemist Justus von Liebig (1803–1873) had already established the use of synthetic fertilizer. Organic chemistry was advancing when Ernst Haeckel (1834–1919) coined the term "ecology" in 1868 in his book "General Morphology of Organisms" (1868). In 1935, when Tansley introduced the term "ecosystem", the very first modelling applications in ecology had just been developed. The starting of ecological modelling occurred during the 1920s. It is quite obvious that we do not report about a canonical field – there are lots of different opinions and views about modelling in ecology. So we just take a glimpse of a transient process, which is not only influenced by achievements arising in ecological science. Modelling techniques and their ecological applications emerge in an intense exchange with scientific advancement in other disciplines.

3.2 Ancestors of Ecological Modelling

Ecological modelling deals with the formalization of dynamic and complex interaction networks, how organisms relate with each other and with their environment. Modelling attempts to uncover implications of understandable relations that are not obvious at first glance when looking at the organisms and the locations where they occur. With model development we hope to identify concealed implications. We attempt to approximate and expand the margins, and the boundaries of what is intelligible. The new insight that a good model provides, goes beyond what can be concluded with direct observation, evaluation and interpretation.

Though the term ecology is relatively young, the assessment of natural processes is much older. We can find the precursors of ecological modelling in natural history and also in other disciplines.

Modelling as Derived from Physics: Ecology as Derived from Natural History

Physics had the role of the paradigmatic, standard-setting science. The establishment of the experimental method as a primary source of rational intelligence was established first for physical relations. Galilei and successors emphasized that a temporally limited experiment, well isolated from the context and carefully arranged, can stand as a prototype for a class of similar phenomena. Knowing one outcome informs about the entire field of identical settings. One case can stand for all – as long as standardization is adequate. This applied to inorganic material, i.e. *res extensa*, as Descartes put it. He exempted *res cogitans*, the domain of intelligence, where mechanistic paradigms would not hold in his opinion. Living beings – though usually considered as plain *res extensa* – posed difficulties to some degree. Discussions sparked about how far man and animals share certain properties, and how the human mind as a domain of free choice and brain as a domain of its physical substrate would relate. In neurology and brain research certain aspects of this controversy continue today (Maasen et al. 2003).

Studying the diversity of life and life forms was the domain of natural history (Mayr 1982). An early impetus of natural history was a theological interest to illustrate the richness of creation (natural theology, e.g. Paley 1803). Natural history was largely descriptive and quantification played a relatively marginal role. This line of tradition was influential in ecology. It remained meaningful and it caused scepticism towards quantitative "physicalistic" descriptions. The founder of ecology, Ernst Haeckel himself, did not emphasize the application of quantitative methods in ecology. He largely used conceptual approaches, verbal descriptions, and graphical representations. The initiative to look at quantitative relations of man and environment did not emerge in the context of natural history or ecology but in physics, in economics, and demography.

Malthus: Basic Ideas in Population Science

Robert Malthus (1766–1834) was one of the first to introduce quantitative considerations in the population context. He considered implications and determinants of the growth of human populations. Mainly operating from an economic perspective, his ideas had subsequent influence for ecological considerations. He linked the growth and well-being of the human population directly with a development of natural resources (Malthus 1798) (Fig. 3.1).

Later, Darwin (1809–1882) considered the malthusian ideas in his development of evolutionary theory. A key idea of Malthus was that population growth tends to exhibit self-similar characteristics: If each of the population member has the same chances of fertility in space and time, the involved growth processes tend to be exponential. With a constant rate of increase, exponential growth accelerates

Fig. 3.1 Robert Malthus
source: wikimedia commons

over time. On the other hand, Malthus argued, resource development would not proceed exponentially but follow a linear dynamic. Malthus saw the discrepancy between these two growth forms of arithmetic versus geometric growth as an inevitable source of tension and instability. This controversy substantially inspired the following scientific debate on well reasoned, quantitative considerations on the human use of natural resources. At that time, Malthus' major impact was not in the field of ecology as he worked in the newly developing field of political economics. Here, in the era of European imperialism, his ideas played an important role in the discussion about how to deal with scarce and limited resources (Claeys 2000).

Verhulst: Early Functional Generalizations

It did not take long until other, successively more elaborated functional forms to describe growth were provided. Pierre Francois Verhulst (1804–1849) (Fig. 3.2) was a Belgian mathematician, who sought a way to describe the modification of growth intensity under the conditions of limited resources. He found a rather simplistic form in 1838. His function is still widely used in ecological modelling under the name of logistic growth (see Chap. 6: Differential Equations (6.7)).

The growth process that this equation describes always tends towards an equilibrium. Interestingly, for quite a long time logistic growth was far less considered, compared to Malthus. It was largely forgotten until Raymond Pearl (1879–1940) rediscovered it when ecological modelling had made its first steps in the 1920s (Pearl 1925). This was when the relevance of quantitative considerations as a foundation of ecology became more and more potent (see Lotka and Volterra, below). In physics, chemistry and other fields of science, advancement had led to

Fig. 3.2 Pierre Francois Verhulst
source: wikimedia commons

a well established mathematical underpinning and to the discovery of more and more quantitative relations. Ecology – terminologically existing as a sub-discipline of biology – was still largely dominated by qualitative assessment.

3.3 Founders of Ecological Modelling

In the first half of the twentieth century, Einstein's relativity theory of 1905 had been successively accepted in physics. Quantum theory was on the way, and David Hilbert discussed infinite dimensional vectors as mathematical objects. With some lag, the relevance of quantitative relations received attention in ecology as well. This started with quite elementary and simple contexts, which did not require elaborated mathematical forms. Differential equations, which describe the change of particular variables over time, played the leading role.

Lotka and Volterra: Setting the Stage for Network Approaches

Independent from each other, Alfred Lotka (1880–1949) and Vito Volterra (1860–1940) developed the same simplistic form to describe the interaction of a predator population and a prey population (Lotka 1925; Volterra 1926). The equations are explained in detail in Chap. 6. In the subsequent time, this model inspired innumerable variations, modifications, and adaptations to specific contexts. To this day, the original works of Lotka and Volterra are among the most frequently cited papers in ecological modelling. The Verhulst-equation, re-discovered by Pearl around the same time, helped to extend the functional repertoire usable in the equations. The Lotka–Volterra (LV) model describes the interaction of two species in the most simplistic way. Actually, any real ecological context is by far more complex than two types of organisms interacting in a constant environment. Interestingly, in a metaphorical sense, it can be stated, that the LV-equations have the same role in ecology as Kepler's laws had in astronomy: the two-body problem can be mathematically solved. In isolation, the mutual gravitational impact of two objects is easy to describe in an equation. But when three or more mass points influence each other, it becomes very difficult to develop a valid model. The three-body problem requires numeric approximation and can be solved only in some special cases. In ecology, it was now possible to write down one, two, or a larger number of equations for the interaction of populations. But the equations could not be solved if their complexity was only a little bit higher than the LV-model. This may have been the reason why ecological modelling still played a peripheral role in the ecological science. This aspect changed with the extent that network interactions could be numerically managed.

von Bertalanffy: System Theoretic Foundation and Generalization

The Austrian biologist and philosopher Ludwig von Bertalanffy (1901–1972) (Fig. 3.3) played an important role in establishing and popularizing the systems perspective in biology in general, and influenced ecology (1949, see also 1969). He developed the concept of a flow equilibrium and emphasised the holistic approach. This encouraged to turn conceptual reasoning about the relations of interacting parts and the whole into practical research agendas. In his time, the transition was

Fig. 3.3 Ludwig von Bertalanffy (courtesy of Bertalanffy Center for the Study of Systems Science, W. Hofkirchner, Vienna http://www.bertalanffy.org)

made where network structures could not only be formulated but also iterated. This allowed to approximate the dynamic processes in complex networks. He, and the following colleagues were influential for network approaches to gain further roots in ecology.

Lindeman, EP and HT Odum, Waddington: Modelling and Ecological Application

There are many scientists who influenced the establishment of quantitative views in ecology. We select a few. Raymond Lindeman (1915–1942) (Fig. 3.4) pioneered the concept of trophic dynamics. In his 1942 work at the University of Minnesota (USA) he elaborated quantitative relations in an ecosystem context. At this time, he still met reservation whether this would bring ecological research forward.

For further description of his work: see Chap. 18 on trophic cascades. In the 1950s, Eugene Odum (1913–2002) published an influential textbook (1953), where he emphasized quantitative relations in a systems context as a means for ecosystem management. With this book, "modern ecology" reached the surface of the scientific mainstream. His brother Howard forwarded the idea to represent relevant relations in ecological systems using energy equivalents incorporated in biomass as a unifying measure. The energetic content of biomass could be used as a basis for homogeneous descriptions applicable for all ecosystem types. Paradigmatic in this context was the Silver Springs Ecosystem study (Odum 1957). Odum and colleagues refined their conceptional ideas in further studies (see Fig. 3.5). Though the concept is generalizable in a formal sense, as any change in ecological system has energetic implications, not all aspects were solved. This was because numerous factors influence the quantitative changes in energy transfer. E.g. limiting factors of

Fig. 3.4 Raymond Lindeman (by courtesy of University of Minnesota Archives)

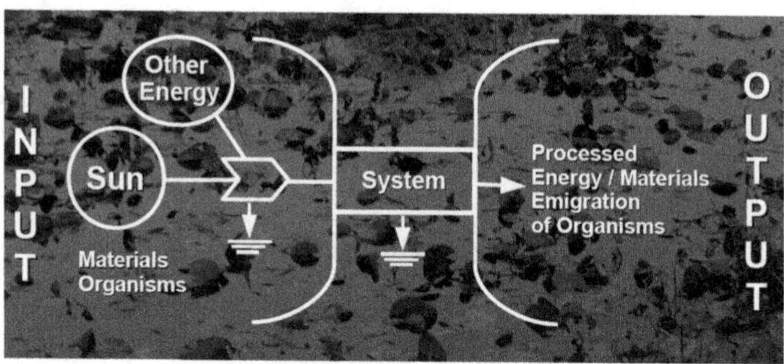

Fig. 3.5 Conceptional ecosystem model with input and output side following Patten (1978) and Odum (1997)

chemical nutrients, and information processing by the organisms. Systems ecology proceeded in a controversial way. The International Biological Programme (IBP), established in 1962, executed 1964–1974, was the first global attempt to provide a quantitative overview across different ecosystem types. The co-ordinator was C.H. Waddington (1905–1975). Representative ecosystems were identified and biotic inventories and quantitative trophic relations were investigated with a comparably large effort (Worthington 1975). For the first time, a large international effort was centred on ecosystems research including co-ordinated modelling efforts. Continuations of the research efforts provided model approaches that were quite important in the understanding of the forest decline, which was a large-scale phenomenon as a consequence of acidification of precipitation through industrial emissions during the 1970s and eighties (Bossel 1986, 1996).

Forrester, Meadows, Patten, Joergensen: Ecological Systems and Interdisciplinary Linkages

The systems approach has been continuously elaborated since the 1950s. Methods were developed that allowed to link knowledge from different disciplines in co-ordinated research frameworks. The interdisciplinary methods are described in further detail in Chap. 4. Origins were in systems philosophy (von Bertalanffy), and in industrial dynamics. Forrester (1961) developed a systems approach that he initially applied to understand complex matter flows in industry and the relating economic flows. The approach was subsequently generalized (Forrester 1969) and applied in a global environmental context (Meadows et al. 1972). Bernhard Patten (1976) exemplified the applicability for systems ecology. The strength of the approach was in facilitating an interdisciplinary understanding, to structure complex research tasks and to organize larger consortia of researchers contributing to an overall goal. With numerous works of Joergensen, e.g. his activity to establish the Journal *Ecological Modelling* in 1975, with the compilation of Textbooks (Joergensen 1986) and handbooks (e.g. Joergensen 1979), and the establishment of the International Society for Ecological Modelling in 1975, the topic was fully established as a conceptionally and methodologically expanding science.

To meet the demands of empirical complexity, it became necessary to involve additional methods. Most of these additional methods were again imported from other disciplines and adapted to ecological requirements and then yielded the repertoire of modelling complex ecological dynamics.

3.4 Diversification and Diversifiers

During the 1980s it became apparent that homogeneous variables had a limited potential to fully capture the complexity of ecological relations. In particular, quantity–quality transitions and inhomogeneous temporal and spatial structures were difficult – at least inconvenient – to be captured in differential equation systems. Criticisms towards modelling as such (e.g. den Boer 1981) would be overcome when modelling methodologies diversified.

The Object Paradigm and Individual-Based Modelling

During the 1960's, Ole Johann Dahl (1931–2002) (Fig. 3.6) and Kristen Nygaard (Fig. 3.7) from the Norwegian Computer Centre in Oslo developed a computer language which went conceptually beyond the established so-called procedural approach: With SIMULA (SIMUlation LAnguage), extending ALGOL (ALGOrithmic Language) they introduced options that allowed a kind of self-structuring during programme execution (Dahl et al. 1968). Reference variables could point to particular addresses of the computer storage to indicate the location of complex data objects. These data objects could be created and deleted when running the programme and utilize references to each other that could be changed during execution. This yielded *object-oriented programming*, which became generally known in computer science with the SMALLTALK-80 programming language

Fig. 3.6 Ole-Jan Dahl, courtesy of Depart. of Informatics, University of Oslo

Fig. 3.7 Kristen Nygaard

with C++ and others during the late 1980s. These developments opened considerable new options in ecological modelling, since object orientation allowed a convenient linkage of structural and functional dynamics, which had been difficult to bring together. In both domains, changes could be synchronously represented – by changing the object structure together with the values of the variables stored in the objects (Fig. 3.8). One of the first, who understood the importance for ecology was Heinrich Kaiser in Aachen (Germany). He used the approach for the development of individual-based models (Kaiser 1976, 1979). A Pioneer in the field was also Paulien Hogeweg in Utrecht (The Netherlands, see Hogeweg and Hesper 1979, 1983). The individual-based approach offered far more options to synchronously represent variability in physiological states of organisms, usage of their behavioural repertoire and responses to time variant structural environmental pattern (Huston et al. 1988; Judson 1994).

The Self-Organization Paradigm

The self-organization paradigm became influential in ecology at about the same time as the individual-based modelling approach. It emerged in the context of physics (Haken 1977), thermodynamics (Glansdorff and Prigogine 1971), and systems theory. The proponents of the self-organization approach argued the following: Science as a whole is grounded on the principle of causality. Any change that occurs has an antecedent cause. Same settings, same causes always produce the same results. In this way, causes and effects were traditionally segregated. What happens if they interact in large numbers and complex networks? With the availability of automated calculation potential, non-trivial effects were discovered in loop-structures where interaction results served as a new input (the effect becomes a new cause that feeds back in iterative cycles). In feedback structures, non-trivial states can emerge that cannot be reduced to plain external impact. Systems can self-generate

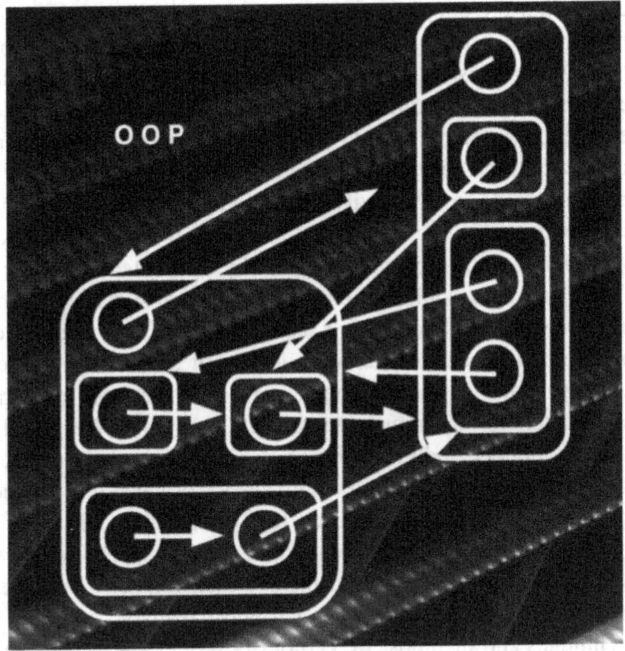

Fig. 3.8 Scheme of an object-oriented programme during execution state. The objects are using pointers to access each other (see Chap. 12)

complexity. This became a very influential idea for interdisciplinary exchange (Jantsch 1980; Prigogine and Stengers 1984). In parallel to ecology, the self-organization discourse influenced biology as a whole, as well as the social sciences, psychology, and other fields of science and philosophy. The Santa Fe Institute (New Mexico, USA) developed the research agenda of Artificial life. Here, formal descriptions of how living entities self-organize were used to study the potential to simulate properties of living systems using physical substrates (Langton 1994). Frequently, object structures were used, but also other approaches like Cellular Automata (see Chap. 8), which had received relevant applications in physics, and thermodynamics. Moreover, fractals were used (Mandelbrot 1982) and new developments in the theory of dynamic systems (Peitgen 1992), all of which had become more widely recognized during the 1980s and 1990s.

3.5 Anything Goes: The Diversity of a Post-Modern Ecology

Where are we now? Ecological modelling is a well established discipline. It is recognized that the understanding of complex phenomena requires modelling and that formal approaches can to a considerable extent capture the quantitative and qualitative understanding we have about biological systems and their environmental interactions. From model representations, we also know that marginal shifts can be amplified and alter directions of development, which allows qualitative understanding but limits prognostic potential. It is clear meanwhile, that not a single methodological approach is equally suitable for all questions and problems. This confronts ecologists with the requirement to select appropriate approaches. Which one to chose depends on the problem to be solved. Nowadays, methods from many different sources are adapted if they are suitable for ecological modelling purposes.

How else could we discover activity pattern and behaviour of marine animals except by combining radio tracking, remote sensing and other technical devices with data evaluation facilities to come up with object oriented behavioural models (see Chap. 12)? How else could we come up with realistic plant shape models without using graphical iteration (see Chap. 11)? How else could we gain an integrated picture of landscape dynamics and how it alters the living conditions for protected organisms, if not using a larger set of methods in parallel with network-like connections of mutual input (see Chap. 21)? We use differential equations, matrix models, individual-based descriptions – as presented in the different chapters of this book as single approaches – or when it is needed, various model components will be coupled together (see Chaps. 20 to 22).

Promising approaches are very welcome for ecological application when it helps and inspires the understanding of organisms and their relations among each other and their environments. Regardless of the currently available, amazing supply of modelling techniques and methods, we ecologists should always stay focussed on one thing: to develop our own model.

Part II
Modelling Techniques and Approaches

Chapter 4
System Analysis and Context Assessment

Broder Breckling, Fred Jopp, and Hauke Reuter

Abstract System analysis is a theory-based approach with a wide range of practical applications for interdisciplinary co-operation that was derived from the General Systems Theory, as introduced by Bertalanffy and others. Formerly being developed as an analytical instrument, it has now become also an integral part of model development. The method starts with the delimitation of the investigated system from the surrounding context, the specification of its compartments, key factors, driving forces and how these interact with each other. To gain a conceptual overview, a cause–effect diagram of a system can be constructed to sketch the influences between the investigated components. Such a graphical examination can be refined in a next step into a flow-diagram (flow chart) that depicts the compartments, connections and controls of the system. This tool is of considerable help when developing model systems further on the basis of differential equations, as it enables to characterize time-dependent quantitative changes of the components as a result of their interactions. Over the recent years, object-oriented systems analysis (OOSA) has been established as a structural extension of the classical systems analysis. By using the object-oriented programming approach, OOSA enables to represent and subsequently simulate dynamic systems which change their structure over time. We provide examples of the resulting complex interaction networks from marine fisheries and functional plant architecture.

4.1 Systems Analysis and Context Assessment: The Starting Point for Model Development

Systems analysis is an essential part of model development. It provides a targeted methodology that helps to identify, represent and connect the important components and relations which are relevant in a given context. System analytical

B. Breckling (✉)
University of Bremen, General and Theoretical Ecology, Center for Environmental Research and Sustainable Technology (UFT), Leobener Str., 28359 Bremen, Germany
e-mail: broder@uni-bremen.de

F. Jopp et al. (eds.), *Modelling Complex Ecological Dynamics*,
DOI 10.1007/978-3-642-05029-9_4,

approaches can be used to structure complex ecological phenomena (Jørgensen and Müller 2000). The methodologies are applicable in practically any discipline, including social, technical and ecological systems. In this chapter we give a short overview on backgrounds and fundamentals in system analysis and frontier views of ecological theory and application. Furthermore, we explain important terminologies and introduce major techniques and strategies used in the system analytical practice as a basis for the development of ecological models.

4.2 Ecological Theory, Systems and Complexity

In former stages of development, systems analysis was seen mainly as an analytical tool which was used to structure and survey smaller, as well as large interaction networks (Forrester 1968). Later, it was used as a scoping approach in coordinating the tasks to be done, in particular to coordinate interdisciplinary research groups where different specializations contribute to a given task and where the facilitation of an interdisciplinary understanding is essential (Patten 1959, 1975). However, it seems necessary to mention that systems analysis is not a standardized procedure. There are many different interpretations of system theory. We largely refer to the approach originating from the works of Bertalanffy (1969, called "General System Theory") and in addition to the more recent Object-Oriented Systems Analysis (Hill 1996) that originated from computer programming.

In ecology, the transition from qualitative consideration to quantitative considerations is largely due to the introduction of a systems perspective (Odum 1953, 1983; Meadows et al. 1972). With the epistemological underpinning developed in the context of Bertalanffy's general system theory (Bertalanffy 1949, 1950, 1969) the concept became widely applied also in biology. Ecological applications frequently related to a systems interpretation as used in cybernetics (see: Patten 1975; Patten and Odum 1981), and elaborated the implications for hierarchical structures relevant in ecological systems (Odum 1971; Allen and Starr 1982; O'Neil et al. 1986). Odum (1977) points out that central concepts of ecosystem theory need to adopt a holistic perspective to deal successfully with the complexity of the system in focus. Applications cover all aspects of analysing and describing ecological systems from short-term physiological processes, adaptive behaviour and reaction to environmental conditions, to long-term evolutionary approaches.

4.3 What Is a System and What Is Systems Analysis

A system is an abstract, hierarchical construct of a set of elements, which is considered to operate as an entity. When referring to any given system, we first need to make a consistent distinction of what exactly belongs to the system and what does not. With this first analytical step, the boundaries of the system are defined. The different elements and their relations are specific for a particular system. The elements can be involved in relations with other elements.

Elements can be grouped into sub-systems which can have relations to other elements or sub-systems. The resulting systems may be hierarchically organized (see e.g. Fig. 2.4). However, in the early days of systems analysis, the hierarchical aspect was followed only in a very limited way, because the representation in differential equations used for dynamic models used to concentrate the interaction on only one focal level (see Chap. 6).

System analysis is a generally applicable strategy to operationalize a given context of any structure and complexity. It allows to check the available information for completeness, consistency, relevance and plausibility, and can be colloquially summarized as an assessment of "how things work together". The strength of the approach is its interdisciplinary applicability which allows to organize and structure workgroups and contents across different scientific specializations, while maintaining the coherence of the subject.

During the first step of model development (see also Chaps. 2 and 3), there are important questions to answer:

- What are the key factors to focus on?
- What are the driving forces behind the investigated phenomena?
- Which interactions are relevant, which can be excluded?
- How to organize and prioritize the analysis?

These decisions have to be made with care since the best model does not help to solve specific research questions if it does not capture the relevant influences. The best way to begin is to provide a communicable structure of what the model intends to deal with. Such a conceptual overview can serve as a decision basis for further model development. The following steps are a short overview of how such a communicable structure can be obtained.

The Development of a Cause–Effect Diagram

An approach serving to delimit a modelling task and identify relevant structural components of the given context is the cause–effect diagram. It is a graphical representation that has the purpose to list the relevant aspects of the considered context and to indicate existing relations and influences. The specification should be made in a way that it prepares the next steps in formalizations:

Every component in the diagram should be selected so that it represents something quantifiable. With regard to interactions it is involved in, it should be possible to specify the component so that it can increase or decrease in quantity. Or to put it another way: Only concepts that allow the specification of an increase or decrease are suitable as a component in analysis. Examples of suitable concepts are e.g. biomass, temperature, concentration or density of a substance, frequency, etc. Less suitable are terms that involve names, tastes, styles, and other concepts that are difficult to interpret what their increase or decrease in quantity could mean. Vague terms that leave room for different ways of interpretation should also be avoided. They should be expressed in other terms or divided further into (sub-) components. The components should have a minimum of internal structure. If a component has a pronounced internal structure it should be considered to disaggregate it further. Disaggregation of the overall context into its components should be continued to approximate a manageable number of components (or elements). Frequently, manageable problems are limited to around ten components. If a higher number seems inevitable, it should be discussed whether the modelling task eventually requires a more specific focusing.

The first step is to write the single components so that relations between them can be drawn in the form of arrows. The arrow between two components means there is an influence of one component on another component concerning quantitative changes. The criterion for where to set an arrow is a specific one: if the component increases in quantity, the other component should either increase as well or decrease in quantity. The influence must not be necessarily in exact proportion, only the direction of increase or decrease should hold. If the influence is mutual, two arrows should be drawn.

For a cause–effect diagram, a relation exists between two components if the quantity of a component changes as a result of a change in another component. A positive influence would exist if component **A** increases, and then, as a result component **B** increases too. Also, a positive relation would be considered if **B** decreases as a consequence of a decrease in **A**.

A negative influence exists, if element **A** increases in quantity, and then element **B** decreases as a result (or **B** increases as a result of a decrease in **A**). The next step is to draw arrows between the elements as relations (influences). This may require shifts of the location of the components so that the overview is optimized and the possible crosses of the arrows are minimized. The quality of the influence, whether it is positive or negative, is indicated by assigning the arrow with an according symbol (+ or –).

The cause–effect diagram (see Fig. 4.1 for an example) facilitates a conceptual overview. Discussing the result can give rise to further refinements and is the crucial step in an identification of the important elements (variables) that are used in further development steps in modelling. Further steps include a quantitative specification of each of the influences as a mathematical function.

The selection of a system analytical interaction network is not as trivial as it may seem at first glance. The selection of each of the components (elements) and influences (relations) requires careful discussion to capture the crucial interaction. It forces the modeller to be explicit about the ideas of how things work together.

The Flow-Diagram (Flow Chart): Compartment, Connections, Controls: The "3 Magic C's"

Cause–effect diagrams provide the first overview. In a following step, they can be refined into flow-diagrams. Flow diagrams are an important tool deriving from the classical systems approach, which deals with the development of differential equation models where time-dependent quantitative changes occur as a result of the

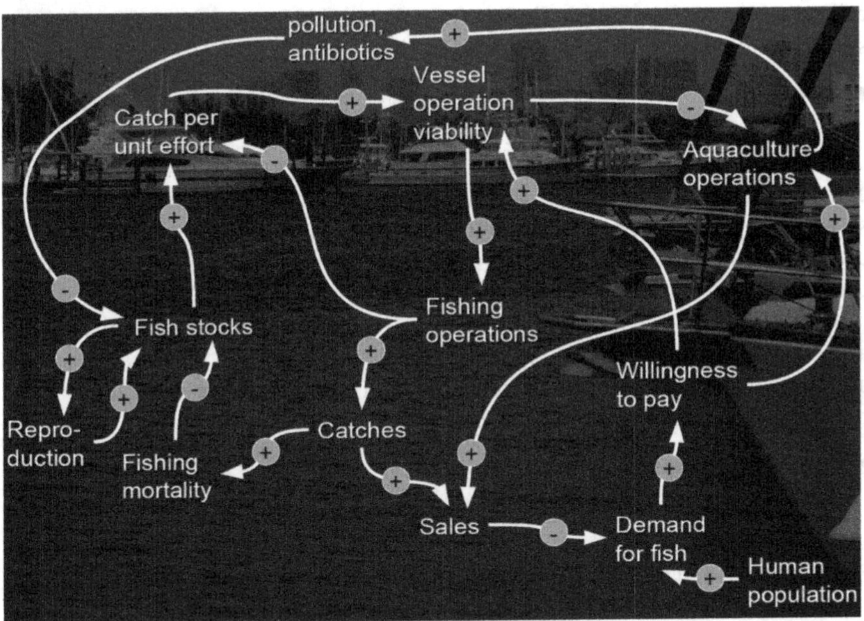

Fig. 4.1 Example of a cause–effect diagram describing relations in coastal fisheries. Explanations of which criteria are used to set-up a cause–effect diagram are given in the text

components interactions. Other modelling approaches go specific ways in successive refinement. The additional achievement of the flow diagram is, that a graphical overview makes it easier to survey larger networks and follow quantitative transitions within the model. The flow chart approach was originally popularized by Forrester (1968) in the context of the simulation package "DynaMo". Later, the approach became modified in different ways and was used also as a graphic user interface in various simulation software (e.g. STELLA, ModelMaker, SIMILE, see http://www.mced-ecology.org). This links the conceptual development of a model with computer implementation. The flow diagram consists of the following major components:

- **Compartments**

Compartments are containers to store material, energy, or any kind of quantity in focus. They represent a variable (element) of a dynamic system. Compartments are drawn as rectangles.

- **Connections**

Connections are drawn as solid arrows. They represent the flow of the quantities between the compartments. A flow always occurs in a dynamic system when the size of a variable changes. An increase is considered as an inflow to the variable, a decrease as an outflow. The flow can come from or go to other compartments, or cross the border of the system [inflow from the outside into the system (i.e. from a source), outflow when it leaves the system (into a sink)].

- **Controls**

Controls are used to specify the extent of flows occurring between different compartments. They are drawn as valves with influences from other parts of the system that are drawn as light or dotted arrows. The dotted arrows represent the causal structure, determining which parts of the system have an impact on other parts.

Compartments, connections, and controls represent the functional units which allow to specify dynamic systems. For convenience, additional graphical elements are sometimes used in systems representation when a direct interfacing to a mathematical formalization is intended.

- **Sources and Sinks**

The system border is indicated only implicitly in the form that flows come from the outside into the system or leave the system to the outside. Sources are used if a flow is represented that does not originate from within the system (i.e. an inflow to a variable that does not originate from another variable). Sinks, usually displayed with the same symbol as a source, are displayed when a flow originates from within the system but does not ends in a systems compartment.

- **Auxiliaries**

In some cases it simplifies the representation by adding a symbol (e.g. a larger circle) to represent expressions, which occur more than once in the specification of different controls.

- **Parameters**

These are constant values that are used to specify controls. They can be visualized by a specific symbol, frequently a small circle, from which influence(s) go to one or more than one control.

The transition from a cause–effect diagram to a flow chart is made by an assignment of the concepts appearing in the cause–effect diagram to either a compartment, connection or control. Arrows from the cause–effect diagram are distinguished whether representing flows or influences. The process to set up a flow diagram is easier when the overall structure has already been considered in terms of causes and effects, rather than starting at the beginning with the set-up of a flow diagram without having systematized the relevant relations to be considered beforehand.

For a correct systems representation, implicit rules emerge for flow charts. Flows can only originate from sources or from compartments and end at sinks or compartments. Influences can originate from parameters, auxiliaries, compartments or controls. They can connect to auxiliaries and controls, but cannot end at compartments. Figure 4.2 represents an example of a Lotka–Volterra flow chart.

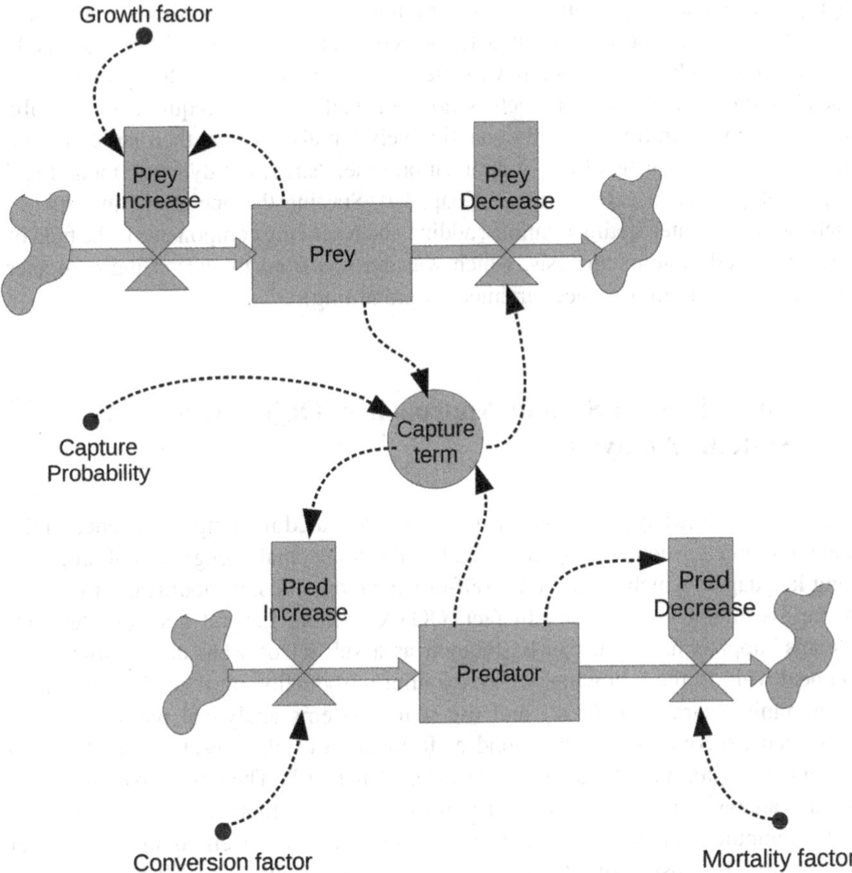

Fig. 4.2 Example of a flow diagram: Flow chart of the Lotka–Volterra equations, see Chap. 6 (6.11). In the classical flow-chart notation the diagram relates to the following differential equation system

d Prey/dt	= Prey_Increase − Prey_Decrease
d Predator/dt	= Pred_Increase − Pred_Decrease
Capture_term	= Capture_Probability ∗ Prey ∗ Predator
Prey_Increase	= Growth_factor ∗ Prey
Prey_Decrease	= Capture_term
Pred_Increase	= Conversion_factor ∗ Capture_term
Pred_Decrease	= Mortality_factor ∗ Predator

Why the General Systems Theory Is Extensible

General Systems Theory, as introduced by von Bertalanffy and others, raised the expectation of general applicability, i.e. being an "all purpose approach". What would be more general than listing what it is that changes, and then assigning how much that would be under certain circumstances. The approach was applied and actually is most appropriate as a developmental basis for (ordinary) differential equation systems. It is less suitable to describe structurally heterogeneous settings, where it is not reasonable to define internally homogeneous variables, where the internal heterogeneity is of crucial interest. Also, the approach is not well suited if it is required to describe systems that vary in time not only quantitatively but also change their structure in a pronounced way. For this kind of application case, "structural dynamic modelling" was developed as an extension (see Chap. 19). Starting the operations in terms of establishing and interrupting relations, adding and removing components is the field of object-oriented systems analysis, which was derived from the according computer programming paradigm (object oriented programming).

4.4 Non-Classical System Approaches: Object-Oriented Systems Analysis

The object-oriented systems analysis (OOSA) originated in computer science in the context of object-oriented programming. It is the conceptual background of programming languages, which allow some self-organization and self-modification features of the executable programme. In fact, OOSA is more general than the "general systems" approach, which can be thought as a sub-set of what object orientation can deal with. In the following, we give a brief introduction into the object-oriented programming paradigm (OOP) and use it for systems analytical purposes. More elaborated descriptions can be found at Rumbaugh et al. (1991) and Hill (1996) and in the chapter on individual-based models (Chap. 12). The term "object-oriented systems analysis" refers to specific features of modern computer programmes.

In computer science, a *class* is a unit that can be specified in a computer programme. It consists of storage reservations for variables as well as executable statements of computer code. The specific feature of a class allows it to be copied ("instantiated") various times during programme execution. Then, the copies are referred to as *objects*. In order to distinguish several copies of a class during programme execution a special type of variable ("reference variable" or "pointer") is required. A pointer contains the address where the object can be found in the computer storage. Hence, using pointers, it is possible to specify relations and also interactions between objects.

4.4.1 Object Oriented Systems Analysis Facilitates Structurally Dynamic System Representations

For ecological applications, classes can be used to describe the set of states and activities of organisms and individuals. Using OOP it is conveniently possible to use one generic description to simulate large numbers of individuals performing their activities independently of each other. OOP allows to handle *structurally variable interaction networks*. This largely extends the range of phenomena that can be modelled, in particular self-organizing spatio-temporal structures on different levels of organization (Breckling et al. 2005; Reuter et al. 2008). The system analytical operations to set up an object-oriented model are explained in Chap. 12, since the approach has its most important application in individual-based modelling.

4.4.2 Application Examples

The object structure of a model can be very closely adapted to quantitative *and qualitative* ecological knowledge and thus allows to investigate implications from

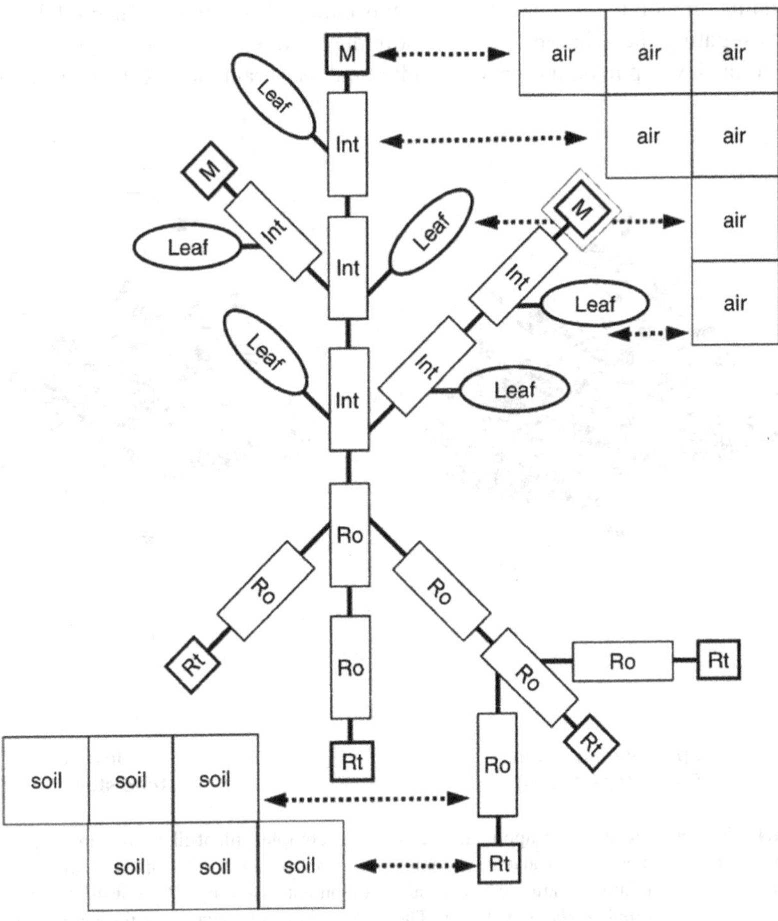

Fig. 4.3 Object network of a structurally dynamic plant model: Meristems, internodes, leaves, roots and root-tips are classes from which objects are instantiated to simulate the plant architectural development according to the physiological processes simulated inside the objects (assimilation, nutrient transport, etc.); see Eschenbach (2005) for details

what is known without being forced by the abstraction concepts to focus only on specific properties. This is shown in the following examples.

Eschenbach (2005) attempted to model structural and physiological implications of plant development jointly. Previous models usually focused on either the structure or on physiological processes. She defined the typical units as classes and specified according methods (rules how to change the variables) within the particular classes. Under certain conditions new class instances are created (e.g. branches, internodes) – or existing ones could be eliminated (e.g. leaves). This allowed to test whether certain assumptions about partitioning that cannot be directly measured, but are important for growth implications were in line with the overall structural development that results after many years of growth. (Figs. 4.3 and 4.4).

The same approach can be used to describe animal interactions. Hölker and Breckling (2002, 2005) applied it to simulate local density heterogeneities of fish depending on individual behaviour and environmental structures. The model aimed at investigating the relation between individual bio-energetic characteristics and population development. It represents individual fish as autonomously acting agents

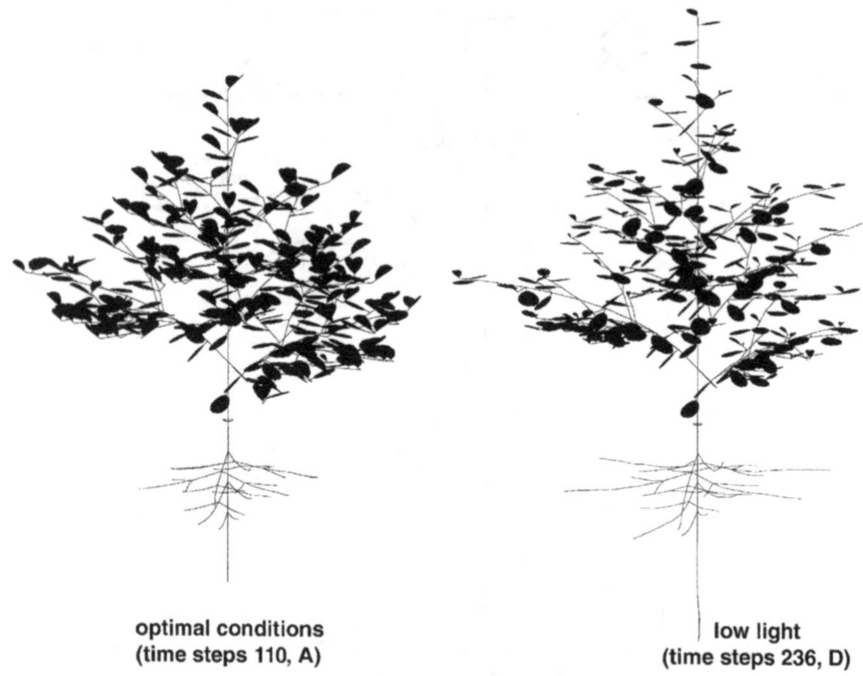

Fig. 4.4 The repeated iterative application can reveal complex plant-like structures with root system, stem, branches, leaves and meristems as shown for the model of alder trees (*Alnus glutinosa*). Two Simulation results of structural development assuming different light intensities are shown (*left*: optimal, *right*: low light). The tree architecture adapts to the environmental conditions according to the response pattern of the single components showing the implications of environmental or physiological differences on the structural development [from Eschenbach (2005)]

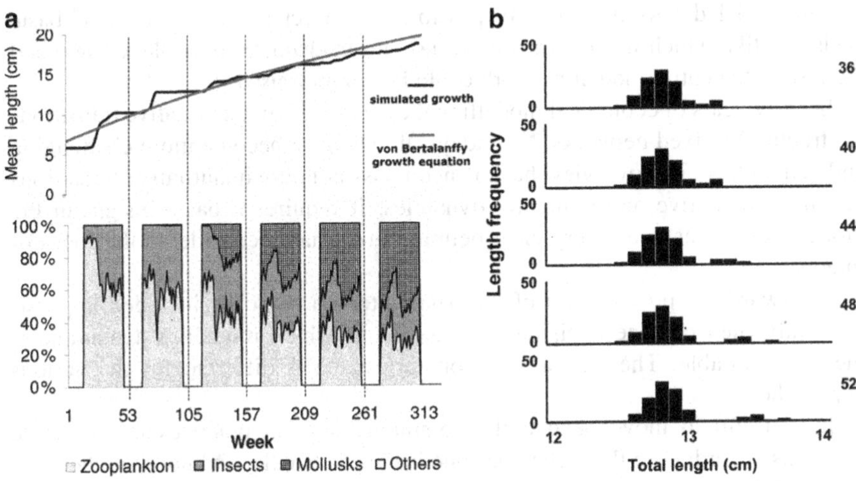

Fig. 4.5 Results from the fish model of Hölker and Breckling (2002, 2005). (**a**) Increase in biomass and size dependent diet of roaches (*Rutilus rutilus*). Large individuals are capable of exploiting bigger food items (molluscs) thus providing a higher nutritional value [Figure adapted from Hölker and Breckling (2005)]. (**b**) Simulation results of the length development of a second year roach cohort. The length frequency illustrates the differentiation in size. Most roaches are below 13 cm and only a few are 0.5–1 cm longer

with an explicit behavioural repertoire to react to its environment and comprises, e.g. food searching, movement and biomass and season dependent reproduction. Bioenergetic processes were implemented in detail to represent a realistic food dependent growth and reproduction, which includes a size dependent exploitation of resources. The model was parameterized for the roach (*Rutilus rutilus*) and simulations were carried out for the abiotic conditions of a Northern German lake of about 1.1 km^2 surface.

The analysis of model results revealed changes in structural relationships. Depending on size the fish were able to exploit further resources (molluscs) providing a higher nutritional value (Fig. 4.5a). Heterogeneities in size structure thus were amplified (Fig. 4.5b) allowing the larger individuals to reproduce 1 year before the rest of the cohort. The model thus allowed to analyze causes for the empirically determined cohort structure and potential reactions to changing environmental conditions.

4.5 System Types and Modelling Approaches: A Modern Diversity

For model development, a structural overview of the relevant relations to be included or left out of consideration is essential. Formulating this overview in an explicit way is the primary step of model development. If there are shortcomings in certain model details it is usually possible to correct them. However, if basic decisions like which model structure to use are inadequate, regardless how good the rest is, the entire modelling work could be compromised.

In early years of ecological modelling the focus was on quantitative transitions in structurally fixed networks. Nowadays, things have become more challenging and demanding. The overview has to encompass not only quantitative transitions but also qualitative and structural dynamics. It requires a basic insight in the biotic interactions and a very clear definition of what is considered the focus of interest.

Afterward a representation of the basic interactions to be included is possible. Only then can the decision be made as to which modelling technique is the most suitable. The way abstractions are made is different for the various approaches.

It is useful to move beyond the commonly applied practices to select the problems according to the techniques one is able to handle, and not the other way around. It is better to choose from a larger range of different tools, according to the state-of-the-art of current ecological modelling techniques. This strategy is not only more efficient and appropriate. It also helps to maintain the quality of the theoretical analysis with the diverse requirements of modern ecological issues.

Chapter 5
Steady State Models of Ecological Systems: EcoPath Approach to Mass-Balanced System Descriptions

Matthias Wolff and Marc Taylor

Abstract We describe the fundamentals and applications of trophic models of ecological systems and show how a simple mass balance approach of the early 1980s was further developed into a very advanced complex software package, freely available on the internet (Ecopath with Ecosim, EwE, http://www.ecopath.org). Through its three decades of evolution, the approach became increasingly popular, with over three hundred Ecopath models being published to date. During its first 10–15 years, the approach was mainly used as a tool to integrate ecological and fisheries data to understand and visualize the trophic flow structure of ecosystems, thereby allowing for the meaningful comparisons between systems. Later (since the mid-1990s) it was increasingly used to explore ecosystem changes under the impact of management or climate impact scenarios. This evolution from a more descriptive mass-balance to a simulation modelling tool was enabled through fundamental changes in the mathematical architecture: the original version of Ecopath was based on linear algebra for input–output analysis to investigate the properties of steady state networks, while in the recent version, coupled differential equations are used for each of the defined system compartments. The new model architecture thus allows the magnitude of flows in and out of compartments to change over time, which makes simulations of changes possible. Foraging arena theory was also taken into account and prey biomass was allowed to vary in its availability (vulnerability) to predators. If the predators consumption is mainly determined by prey availability (bottom-up control, low vulnerability), predator biomass would greatly respond to changes in prey biomass, while under a situation of strong predator (top-down) control, changes in predator biomass would greatly impact its prey (high vulnerability). Since EwE also allows for specifying different resource use types (e.g. types of fisheries or other resource uses) and economic variables associated to them (operational costs, number of people employed etc.), different management regimes can also be explored in terms of their socio-economic

M. Wolff (✉)
Department of Ecological Modelling, Leibniz Center for Tropical Marine Ecology GmbH (ZMT), Fahrenheitstr. 6, 28359 Bremen, Germany
e-mail: matthias.wolf@zmt-bremen.de

F. Jopp et al. (eds.), *Modelling Complex Ecological Dynamics*,
DOI 10.1007/978-3-642-05029-9_5, © Springer-Verlag Berlin Heidelberg 2011

outcome. If spatially explicit data on biomass distribution of model groups and dispersal rate values are available, the Ecospace module of the EwE package may be used to derive at a spatially explicit, dynamic trophic model able to simulate how the management of one part of the system area may affect other parts. The incorporation of these spatial dynamics is of particular interest in the exploration of management questions, such as in the size and placement of Marine Protected Areas (MPAs) (this application is not further described in this chapter).

5.1 Introduction

The steady state trophic modelling approach came into focus in the late 1970s, when the evidence was increasing that any strong single species fishery (like the Cod fishery in the North Atlantic) substantially affects other parts of the ecosystem that are trophically linked with the target resource. People sought to quantify trophic interactions within aquatic ecosystems and different multispecies approaches and models evolved. Among the pioneering work was the North Sea Model of Anderson and Ursin (AU model, 1977), which includes the dynamics of nutrients, phytoplankton, zooplankton and age-structured Beverton–Holt yield models for the main fish groups of the system. The full model was based on 308 differential equations and required an enormous amount of biological information for model formulation. The Multispecies Virtual Population Analysis (MSVPA) followed shortly in the early 1980s and the AU model also represents a multispecies extension of traditional single species fisheries stock assessment models. It is based on values of predation mortality based on catch-at-age data; predator ration and predator diet information and allows estimating the predation mortalities produced by predators on prey species and the annual consumption of prey by predators.

While the approaches mentioned, and other related ones, were developed from age structured population models, the ECOPATH trophic modelling concept, developed during the early 1980s by Polovina (1984), followed a different, more holistic approach: it uses biomass pools of functional groups, which are connected through a predator–prey diet matrix. Energy flow through the food web is balanced by equating the biomass production with internal consumption and exports to the fishery of the different compartments. This approach allows integrating a large number of species into ecological functional groups, requires much less biological information and is thus applicable for data – sparse situations that are often found in tropical waters.

During the first decade of its genesis, ECOPATH was greatly enriched through the input of Ulanowicz's (1986, 1997) ideas of network growth and development of natural systems and through substantial contributions by Christensen and Pauly (1992), which were all integrated into the ECOPATH II software. In the mid-1990s Carl Walters joined the team and developed a simulation software, then called ECOSIM (Walters et al. 1997). It was based on ECOPATH II, but replaced the simple linear algebraic equations by coupled differential equations, thus allowing compartment biomasses to change over time as a response to changes in mortality or consumption or environmental forcing. Thus, a tool was created that allows, by using a balanced ECOPATH model as a reference, to explore ecosystem response of different fishing policies and to follow ecosystem changes over time by fitting time series of observed biomasses and catches to fishing effort and/or environmental parameters.

Application Range

ECOPATH requires the identification of main functional groups within a system, and allows to quantify and visualize biomass flows, to identify key compartments of high biomasses and productivities and to understand the relative importance of the biomass export to the fishery for the energy cycling of the system.

Relevant research questions are related to (a) the relative impact of the fishery and natural predation on the key resources of the system; (b) the overall functional structure (i.e. the ratio of benthic to pelagic compartments; relevance of benthic versus pelagic primary productivity etc.); (c) the trophic efficiency between trophic levels; (d) network flow characteristics (such as degree of connectivity, amount of internal cycling, role of detritus, etc.); (e) mean trophic level of the catch, Primary Production Required (PPR) to sustain the catches and many more summary statistics that can be derived from this holistic approach. It is frequently used for comparing network structure, overall degree of development/disturbance and the relative role of certain functional groups of aquatic ecosystems. Hundreds of aquatic ecosystems have been modelled with this approach and the database for comparative research has grown substantially over the past two decades. Following the Large Marine Ecosystem (LMEs) initiative of Sherman et al. (2003) ECOPATH models were and are being constructed for each of the LMEs. A substantial number is already accessible on the web (see: http://www.seaaroundus.org).

5.2 Conceptual and Mathematical Basis of Approach

5.2.1 Mass-Balance Modelling: Ecopath

The core routine is based on the assumption of mass balance over a given time period (usually 1 year, but any other interval may be chosen). In its present form Ecopath is based on two master equations, one to describe the production term (5.1) and one for the energy balance for each group (5.4).

Master equation one:

$$P_i = Y_i + M2_i * B_i + E_i + BA_i + M0_i * B_i, \tag{5.1}$$

where Y_i is the total fishery catch rate of i, $M2_i$ is the instantaneous predation rate for group i, E_i the net migration rate (emigration–immigration), BA_i is the biomass accumulation rate for i, while $M0_i$ is the "other mortality" rate for i. P_i is calculated as the product of B_i the biomass of i and $(P/B)_i$, the production/biomass ratio for i. The $(P/B)_i$ rate under most conditions corresponds to the total mortality rate, Z (see Allen 1971), commonly estimated as part of fishery stock assessments. The "other mortality" is all mortality not included elsewhere, and is internally computed from (5.2)

$$M0_i = \frac{P_i * (1 - EE_i)}{B_i}, \tag{5.2}$$

where EE_i is called the "ecotrophic efficiency" of i, and can be described as the proportion of the production that is utilized in the system. The predation term, $M2$, in (5.1) serves to link predators and prey as

$$M2_i = \sum_{j=1}^{n} \frac{Q_j * DC_{ji}}{B_i}, \tag{5.3}$$

where the summation is over all n predator groups j feeding on group i, Q_j is the total consumption rate for group j, and DC_{ji} is the fraction of predator j's diet contributed by prey i. Q_j is calculated as the product of B_j, the biomass of group j and $(Q/B)_j$, the consumption/biomass ratio for group j. For parameterization Ecopath sets up a system with (at least in principle) as many linear equations as there are groups in a system, and it solves the set for *one* of the following parameters for each group, biomass, production/biomass ratio, consumption/biomass ratio, or ecotrophic efficiency. The other three parameters along with the following parameters must be entered for all groups, catch rate, net migration rate, biomass accumulation rate, assimilation rate and diet compositions (for more details see Christensen and Walters 2004).

Within each group energy balance is ensured using the following master equation two:

Master equation two:

$$\text{Consumption} = \text{production} + \text{respiration} + \text{unassimilated food.} \tag{5.4}$$

This equation is in line with Winberg (1956), who defined consumption as the sum of somatic and gonadal growth, metabolic costs and waste products. In Ecopath it was chosen to estimate respiration from the difference between consumption and the production and unassimilated food terms. This reflects the focus on application for fisheries analysis, where respiration rarely is measured while the other terms are more readily available. Besides units of wet weight biomass, Ecopath models can be constructed using energy as well as with nutrient related currencies. If a nutrient-based currency is used in Ecopath the respiration term is excluded from the above equation (as nutrients are not respired), and the unassimilated food term is estimated as the difference between consumption and production.

Addressing Uncertainty

A resampling routine, Ecoranger, has been included to accept input probability distributions for the biomasses, consumption and production rates, ecotrophic efficiencies, catch rates, and diet compositions. Using a Monte Carlo approach, a set of random input variables is drawn from user selected frequency distributions and the resulting model is evaluated based on user-defined criteria and physiological and mass balance constraints.

To facilitate this task of describing probability distributions for all input parameters (including the diet compositions matrices) and to make the process more transparent a "Pedigree" routine was implemented (Pauly et al. 2000) that allows the user to mark the data origin using a pre-defined table for each type of input parameters (from in-situ sampling, taken from other models etc.). The confidence interval around the input parameter is smallest (±10%) for data derived from sampling the same system and largest for those derived from other models (±80%).

The Pedigree index values for input data scale from 0 for data that is not rooted in local data up to a value of 1 for data that are fully rooted in local data. Based on the individual index value an overall "pedigree index", τ, is calculated as the average of the individual pedigree value based on

$$\tau = \sum_{i=1}^{n} \frac{\tau_{i,p}}{n}, \tag{5.5}$$

where $\tau_{i,p}$ is the pedigree index value for group i and input parameter p for each of the n living groups in the ecosystem; p can represent either B, P/B, Q/B, Y or the diet composition, DC. To scale based on the number of living groups in the system (n), an overall measure of fit, t^* is calculated (using an equation based on how the t-value for a regression is calculated) as

$$t^* = \tau * \frac{\sqrt{n-2}}{\sqrt{1-\tau^2}} \tag{5.6}$$

This measure of fit describes how well rooted a given model is in local data.

Mass-balancing an ECOPATH model is usually achieved by manually adjusting biomasses, mortality rates, diets, etc., searching for data inconsistencies and gradually obtaining a balanced model. An iterative method for obtaining mass-balance has been added to EwE, offering a well defined, reproducible approach, while also allowing exploration of alternative solutions based on parameter confidence intervals as explained above. This routine, called *automated mass balance*, is further explained in Christensen and Walters (2004).

Ecosystem Indicators

A selection of ecosystem indicators was included in EwE to describe the state of the system. Following Odum (1969, 1971) it is assumed that an undisturbed ecosystem is mature and that in a more mature system most niches are filled; that a larger part of the energy flows should be through detritus-based food webs; that primary production should be more efficiently utilized; that the total system biomass/energy throughput ratio should be higher, etc. As shown by Christensen and Pauly (1998) (and many others) the use of a composite of ecosystem indices may allow to describe the state of a given system and how it may have changed over time. A selection of relevant ecosystem "health" indices included in EwE are given in the following Table 5.1.

5.2.2 Time-Dynamic Simulation: Ecosim

The basics of Ecosim are described in detail by Walters et al. (1997, 2000) and will only be overviewed here. Ecosim consists of biomass dynamics expressed through a series of coupled differential equations. The equations are derived from the Ecopath master (5.1), and take the form

$$\frac{\mathrm{d}B_i}{\mathrm{d}t} = g_i \sum_j Q_{ij} - \sum_j Q_{ij} + I_i - (M0_i + F_i + e_i) * B_i, \tag{5.7}$$

where $\mathrm{d}B_i/dt$ represents the growth rate during the time interval $\mathrm{d}t$ of group i in terms of its biomass, B_i; g_i is the net growth efficiency (production/consumption); $M0_i$ the non-predation ("other") natural mortality rate estimated from the ecotrophic efficiency, F_i is fishing mortality rate, e_i is emigration rate, I_i is immigration rate (assumed constant over time, and hence independent of events in the ecosystem modelled), and $e_i \times B_i - I_i$ is the net migration rate of (5.1). The two summations estimate consumption rates, the first expressing the total consumption by group i, and the second the predation by all predators on the same group i. The consumption rates, Q_{ji}, are calculated based on the "foraging arena" concept, where B_i's are divided into vulnerable and invulnerable components (Walters et al. 1997, Fig. 5.1), and it is the transfer rate (v_{ij}) between these two components that determines if control is top-down (i.e. Lotka–Volterra), bottom-up (i.e. donor-driven), or of an intermediate type.

Ecosim bases the crucial assumption for prediction of consumption rates on a simple Lotka–Volterra or "mass-action" assumption, modified to consider "foraging arena" properties. Following this, prey can be in states that are either vulnerable or un-vulnerable to predation, for instance by hiding (e.g. in crevices of coral reefs or inside a school), when not feeding, and only being subject to predation when having left their shelter to feed (Fig. 5.1). In the Ecosim formulation (Walters et al. 1997, 2000) the consumption rate for a given predator feeding on a prey was thus predicted from the effective search rate for predator–prey specific interactions, base

Table 5.1 Ecosystem indices given in the EwE software

Name of index	Meaning
Cycling index	Fraction of an ecosystem's throughput that is recycled
Predatory cycling index	Corresponds to the cycling index but is excluding detritus groups
Cycles and pathways	A routine that presents the numerous cycles and pathways that are defined by the food web representing an ecosystem based approach suggested by Ulanowicz (1986)
Connectance index	The ratio of the number of actual links to the number of possible links. Feeding on detritus (by detritivores) is included in the count, but the opposite links (i.e. detritus "feeding" on other groups) are disregarded
System omnivory index	The average omnivory index of all consumers weighted by the logarithm of each consumer's food intake. It is a measure of how the feeding interactions are distributed between trophic levels. An omnivory index is also calculated for each consumer group, which is a measure of the variance of the trophic level estimate for the group
Trophic level decomposition	Aggregates the system into discrete trophic levels *sensu* Lindeman. The routine reverses the routine for calculation of fractional trophic levels.
Trophic transfer efficiencies	Calculated for a given trophic level as the ratio between the sum of the exports plus the flow that is transferred from one trophic level to the next, and the throughput on the trophic level. The transfer efficiencies are used for construction of trophic pyramids, and others
Primary production required (PPR)	To estimate the PPR to sustain the catches and the consumption by the trophic groups in an ecosystem the following procedure is used: all cycles are removed from the diet compositions, and all pathways in the flow network are identified using the method suggested by Ulanowicz (1995). For each pathway the flows are then raised to primary production equivalents using the product of the catch, the consumption/production ratio of each path element times the proportion the next element of the path contributes to the diet of the given path element
Mixed trophic impact (MTI)	Leontief (1951) developed a method for input–output analysis to assess the direct and indirect interactions in the economy of the USA, using what has since been called the "Leontief matrix". A modified input–output analysis based on the procedure described by Ulanowicz and Puccia (1990) is implemented in EwE. The MTI describes how any group (including fishing fleets) impacts all other groups in an ecosystem trophically. It includes both direct and indirect impact, i.e. both predatory and competitive interactions.
Ascendency	EwE includes a number of indices related to the ascendency measure described in detail by Ulanowicz (1986). Ascendency is seen as a measure of ecosystem growth and development

vulnerabilities expressing the rate with which prey move between being vulnerable and not vulnerable, prey biomass, predator abundance. The model as implemented implies that "top-down versus bottom-up" control is in fact a continuum, where low v's implies bottom-up and high v'stop-down control. EwE has incorporated numerous further computational routines; to force time series, to handle complex

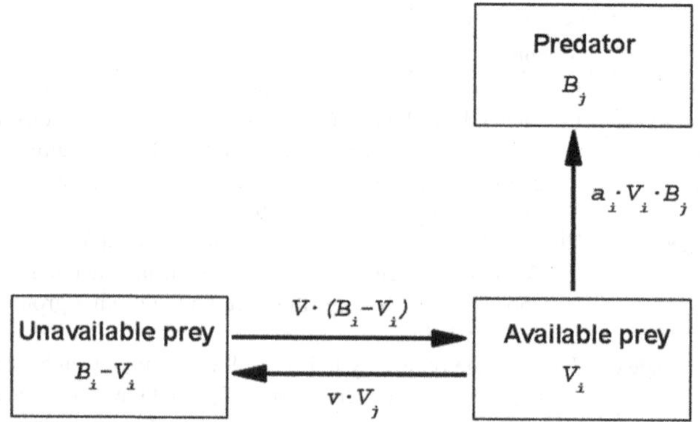

Fig. 5.1 Simulation of flow between available (V_i) and unavailable (B_i–V_i) prey biomass in Ecosim. a_i is the predator search rate for prey i, v is the exchange rate between the vulnerable and un-vulnerable state. Fast equilibrium between the two prey states implies $V_i = vB_i / (2v + aB_j)$ (based on Walters et al. 1997)

life histories, to trace nutrients and pollutants through the food web and several others (for more details see Christensen and Walters 2004).

5.3 Application Examples

5.3.1 Ecopath Approach

The following example is based on Wolff (2006), who compared Ecopath II models of two mangrove fringed estuaries in Costa Rica (Gulf of Nicoya, at the Pacific shore) and Brazil (Caeté estuary, NE of Belem) with respect to biomass and energy flow distributions, productivity and fisheries potential, with the added objective to obtain guidelines for conservation and management of these systems.

Figure 5.2 shows the flow charts of the two models derived from the ECOPATH programme.

Figure 5.3 shows a routine of ECOPATH, which allows the visualization of flows that enter (as food) and leave (to the predators and the fishery) a compartment. In the example below the central mangrove consumers in both systems (land crab in the Caeté system and shrimps in the Nicoya Gulf) are shown. A biomass pyramid of both systems is also included.

The input data matrix (see additional material at http://www.mced-ecology.org) needs three input parameters for each compartment: biomass (B), turnover rate (P/B) and consumption rate (Q/B). The trophic level of each group is calculated by the program based on the diet matrix, which connects the model compartments.

Differences in biotic structure, energy flow and resource productivity's between both systems proved to be substantial as seen by the summary statistics of the model calculated (but not included here). These are largely due to differences in topography,

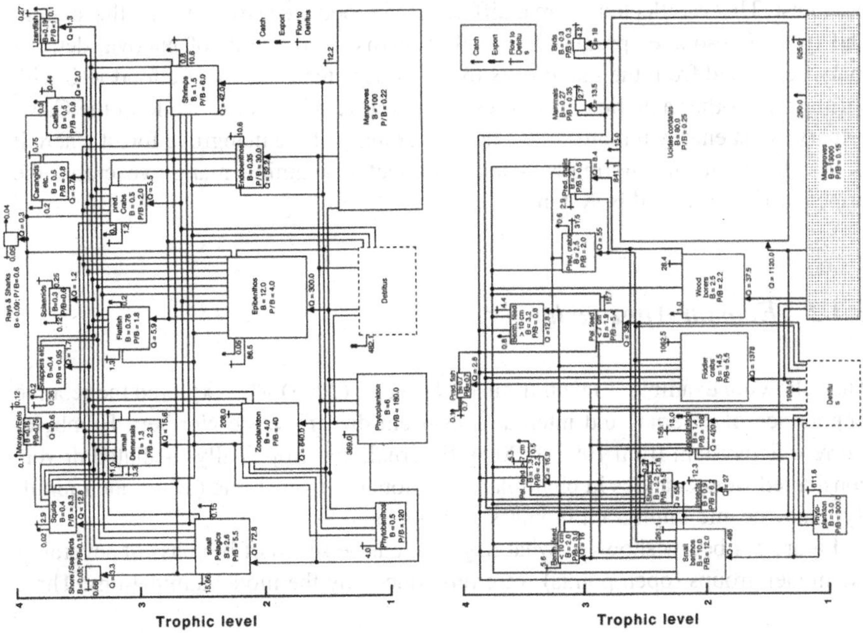

Fig. 5.2 Flow charts of the Gulf of Nicoya (*left*) and of the Caeté estuary (*right*); box size is proportional to square root of compartment biomass (except for mangrove compartment); Q-values represent total amount of biomass entering the compartment; flows are given in gm^{-2} (wet mass)

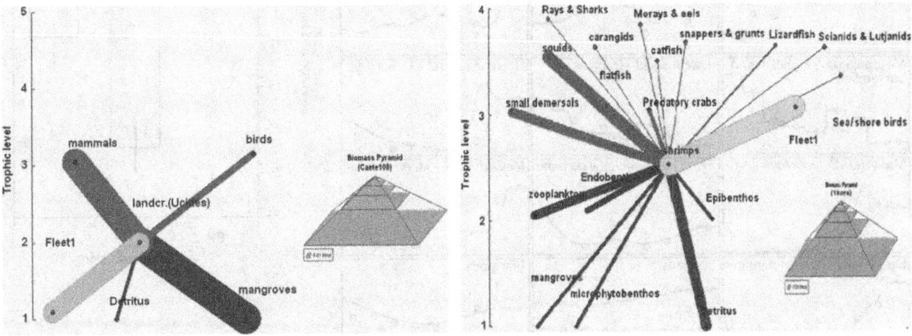

Fig. 5.3 Caeté estuary (*left*) and Gulf of Nicoya (*right*). Role of land crabs (*Ucides cordatus*) and shrimps (*Penaeus* spp.) in the ecosystem and system biomass pyramid; food biomass from prey compartments (*black bars*); food biomass to predators (*dark grey bars*); catches (*light grey bar*) (widths of bars proportional to amounts transferred) (from Wolff 2006)

tidal regime and mangrove cover between both systems. The Gulf of Nicoya is exposed to semidiurnal tides and an efficient daily water exchange between the mangrove stands and the gulf and thus to a strong mangrove matter export to the gulf water. In comparison, the mangrove forest of the Caeté estuary is only flushed each fortnight and the largest part of the mangrove production thus remains within the forest. This is reflected in great differences in food web structure and the amount and type of resources produced in both systems. In the Gulf of Nicoya, detritus matter exported from the mangroves to the estuary feeds an aquatic food web with shrimps and other aquatic detritivores in the centre of the web, while in the Caeté estuary, most energy remains in the benthic domain of the mangrove forest where it is transferred to an enormous biomass of leaf consuming mangrove crabs, the principal resource of this system.

5.3.2 *Ecosim Approach*

The following example is taken from Taylor et al. (2008) who explored the relative importance of external and internal ecosystem drivers in the Northern Humboldt Current Ecosystem from 1995 to 2004. External, non-trophically-mediated drivers considered were changes in phytoplankton biomass, fishing rate (effort and mortality), and oceanic immigrant biomass (mesopelagic fish).

Figure 5.4 below shows to what degree the time series of biomass of several of the model groups (open points) were reproduced by the model simulations. Three

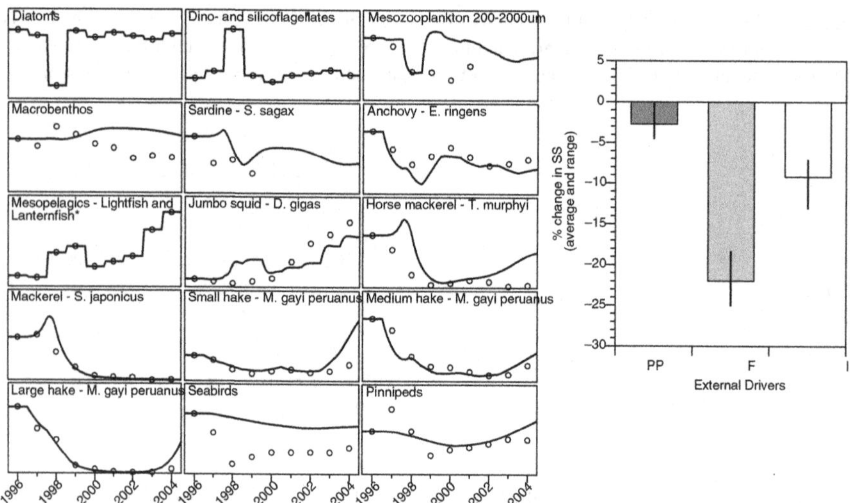

Fig. 5.4 *left*: Time-series trends of biomass changes from the data sets (*dots*) and Ecosim simulations (*lines*). Presented is the best-fit simulation (i.e. lowest SS), using all drivers (phytoplankton, PP; fishery, F; and immigrants, I) followed by a "fit-to-time-series" routine. Yearly data points represent "biological years" (i.e. July–June of the following year). *Asterisks* indicate artificially-forced functional groups (diatoms, dino- and silicoflagellates, and mesopelagics); *right*: Sum of square (SS) changes under the application of external drivers in all possible sequences and combinations. All simulations use intermediate, default control settings (i.e. all predator–prey vulnerabilities equal 2.0). Negative values (i.e. decrease in SS) indicate an improvement in fit (from Taylor et al. 2008)

model groups (diatoms, dinoflagellates and myctophic fish) were fixed during the model runs. As shown by the figure, anchovy dynamics are well simulated. The initial decrease in anchovy biomass during 1997–1998 is mainly reproduced by forcing phytoplankton abundance downwards; specifically, a decrease in diatom biomass and subsequently, a decrease in the second most important food item, mesozooplankton.

The application of fishing as an external driver improved the fit of the simulation and helped to explain the long-term dynamics of several main target species. The fishing driver decreased long-term variance by 22% (Fig. 5.4) pointing to the important role of fishery in shaping the trophic flows and biomass distribution in the system.

5.4 Outlook

Through the integration of the biomass budget (balanced flows) approach of Ecopath with the time-dynamic approach of Ecosim, a modelling software package (EwE) was created, which can be used for a great variety of data situations and purposes. In the most data sparse situation, Ecopath can be used to integrate available biological, ecological and fisheries information of the ecosystem to obtain a first holistic view and understanding of the system under study. The process of model construction is most instructive since questions of ecosystem functioning and process rates are addressed. Since model construction requires the knowledge input of different experts (for the different model groups studied), the process of model construction may help to integrate groups of researchers working in the same

Table 5.2 Further application examples using the EwE-software

Application	EwE package	Author(s)
Comparison across ecosystems	Ecopath	Christiansen and Pauly (1993), Moloney et al. (2005)
Comparison of ecosystem states	Ecopath	Shannon et al. (2003), Tam et al. (2008), Taylor et al. (2008c), Wolff (2006)
Ecosystem phase shift exploration	Ecosim	Shannon et al. (2004)
Optimal harvest strategies	Ecosim	Arreguín-Sánchez et al. (2004a, b)
Mediation of trophic controls with actual environmental time series	Ecosim	Field et al. (2006)
Mediation of trophic controls by a third functional group	Ecosim	Guenette et al. (2008)
Fitting of trophic mediation time series with post-hoc comparison to actual time series	Ecosim	Shannon et al. (2008), Coll et al. (2009)
Exploration of trophic and environmental drivers	Ecosim	Taylor et al. (2008a–c)
Marine protected area exploration	Ecospace	Okey et al. (2004)

system. If the available data base is rich, the model architecture may be more complex, with more functional compartments defined and meaningful input parameters obtained. If time series of compartment biomasses, environmental parameters, fisheries effort or catch are available, the mechanisms causing observed ecosystem changes can be explored over any relevant time period. EwE can thus also be applied as a complex modelling tool for simulating scenarios of ecosystem change under varying boundary conditions. As mentioned in Table 5.2 the spatial dynamics of the system can also be modelled if spatially explicit data for the model compartments are available.

Further Reading

For Beginners

Christensen V, Pauly D (1995) Fish production, catches and the carrying capacity of the world oceans. ICLARM Q 18(3):34–40

Polovina J (1984) Model of a coral reef ecosystem I. The ECOPATH model and its application to French frigate shoals. Coral Reefs 3:1–11

Wolff M (2002) Concepts and approaches for marine ecosystem research with reference to the tropics. Rev Biol Trop 50(2):395–414

For Advanced Scientists

Christensen V, Walters CJ (2004) Ecopath with ecosim: methods, capabilities and limitations. Ecol Model 172:109–139

Pauly D, Christensen V, Walters C (2000) Ecopath, ecosim, and ecospace as tools for evaluating ecosystem impact of fisheries. ICES J Mar Sci 57:697–706

Ulanowicz RE (1986) Growth and development: ecosystems phenomenology. Springer, New York, 203 pp

Chapter 6
Ordinary Differential Equations

Broder Breckling, Fred Jopp, and Hauke Reuter

Abstract Differential equations represent a centrally important ecological modelling approach. Originally developed to describe quantitative changes of one or more variables in physics, the approach was imported to model ecological processes, in particular population dynamic phenomena. The chapter describes the conceptual background of ordinary differential equations and introduces the different types of dynamic phenomena which can be modelled using ordinary differential equations. These are in particular different forms of increase and decline, stable and unstable equilibria, limit cycles and chaos. Example equations are given and explained. The Lotka–Volterra model for predator–prey interaction is introduced along with basic concepts (e.g. direction field, zero growth isoclines, trajectory and phase space) which help to understand dynamic processes. Knowing basic characteristics, it is possible for a modeller to construct equation systems with specific properties. This is exemplified for multiple stability and hysteresis (a sudden shift of the models state when certain stability conditions come to a limit). Only very few non-linear ecological models can be solved analytically. Most of the relevant models require numeric approximation using a simulation tool.

6.1 Background and Purpose of the Chapter

Differential equations play a highly relevant role in the history of modern ecology. The introduction of differential equation-based modelling was an important achievement in the paradigm shift from of a previously more qualitatively oriented science to a leading role of quantitative approaches. The concept of differential equations originated in classical mechanics. It was developed to describe the motion of mass points, acceleration, and other time dependent processes. Early last century, differential equations were adapted by a few ecologists, who focused on quantitative

B. Breckling (✉)
University of Bremen, General and Theoretical Ecology, Center for Environmental Research and Sustainable Technology (UFT), Leobener Str., 28359 Bremen, Germany
e-mail: broder@uni-bremen.de

F. Jopp et al. (eds.), *Modelling Complex Ecological Dynamics*,
DOI 10.1007/978-3-642-05029-9_6, © Springer-Verlag Berlin Heidelberg 2011

considerations varying over time (Kingsland 1995). The introduction of differential equations successively refined analysis of ecological relations and helped to assess the backgrounds and driving forces of quantitative changes in natural systems. Differential equations describe the change of one or more variables over time. The change can be influenced by the quantity of the variable itself, by external impact or by responses to other variables. Differential equations can be used to describe single elements as well as complex networks of dynamic systems (Bertalanffy 1976).

This chapter introduces the use of differential equations in ecological models. It does not provide a complete overview of the mathematical theory. Here, those aspects are selected and explained, that are most important to understand the contribution of differential equations in ecological theory and its applications. We facilitate an understanding of which kind of dynamic representations have particular relevance for describing ecological processes. The chapter addresses the central terms and topics that are required for model construction and that are useful for understanding the scope and the limitations of the approach. For this purpose, a selection was made that reduces the difficulty of mathematical formalism. The selection of topics builds on lecture experiences at the University of Bremen, and feedback from a large number of students over the years. Not only are the mechanisms of increase and decrease of different variables explained, but also the phenomena of multiple equilibria, hysteresis and deterministic chaos.

6.2 What Are Differential Equations?

Differential equations represent a concept of abstraction with very specific requirements. Though the approach is used in a modelling strategy called "general systems theory" (Bertalanffy 1976) differential equations can be successfully applied only when specific preconditions hold:

- They deal with homogeneous quantities. They do not describe the internal heterogeneities that may lie behind particular variables. The focus of interest, is on "how much?", assuming that the internal *quality* of what the variable represents, is invariant and structurally homogeneous.
- Differential equations are deterministic and functional. A given state of the system always determines precisely the subsequent states. Stochastic influences are excluded.
- Differential equations are continuous. They describe the succession of states in infinitely small intervals, i.e. for any point in time of the considered simulation interval.

These specifications can best be visualized by a pool metaphor: a pool – to be filled with water (or whatever imaginary liquid), having an inflow and an outflow (Fig. 6.1). Together with the initial filling level of the pool, the regulation of inflow and outflow determines the filling state (i.e. the value of the variable), which we can also call the compartment size. In this metaphor, the equations describe the operation

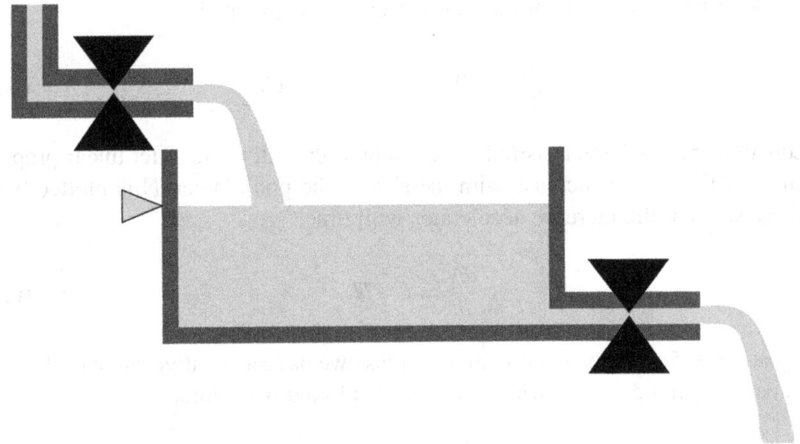

Fig. 6.1 The pool metaphor – how differential equation models describe quantitative relations. Any change of the pool size (the size of the variable) implies a flow. How much of a flow per unit of time occurs, is considered to be regulated (by valves). This regulation is described as a function, which can depend on the size of the variable itself, on other variables or flows, or on external conditions

of the valves that regulate the amount of inflow and outflow per unit of time. As functions, inflow and outflow can depend on specific constants, including the pool size itself (or other pools), and other (externally determined) functions. For ecological applications, variables typically represent populations in relatively homogeneous and constant environments.

With some elementary examples we can see how the metaphor and the formulation of equations relate. First, we look at the simplest equations, combine and expand them to model successively more complex dynamic phenomena. A constant inflow only (i.e. with no outflow) would lead to a constant increase of the pool size per unit of time. Constant inflow and outflow would lead to a net change rate – either positive or negative, depending on the relative flows (or no change if the inflow equals the outflow). If the state variable size feeds back to influence inflow or forward to influence outflow in exact proportion to its current pool size, exponential growth or decline will be modelled. In (6.1)–(6.3), N denotes the pool size state, C is a constant >0, and t is the time.

Equation (6.1): Constant rate of increase of a pool (variable) per unit of time

$$\frac{dN}{dt} = C. \tag{6.1}$$

Equation (6.2): Constant decrease of a pool (variable) per unit of time and

$$\frac{dN}{dt} = -C. \tag{6.2}$$

Equation (6.3): Several constant factors can be aggregated

$$\frac{dN}{dt} = (C1 + C2 - C3) = C. \tag{6.3}$$

Equation (6.4): Characteristic for exponential growth is the strict linear proportionality of the rate of increase with the size of the pool. When N is plotted over time, we see that the increase accelerates with time.

$$\frac{dN}{dt} = C * N. \tag{6.4}$$

Equation (6.5): For the exponential decrease we have a negative slope of the rate and a decelerating decrease when pool size is plotted over time.

$$\frac{dN}{dt} = -C * N. \tag{6.5}$$

Equations (6.4) and (6.5) describe flows which depend only on the pool size and a constant. As a result, exponential growth or exponential decline, respectively, will occur.

Now, we extend the functions used in the equations to successively greater complexity and discuss the dynamic results. In (6.6) constant increase and exponential increase are combined. In principle, we can use any function which allows us to determine the rate of change – with one or more variables interacting, for example exponential growth and negative quadratic decrease (6.7).

Equation (6.6): Exponential growth together with constant increase

$$\frac{dN}{dt} = C * N + C. \tag{6.6}$$

Equation (6.7): Exponential growth with negative quadratic decrease. This is the so-called logistic growth function (see below)

$$\frac{dN}{dt} = C1 * N - C2 * N^2. \tag{6.7}$$

We can simultaneously model several variables that can be either independent of each other, or coupled; that is, mutually influence each other. One of the simplest pairs of coupled equations, that is, having two interacting variables and interesting dynamics, is

$$\frac{dN1}{dt} = C1 * N2 \tag{6.8}$$

$$\frac{dN2}{dt} = -C2 * N1.$$

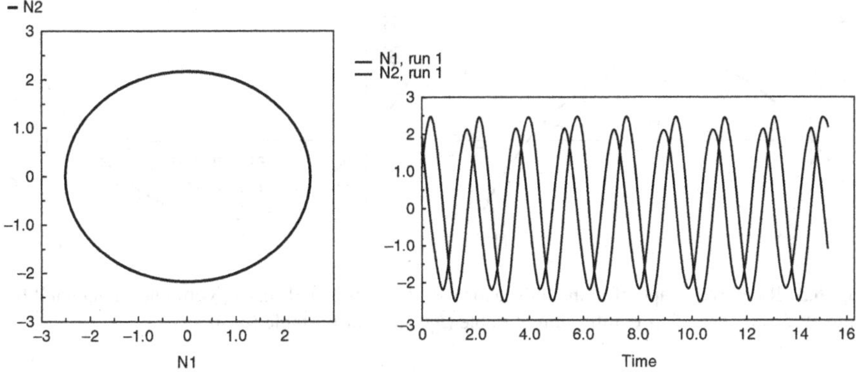

Fig. 6.2 Simulation of (6.8). *Left*: $N2$ over $N1$, *Right*: $N1$ and $N2$ over time. Initial conditions: $N10 = 1.0$, $N20 = 2.0$ Parameter: $C1 = 3.0$, $C2 = 4.0$

$C1$ and $C2$ are positive constants. The variable $N1$ increases exponentially in proportion to the variable $N2$. The change of $N2$ depends negatively on the size of $N1$. The resulting dynamics is an oscillation, depending on the initial conditions and the parameter size (Fig. 6.2). For $C1 = C2 = 1.0$, a sine curve will result when plotted over time – or a circle when one variable is plotted over the other.

Formal structures like this are used to capture some of the dynamic phenomena of ecological processes.

6.3 Differential Equations as a Modelling Approach for Dynamical Systems

When experimenting with differential equations, it is a frequent experience that explosions (rapid exponential growth) or a collapse (approximation of zero) occur unintentionally, if the model was not carefully designed. The art of modelling is to select quantitative relations in a meaningful way so that they capture relevant and dominating elements of observable (physical, biological) phenomena. Simplifications are inevitable, which always brings the possibility of interesting discoveries as well as irrelevant or trivial results. In cases where only one equation is used, relevant information about the resulting dynamics can be obtained from a graphical representation that plots the rate of the change as a function of the size of the variable. This representation is called *rate level graph*. For one-dimensional systems, which are described by only one differential equation, the general type of dynamics can be directly deduced. All intersections of the plot of the rate of change with the zero line are equilibria. This is because, at that particular variable value, the changes are zero. If the plot of the rate of change of a variable N first decreases with increasing N, then becomes negative for larger N, the intersection with the zero-line is a *stable equilibrium*. This means, if a system in an equilibrium state is

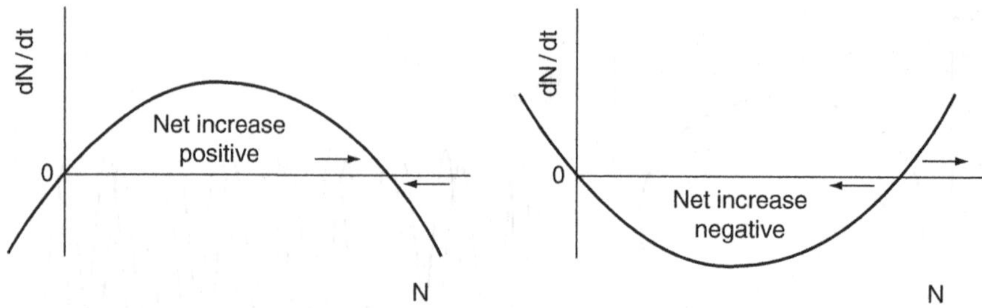

Fig. 6.3 Rate level graph of a one-dimensional system [(6.9) and (6.10)] with one trivial equilibrium at $N = 0$. The other equilibrium is stable (*left side*) or unstable (*right side*)

slightly disturbed (i.e. is shifted to the vicinity of the equilibrium), it will move back. The variable increases in the case that N is smaller than the equilibrium and decreases for N larger than the equilibrium. In the opposite case we have an *unstable equilibrium*. Then the variable decreases for N smaller than the equilibrium and increases for N larger than the equilibrium – it successively moves away from the equilibrium – regardless how close it is to the equilibrium point, as long as a difference exists [(6.9), (6.10), Fig. 6.3].

$$\frac{dN}{dt} = C1 * N - C2 * N^2 \text{ with stable equilibrium at } C1/C2 \tag{6.9}$$

$$\frac{dN}{dt} = -C1 * N + C2 * N^2 \text{ with unstable equilibrium at } C2/C1 \tag{6.10}$$

It is important to note that in a rate level graph only the rate of change for different N is shown, not the change over time. Rate over stock is different from stock over time. Figure 6.4a, b show the respective examples with N plotted over time.

Using this kind of functional approach, the change of animal-, plant- or microbial populations can be approximated by treating the population sizes as pools. With the concept outlined so far, we can now look at a frequently considered starting point for quantitative population ecology, the Lotka Volterra model.

6.4 Lotka–Volterra Equations as a Starting Point for Ecological Modelling

The Lotka–Volterra model is the simplest way to describe the interaction of a predator population and a prey population. It was proposed independently by Alfred Lotka (1925) and Vito Volterra (1926). It is extremely simplified and thus not very realistic; however, this simplicity is what makes it interesting. Frequently, models are started with a by far too simple approach and then refined in a step-wise process.

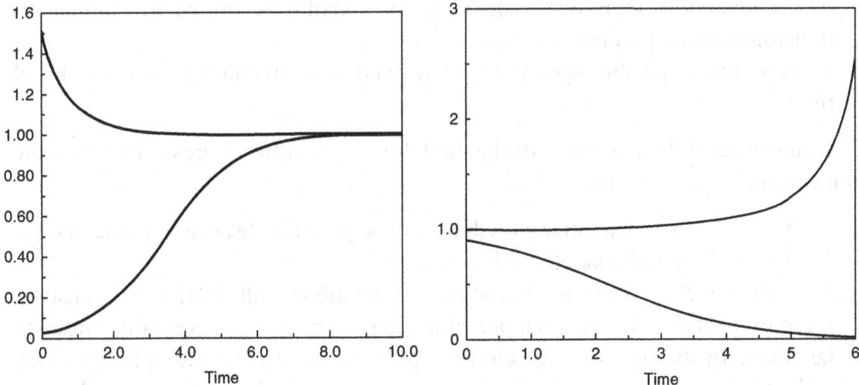

Fig. 6.4 *Left*: Temporal dynamics of (6.9) (initial condition: $N = 1.5$ for the *upper curve* and $N = 0.05$ for the *lower one*). The stable equilibrium is at $N = 1.0$. *Right*: Temporal dynamics of (6.10) (initial condition: $N = 1.0015$ for the upper curve and $N = 0.95$ for the lower one). The unstable equilibrium is at $N = 1.0$

At all intermediate steps, model behaviour is tested to make sure that even in complex situations the overview does not get lost. This approach is called rapid prototyping and is frequently applied in various modelling approaches. The Lotka–Volterra equations can be considered as a simple prototype of a predator–prey system. The equations are:

$$\frac{dPrey}{dt} = C1 * Prey - C2 * Prey * Pred \tag{6.11}$$

$$\frac{dPred}{dt} = C2 * b * Prey * Pred - C3 * Pred$$

To understand the model, we first look at the different components of the equations.

Prey represents the size of the prey population.

Pred represents the size of the predator population.

In the literature, predator and prey are frequently denoted as $N1$ and $N2$, which we will use also below.

$dPrey/dt$ represents the extent of change in the prey population at each point in time.

$dPred/dt$ represents the extent of change in the predator (pred) population at each point in time.

- $C1$, $C2$, $C3$ and b are positive constants. In typical cases, they have a small value below 1.0.
- $C1$ is the rate of increase per unit of time (growth rate) for the prey population.
- $C2$ is a predation factor, specifying what fraction of prey will be caught per unit of time depending on the size of the predator and prey population.

- *b* is a conversion factor. It specifies what fraction of the captured prey biomass is transformed into predator biomass.
- *C*3 is a mortality factor specifying what fraction of predators is lost per unit of time.

To understand the dynamics of the model we look at the expressions which are summed up in the function:

- *C*1 * *Prey* is an exponential growth term. The prey (in absence of predators, i.e. *Pred* = 0) would exponentially increase
- *C*2 * *Prey* * *Pred* is the product of both populations multiplied with a constant. It represents the frequency that predator and prey meet. *C*2 is typically chosen to be small. In the predator equation we find the same term multiplied with an additional constant *b*. This represents the amount of caught prey which is converted to predator biomass.
- *C*3 * *Pred* describes an exponential decay. Per unit of time a certain fraction of the predator population is lost. This is how the model describes death on the population level. We see the implication of the pool approach: The age of individuals does not play a role. Depending on the size of the pool always the same proportion is subtracted.

Now we look at the equations as a whole and can conclude which properties are represented in the equations and which typical properties of real organisms are ignored. We can see that, without predators, prey would grow infinitely. There is no capacity limitation. Without prey, on the other hand, the predators would decline at a constant rate. The implication is that the smaller the population becomes, the longer each remaining predator will survive. There is no limitation of life span. Another unrealistic feature of the model is that an increase in the amount of prey will always lead to an increase in the rate of prey capture and rate of predator increase. The predators have no saturation limit and thus no limit how fast they can grow. These aspects could easily be modelled more realistically, but this would make the model less simple. Now we take a look at its dynamic features and then add model properties to show how the equations can be extended.

An overview of the system dynamics can be obtained if we display the system using one axis to show the size of the prey population and the other axis for the predator population. Each combination of predator and prey values is marked by a point in the plane. This is why the number of the variables is said to indicate the *dimension of the system*. A two-dimensional system can be easily displayed in the plane. This form of display is called *state space* or *phase space*. We can get a coarse overview of what is happening in the phase space by calculating the resulting dynamics for a set of grid points and draw them as vectors. Such a graphic is called *direction field* (see Fig. 6.5).

Any calculation of the dynamics has to start with initial values for the predator and prey populations (*initial conditions*). Displayed in the phase space, the combination of the two initial values will appear as a point. The differential equations describe the successive fate of systems states emerging from this starting point.

It yields a continuous line connecting successive states. Such a line is called a *trajectory*. A trajectory is a line in the phase space showing the fate of a system, i.e. the successive sizes of the variables. The starting point is at the initial condition. Can two-dimensional dynamic systems describe oscillatory processes? Yes, and this implies that there are domains of the phase space where an increase of a variable dominates and others where a decrease dominates. These domains can be separated by a line where increase and decrease are balanced, i.e., the change of that variable is zero for an infinitely small moment. This occurs whenever the size of a variable transits a maximum or minimum value. If we consider all possible points of the phase space on which zero growth for one of the variables occurs, we obtain a line which is called the *zero growth isocline*, or just *isocline*, for that variable. There is an isocline for each variable. An intersection of the isoclines represents an equilibrium point of the system (where the rates of change of both variables are zero). The area of the phase space from which an equilibrium point is reached is called a *domain of attraction* or a *basin*. Points or areas (!) in the phase space that are approached during the system dynamics are also referred to as *attractors*.

Before we proceed to a simulation of the Lotka–Volterra equations, we employ the introduced concepts to anticipate some aspects of the dynamics of the system. We first find equations for the isoclines. To do that we only need to set the rate of change of each variable to zero and solve the resulting equation for one variable in terms of the other. Using this procedure, the prey isocline is: (see 1.243)

$$0 = C1 * Prey - C2 * Prey * Pred, \tag{6.12}$$

or

$$Pred = \frac{C1}{C2}.$$

When the size of the predator population equals $C1/C2$, the momentary change of the prey population is zero (i.e. the prey population dynamics transits from increase to decrease – or decrease to increase). The predator isocline is: (see l. 243) calculated by the same procedure to obtain:

$$0 = C2 * b * Prey * Pred - C3 * Pred, \tag{6.13}$$

or

$$Prey = \frac{C3}{(b * C2)}$$

It is apparent, that the isoclines for the Lotka–Volterra equations are both constant. The change of all prey population sizes is zero when the size of the predator population has a specific value. And the change of all predator values is zero for a specific prey population size. The intersection of these lines is an equilibrium point where no changes in population size occur. The coordinates of the equilibrium point are defined by (6.12) and (6.13).

At all other points of the phase space both variables of the system change over time. The isoclines separate the phase space into four sectors: One where both populations grow, one where both decline, and two other sectors where either prey or predator increase while the other decreases (Fig. 6.5).

The analytical result obtained in Fig. 6.5a can be refined, if the values are calculated grid-wise and the resulting direction of prey and predator vector is drawn. This yields the example of a direction field in Fig. 6.5b. It can be seen, that the system oscillates. The oscillation occurs for all initial combinations of predator and prey population sizes; however, the amplitude depends on the starting value (Fig. 6.5, Fig. 6.6). Only for a starting point precisely at the equilibrium, there would be no subsequent change in the values of the variables.

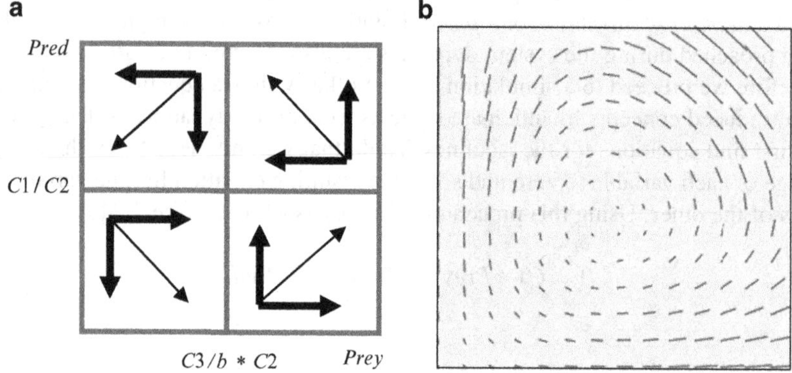

Fig. 6.5 (**a**) Lotka–Volterra isoclines. The vectors indicate in which direction prey and predator populations develop in the phase space. (**b**) Example of a direction field for the Lotka–Volterra model (6.11). The following values were used: $C1 = 0.1$, $C2 = 0.001$, $C3 = 0.1$, $b = 0.1$. The model equilibrium occurs at $Pred = 100$ and $Prey = 1{,}000$, x-axis: 0.0 ... 2,000, y-axis: 0.0 ... 200

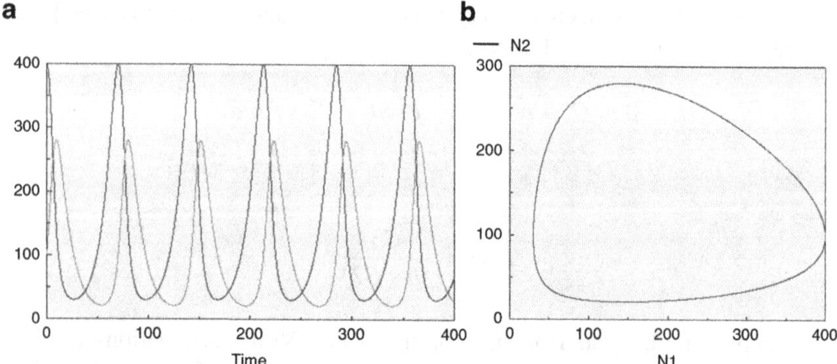

Fig. 6.6 Simulation the Lotka–Volterra system (6.11). (**a**) Display of Pred and Prey over time, (**b**) Display of Pred over Prey (trajectory drawn in the phase space). Initial conditions: $Prey = 400$, $Pred = 100$, Parameter: $C1 = 0.1$, $C2 = 0.001$, $C3 = 0.1$, $b = 0.7$

Now we have the main conceptual ideas at hand to deal with more complex cases. A strategic approach to represent more complex situations can move in two directions – to increase the number of variables, and to replace constants by particular algebraic expressions involving the variables; i.e. to introduce nonlinearities. While the Lotka–Volterra equations can be solved analytically, equations with higher nonlinearities usually require numeric approximation (simulation).

6.5 The Few Basic Types of Long-Term Behaviour in Deterministic Dynamical Systems

The determinism implied in differential equations always results in the same trajectory when identical functions and identical initial conditions are used. From this general condition we can derive relevant restrictions concerning the types of dynamics that can occur in these systems. The restrictions apply to systems which do not receive external inputs and are completely described by the equations. Such a model is called an *autonomous system*. The knowledge about these restrictions can be used to determine the minimum dimensions for certain types of dynamic behaviour to occur. In the following we introduce the most important types of dynamics in differential equations. If the description of a system's behaviour requires the use of external forcing functions, it is obvious that the system equations themselves capture only a part of what determines the dynamics of the considered variables. This usually invokes consideration of whether a more complete description could be achieved in further studies.

6.5.1 Dynamic Properties in One-Dimensional Systems

One-dimensional systems are the most restricted concerning the potential dynamics. The only possibilities are to (a) approach zero, (b) approach infinity or (c) approach a steady state (steady state equilibrium).

There can be more than one equilibrium point in a particular equation if nonlinearities are involved. If there is more than one equilibrium point, the initial conditions are crucial which of the alternative equilibria is approximated (see below: domain of attraction).

Collapse or Explosion

A simplistic way to describe collapsing or exploding dynamics is by exponential increase or decline as the only possible alternatives of the dynamics. An example was presented in (6.8) with $N_{\text{initial}} > 0$ and $c > 0$ for explosion, $c < 0$ for collapse, and marginal stability for $c = 0$. This type of behaviour can also occur in systems with a higher number of variables.

Stationary States (Single or Multiple Equilibria)

The third alternative, that a one-dimensional system approaches a nontrivial equilibrium (equilibrium at a finite value of N), requires in a minimum setting the combination of linear increase and exponential decrease (6.14). The introduction of higher nonlinearities can lead to multiple equilibria

$$\frac{dN}{dt} = C1 - C2 * N \tag{6.14}$$

with $C1$ and $C2 > 0$. A stable dynamic equilibrium exists for $N = C1/C2$. The logistic equation [see (6.9)] has also a stable equilibrium.

6.5.2 Dynamic Properties in Two-Dimensional Systems

In two dimensional systems, the same types of dynamic behaviour can occur as in one-dimensional systems. Oscillations are an additional type of dynamics that are not found in 1D systems. There are different types of oscillations. The simplest oscillator results from a positive and negative coupling of two variables, as we saw in (6.8) and Fig. 6.2.

In case of models that include nonlinearities, the oscillations can either increase or decrease in amplitude over time. At the transition point between both types there are so-called *marginal stable oscillations*, which neither increase nor decrease in amplitude, but maintain the amplitude set by the initial conditions. This is the dynamic behaviour of the Lotka–Volterra equations (Fig. 6.6). An additional type of oscillation occurs, if a dynamic behaviour which leads to increasing oscillation in a certain part of the phase space is limited by a region of the phase space where a decreasing amplitude prevails. Over the long term, a specific cycle results, independent of the initial conditions. This is called a stable *limit cycle*. This dynamic type will be presented in a simple model example below. An unstable limit cycle is also possible. It would be more difficult to observe, since the slightest, infinitesimal deviation would induce a transit to either one of two different alternative states, which could be explosion, collapse or another stationary state.

Oscillations with Damped Amplitude (Approximating a Steady State)

The following system (6.15) oscillates with decreasing amplitude and approaches an equilibrium point (stationary steady state). Over time, the oscillations decay. To obtain this behaviour, the Lotka–Volterra equations (6.11) are used and it is additionally assumed that the growth capacity of the prey is limited.

$$\frac{dPrey}{dt} = C1 * Prey\frac{(K - Prey)}{K} - C2 * Prey * Pred \tag{6.15}$$

$$\frac{dPred}{dt} = C2 * b * Prey * Pred - C3 * Pred$$

The form used in the prey equation is equivalent to the logistic curve, as given in (6.9), which can be verified by multiplying and replacing $C1/K$ by $C2$. Since the behaviour is more easily understood if the size of the environmental capacity can be directly specified as a constant, the form in (6.15) is frequently used in models. Figure 6.7 presents a simulation result of (6.15).

Oscillations with Increasing Amplitude

The system described in (6.16) exhibits oscillations with increasing amplitude, successively moving away from an unstable stationary state. Since the amplitude grows exponentially, the effect can be seen best by starting the system close to the equilibrium (Fig. 6.8). In this case, the Lotka–Volterra equations are extended by a saturation term. This term describes increasing predator growth rate with increasing prey population, up to a saturation level. The same form is frequently used to describe enzyme kinetics (Michaelis–Menten equation). One constant specifies the maximum achievable rate (here: $C11$) and the other one the half saturation concentration (here: population size at which half of the maximum rate occurs, CS).

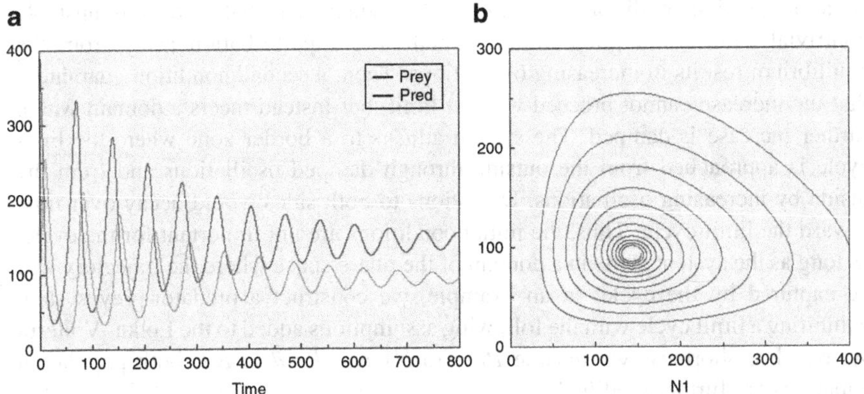

Fig. 6.7 Damped oscillations obtained from a simulation of (6.15). (**a**) Display of Pred and Prey over time. (**b**) Display of Pred over Prey. Initial conditions: $Prey = 400$, $Pred = 100$, Parameter: $C1 = 0.1$, $C2 = 0.001$, $C3 = 0.1$, $b = 0.7$, $K = 2{,}000$

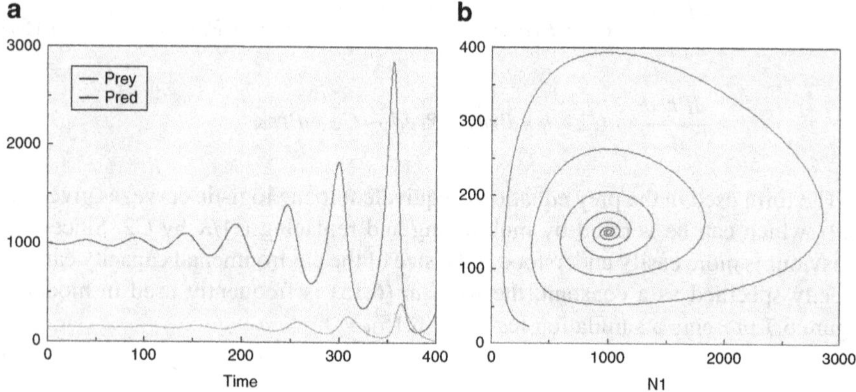

Fig. 6.8 Increasing oscillations obtained from a simulation of (6.16). (**a**) Display of Pred and Prey over time. (**b**) Display of Pred over Prey. Initial conditions: $Prey = 1{,}000$, $Pred = 152$. The unstable equilibrium is at $Prey = 1000$, $Pred = 150$. Parameter: $C1 = 0.1$, $C11 = 0.1$, $CS = 1{,}000$, $C2 = 0.001$, $C3 = 0.1$, $b = 0.7$

$$\frac{dPrey}{dt} = \left(C1 + \left(\frac{C11 * Prey}{CS + Prey}\right) * Prey\right) - C2 * Pred * Prey \tag{6.16}$$

$$\frac{dPred}{dt} = C2 * b * Prey * Pred - C3 * Pred$$

Limit Cycles (Periodic Equilibria)

A limit cycle can be obtained when additional properties are introduced to the equations. A first condition for a limit cycle frequently is that there is an unstable nontrivial equilibrium point of the system, i.e. a perturbation away from the equilibrium results in increasing oscillations. Then, a second condition guarantees that the increase cannot proceed without limit, but instead meets a domain where further increase is damped. The system adjusts to a border zone where the limit cycle is approached from the outside through damped oscillations and from the inside by increasing oscillations. Deviations to both sides would decay over time toward the limit cycle. Thus, the initial conditions are not important for the cycle, as long as the system starts in a domain of the phase space where the trajectory will be captured by the cycle. In an example, we construct a predator-prey system exhibiting a limit cycle with the following assumptions added to the Lotka–Volterra system. For shortness we rename *Prey* to *N*1 and *Pred* to *N*2; both population equations are further modified.

A hyperbolic function strongly decreases predation probabilities when the prey population is small. This can be due to the effect of a refuge with limited capacity where the prey would be relatively safe. To represent this, the predation function

$C2 * N1 * N2$ is modified (6.17) and added to the predator as well as the prey equation

$$\frac{C2 * N1 * N2}{\frac{A+N1}{N1}} \tag{6.17}$$

At $N1 = A$ the predation efficiency is reduced to 50%. This rate decreases for smaller $N1$. Further modifications are required: (a) the introduction of a saturation function into the growth term. An appropriate selection of constants can ensure that the oscillations close to the equilibrium point are unstable (this can lead to an increasing amplitude around the equilibrium, as was seen in (6.16) and Fig. 6.8); (b) the introduction of a logistic term (capacity limitation), preventing an infinite increase of the oscillations, as it was seen in (6.15) and Fig. 6.7. Then, the following differential equation system results (6.18), where increasing and decreasing oscillations can be observed (Fig. 6.9).

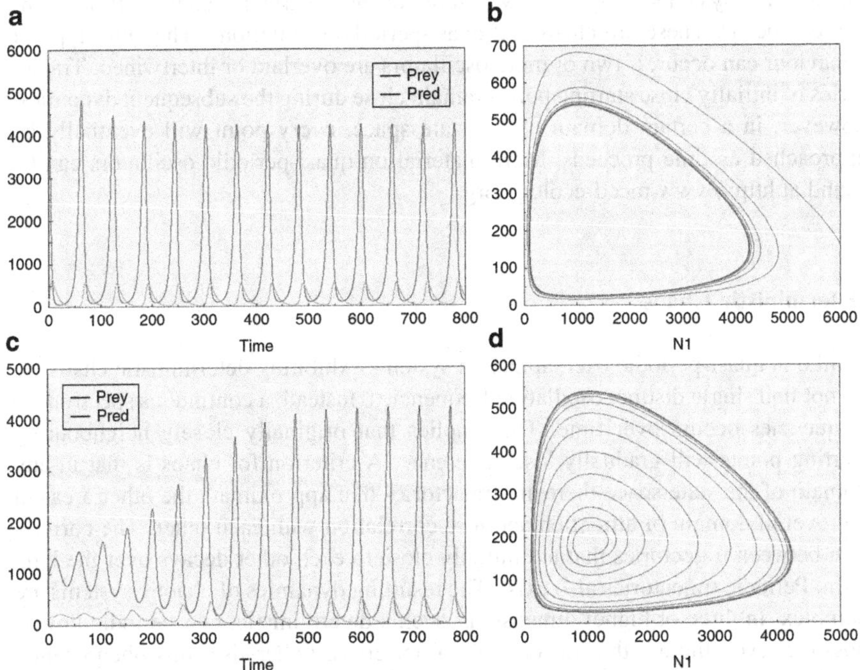

Fig. 6.9 Simulation results for (6.18). *Left*: (**a**) and (**c**) show the display of predator ($N2$) and prey ($N1$) over time. *Right*: (**b**) and (**d**) show the trajectories in the phase space, Predator ($N2$) over prey ($N1$). If we use the initial condition $N1 = 6{,}000$, $N2 = 160$, the amplitude decreases towards a limit cycle (*top*: **a** and **b**). If we use the initial condition $N1 = 1{,}000$, $N2 = 160$, the amplitude increases towards a limit cycle (*bottom*: **c** and **d**)

$$\frac{dN1}{dt} = \left(C1 + \frac{C11 * N1}{Cs + N1}\right) * N1 * \left(\frac{K - N1}{K}\right) - \frac{(C2 * N1 * N2)}{\frac{(A+N1)}{N1}} \tag{6.18}$$

$$\frac{dN2}{dt} = \frac{(b * C2 * N1 * N2)}{\frac{(A+N1)}{N1}} - C3 * N2$$

For the following parameter values (6.18) exhibits a globally stable limit cycle: $C1 = 0.1, C11 = 0.2, Cs = 1{,}000, K = 10{,}000, C2 = 0.001, A = 50, b = 0.1, C3 = 0.1$.

6.5.3 Dynamic Properties in Three-Dimensional or Higher Dimensional Systems

There are dynamic phenomena that cannot be observed in one-dimensional or two-dimensional systems. In systems with three or more variables additional phenomena can occur. These are chaos and quasi-periodic oscillations. The latter type of behaviour can occur, if two or more oscillators are overlaid or intertwined. Trajectories of initially close starting points remain close during the subsequent dynamics; however, in a certain domain of the state space, every point will eventually be approached as time proceeds. More material on quasi-periodic oscillators can be found at http://www.mced-ecology.org.

Deterministic Chaos

Unlike in quasi-periodic oscillations, in systems exhibiting deterministic chaos we do not find single distinct oscillation frequencies. Instead, a continuous spectrum of frequencies occurs over time. This implies that originally closely neighbouring starting points will gradually lose coherence. A criterion for chaos is that in any domain of the state space there are trajectories that approximate the other areas in the overall domain of attraction and lose correlation with each other. The correlation between trajectories that are initially close to each other decays over the long term. Periodic trajectories also exist. The resulting dynamics of chaotic systems are complex. In three or higher dimensions, such a deterministic, non-periodic flow is possible, even though the trajectories do not cross. Otherwise, this phenomenon would not be possible in deterministic systems. An example of a chaotic system inherent in the equations of a predator–prey system was discovered by Gilpin 1979. It was published some time after Lorenz (1963), who had discovered chaotic behaviour in dynamic systems for the first time when modelling turbulent atmospheric processes.

$$\frac{dN1}{dt} = N1(r1 - a11 * N1 - a12 * N2 - c1 * N3) \tag{6.19}$$

$$\frac{dN2}{dt} = N2(r2 - a21 * N1 - a22 * N2 - c2 * N3)$$

$$\frac{dN3}{dt} = N3(b(c1 * N1 + c2 * N2) - d)$$

with $r1 = 1$; $r2 = 1$; $a11 = 0.001$; $a12 = 0.001$; $a21 = 0.0015$; $a22 = 0.001$; $c1 = 0.01$; $c2 = 0.001$; $b = 0.5$; $d = 1$ and the initial conditions $N1 = N2 = N3 = 50$.

The Gilpin equations (6.19) describe a predator population ($N3$) and two competing prey populations ($N1$, $N2$). The only new aspect to what we have discussed so far is the inclusion of competition. It relates to logistic growth, in which growth of a population is limited to a finite total carrying capacity, where the increase of each competing population is limited by both its own size, and the size of the competing population, in terms of a competition coefficient. Without predators, in Gilpin's model one prey population would be outcompeted and go extinct. Both populations persist through the predator's influence, which modulates the competition effect. This model was used as a default for the "interaction engine", a simple differential equations integrator of the POPULUS software (Alstad 2007). Figure 6.10 shows the simulation of the three variables over time. In Fig. 6.11 it looks "as if" the trajectories would cross, but this is only because a projection of only two of the three variables in a plane was shown ($N2$ over $N1$). There are also other ways in which deterministic chaos can occur in the interaction of three

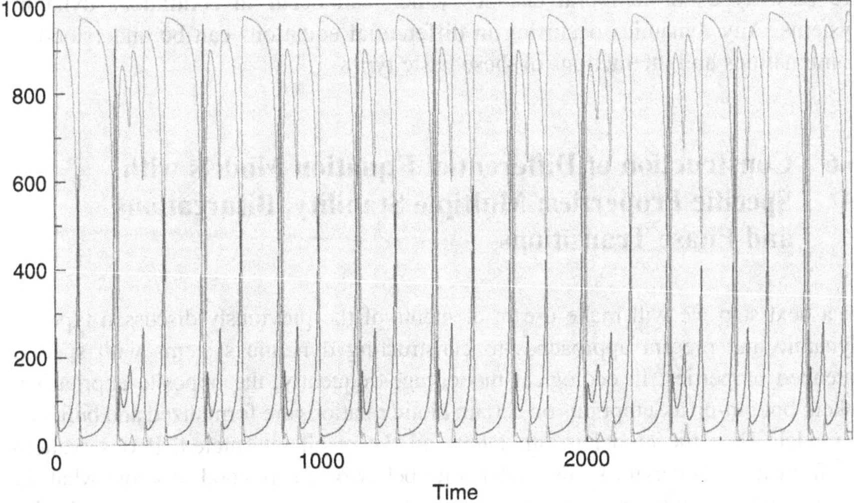

Fig. 6.10 Gilpin's Spiral Chaos Attractor. The simulation results of (6.19) with the initial conditions $N1 = N2 = N3 = 50$. $N1$, $N2$ and $N3$ are plotted over time (3,000 time steps)

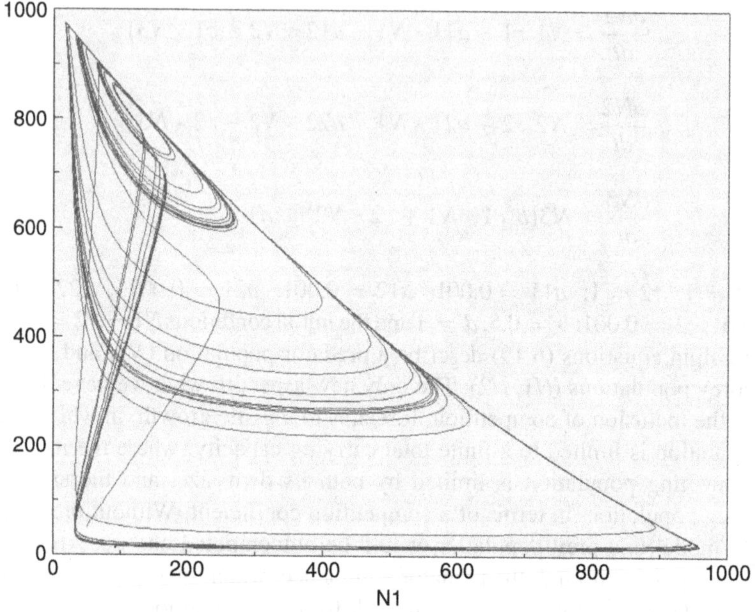

Fig. 6.11 2D-view of Gilpins Spiral Chaos Attractor. The simulation results of (6.19) with the initial conditions $N1 = N2 = N3 = 50$. $N2$ is plotted versus $N1$ (3,000 time steps)

variables. Hastings and Powell (1991) described a three species food chain model exhibiting chaotic dynamics.

Collapse, explosion, single or multiple stationary states, oscillations and chaos are basic types of temporal changes which can occur in continuous dynamic systems. Any dynamics occurring in differential equations can be understood as combinations and interactions of these basic types.

6.6 Construction of Differential Equation Models with Specific Properties: Multiple Stability, Bifurcations and Phase Transitions

In a next step we will make use of elements of the previously discussed types of dynamic and present approaches to constructing dynamic systems with specific intended properties. In ecological modelling, frequently, the opposite approach is taken: bottom-up assumptions on certain biotic relations are formalized and built into a model. Then the model is simulated and the result interpreted. If observations confront us with certain kinds of dynamic behaviour, it is good to know what are the minimum conditions for particular properties to occur in a model, and what specific structures are crucial for a certain dynamic behaviour. When bottom-up and top-down elements can be combined, it will facilitate a more directed development. Especially for two-dimensional systems, the isoclines (see Sect. 6.4) are of particular importance in model construction. They allow one to see when small changes in the system, e.g. the size of a single parameter, can give rise to basic changes of the overall system behaviour. Intersections of the isoclines are the points where the temporal changes in the system are zero. If such an intersection emerges or disappears as a function of changes in parameter values, this represents an important change in system behaviour. The possibilities of isoclines intersecting are limited if isoclines are straight lines. Nonlinearities in the system can give rise to curved isoclines. This can bring various kinds of interesting dynamic properties.

6.6.1 Logistic Growth

An interesting nonlinearity consists in the introduction of a negative quadratic term. As we saw above, when this is added to a simple exponential equation, it yielded the logistic curve (6.9). Simulating the equation, we saw, that starting with very small N, a rapid increase occurred that transitioned towards a stationary state. Starting with a very large N, we observe a declining trend, which stabilizes to the same steady state (Fig. 6.3, logistic growth). Logistic terms are frequently used in ecological models to simulate limited carrying capacities.

6.6.2 Multiple Equilibria States and Hysteresis

The mathematical term "catastrophe" refers to a rapid transition of a system equilibrium point from one state to another. It can occur, if the change of a condition causes the system to leave a stable domain beyond a critical point, after which a *bifurcation* occurs. Then, the system shifts towards another alternative stable state. Such a situation is shown in Fig. 6.12. Here the upper and the lower branch of the graph (solid lines) are the stable state regions of the system. Once the induced change in conditions forces the system to cross a bifurcation point (either of the two black dots in the graph), the system shifts towards another alternative state. The dashed line in the graph represents the unstable region between the two alternative system states. One consequence is that such a system can reach multiple equilibrium points, only depending on the impact of the driving forces (see Fig. 6.12). These dynamics can result in *hysteresis* behaviour, and they have been intensively studied in lake ecosystems. Scheffer et al. (1993) found such concurrent alternative equilibrium states exist in shallow lakes, which tend to be either algae or macrophyte dominated. The transitions between these alternative states can occur through changes in the nutrient load. Scheffer et al. found, also, that the stability of a given stable state (see the upper and lower branches in Fig. 6.12) prevents an

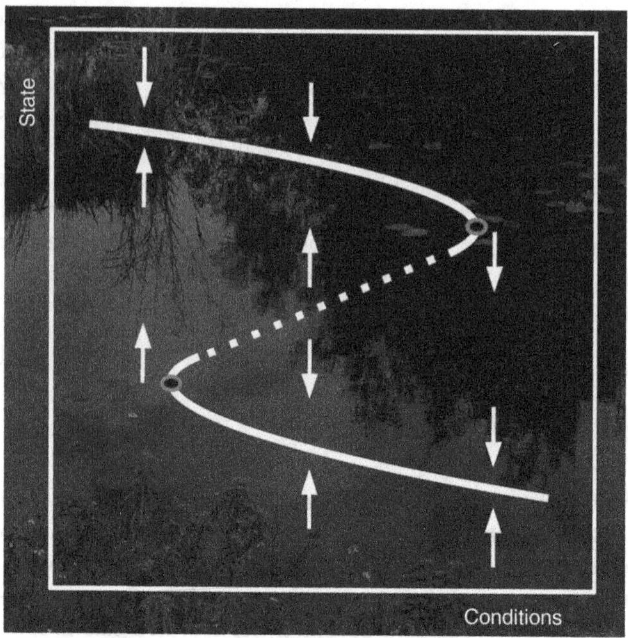

Fig. 6.12 Dynamics of alternative ecosystem states, sensu Scheffer et al. (1993) (further explanations in the text)

immediate return to the former stable state. For example, if the upper branch represents a macrophyte-dominated lake and the lower branch an algal-dominated lake, and nutrient input increases along the *x*-axis, then the nutrient input to the lake must be reduced to a much lower level (left black dot) to cause a transition from algal- to macrophyte dominated than the nutrient level (right black dot) at which the reverse transition occurred. Dong et al. (2002) found the same dynamics for a system of algae and periphyton in a model which describes functional group interactions in the Everglades marshland. Currently this kind of dynamics is intensively discussed for transitions in reefs systems from coral dominated to algae dominated states (see also Chap. 17).

In the following, we use an abstract model to investigate the relations and conditions of hysteresis. The calculation of the isoclines will facilitate the overview. Equation (6.20) represents a predator-prey model with nonlinear interactions for the prey and a logistic growth of the predators. In (6.20), the predator death rate b plays the role of the critical parameter leading to the transition between different equilibria points.

$$\frac{dN1}{dTime} = \left(A * \left(-P3 * N1^3 + P2 * N1^2 - P1 * N1 + P01\right) + P02\right) * N1 - N1 * N2$$

$$\frac{dN2}{dTime} = N1 * N2 - b * N2^2 \qquad (6.20)$$

The zero growth isocline for $N1$ is

$$N2 = A * \left(-P3 * N1^3 + P2 * N1^2 - P1 * N1 + P01\right) + P02 \tag{6.21}$$

and for $N2$ it is

$$N1 = b * N2 \tag{6.22}$$

with $A = 2.0$; $P3 = 4.0$; $P2 = 6.0$; $P1 = 2.5$; $P01 = P02 = 0.25$; b between 1.0 and 3.0

The parameter b influences the slope of the $N2$-isocline (6.22). There is a domain with one equilibrium where prey numbers, $N1$, are low (Fig. 6.13a), which is a stable attractor. In Fig. 6.13b, where b has been increased, we see that there can be three intersections of the isoclines, of which the two outer ones represent stable equilibrium points. The system will remain in the original stable equilibrium until b increases further and the middle intersection points disappear. This situation is shown in Fig. 6.13c, where also the state of the system transitions to the other stable equilibrium, with higher $N1$, can be seen. Figure 6.13d shows the results of increasing and decreasing b, which leads to an array of different transition points.

In the next model we construct a situation, with three concurrent equilibria and four transition paths between them (6.23).

$$\frac{dN1}{dTime} = N1 * \left(A * \left(-N1^5 + P4 * N1^4 - P3 * N1^3 + P2 * N1^2 - P1 * N1 + P0\right) + N1\right) - N1 * N2$$

$$\frac{dN2}{dTime} = N1 * N2 - b * N2^2 \tag{6.23}$$

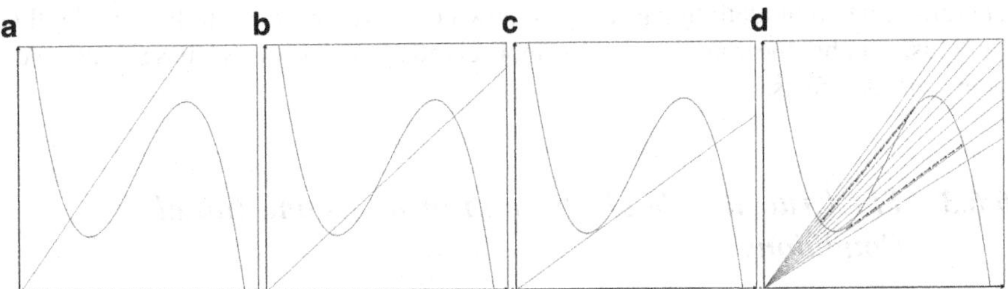

Fig. 6.13 Three different situations for different values of b (6.20). The parameter b determines the slope of the isocline for $N2$ (6.22). (**a**) Only one equilibrium exists where $N1$ has a low level; (**b**) two alternative equilibria exist. Depending on the initial conditions, either one of the equilibria is approached. (**c**) only one equilibrium exists where $N1$ has a high level. (**d**) when the system is in the low level equilibrium state, it moves to the upper one only, if the isocline is beyond the lower dotted line. Being in the upper equilibrium, it would move back only if the isocline is above the *upper dotted line*. The displayed area is $x = 0.0 \ldots 1.0$; $y = 0.0 \ldots 0.5$

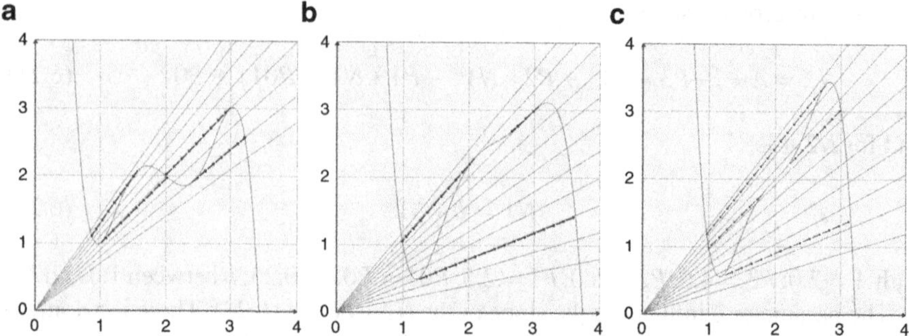

Fig. 6.14 The isoclines were obtained for (6.23) with different parameter. (**a**) A system with three alternative equilibria and four transitions between them (A = 2.0, P4 = 10.0, P3 = 38.0975, P2 = 68.585, P1 = 58.0725, P0 = 18.585). *B* should range upwards and downwards between 0.5 and 2.0. (**b**) A system with three alternative equilibria but only three transitions between them was obtained for A = 1.2, P4 = 11.5, P3 = 51.38, P2 = 110.636, P1 = 113.604, P0 = 43.848. *B* should range upwards and downwards between 0.5 and 3.0 In one direction, there is one transition, the reverse goes in two distinct steps. (**c**) A system with three alternative equilibria but only two transitions were obtained using the parameter A = 3.5, P4 = 10.0, P3 = 38.99, P2 = 73.94, P1 = 67.89, P0 = 23.94. B should range upwards and downwards between 0.5 and 3.0. The trajectory starting at the initial conditions $N1 = N2 = 2.04$ exhibits a damped oscillation approaching an equilibrium at $N1 = N2 = 2.0$

Figure 6.14a shows a parameterization where four transitions occur. Figure 6.14b demonstrates a parameterization of (6.23) where in one direction there are two transitions between different equilibria, whereas in the other direction there occurs only one. Figure 6.14c shows a situation where a central stable branch is fully masked as the system jumps over it when transiting between alternative states. This central "hidden" stable branch would become apparent only when using initial conditions which are close to this equilibrium when b is set to 1.0.

When more complex nonlinearities are involved, we have seen, that hysteresis effects are not only limited to transitions between two alternative equilibria. Therefore, when modelling partially unknown contexts, we need to be sufficiently cautious, whether there are previously unobserved nonlinearities as these could lead to hysteresis effects.

6.6.3 The Crucial Role of Phase Transitions and Initial Conditions

If there are multiple stable states in a model, normally only one becomes directly apparent. The part of the phase space from which a particular equilibrium is approached, is called *domain of attraction*, or sometimes also *basin of attraction*.

In its basic form, the Lotka–Volterra system is a marginal case. As long as positive parameters are used, in principle the type of behaviour is always the same: marginally stable oscillations. This holds even though the speed of change can become so extremely rapid (or slow), that standard numeric approximations fail. From what has been previously discussed, we can derive conclusions for *phase transitions*. In nonlinear systems with multiple equilibria, parameter changes can lead to the emergence of new alternative equilibria or induce the destabilization of previously stable states. In particular, models with more complex nonlinearities facilitate the occurrence of complex combinations. A still relatively simple type of transition is the breaking up of a stable equilibrium and the emergence of a limit cycle. This transition was named after the Austrian mathematician Eberhard Hopf. For (6.18) we can observe such a Hopf-bifurcation when changing the parameter $C3$. For small and large parameter values the limit cycle vanishes and the system approaches a steady state instead of a limit cycle (Figs. 6.9 and 6.15).

To be sufficiently careful in the interpretation of a model, the modeller needs to have an overview of the potential range of dynamic behaviour. Since ecological models are always a simplification focussing on a limited set of interactions and conditions ("*ceteris paribus*" – i.e. all other conditions remain the same), it is useful to know about potential implications. This applies not only to the field work, where the empirical background for structural and functional model specifications are conceptualized – the number of variables, the quantitative relations and the parameter ranges. It is also necessary to have an intuition about the potential properties of dynamic systems representations.

For a modeller it is not enough to be able to write down equations and let the computer evaluate them numerically. An appropriate interpretation of simulation results should also take into account that slight parameter changes could lead to phase transitions, which would alter the overall dynamics. Without knowing the potential effects of the phase transitions between alternative equilibria (which are quite difficult to experimentally investigate in the field, if the system state cannot be arbitrarily manipulated), adequate interpretation of the model results might be difficult. This leaves much potential for uncertainty. However, other difficulties can be managed by understanding the equations and protecting against simulation artefacts. Some of these strategies to tackle "standard" uncertainties are compiled at http://www.mced-ecology.org and in Chaps. 2 and 23.

6.7 Solving Differential Equations Analytically and Numerically

For most ecological questions where dynamic models are developed, a mathematical solution is not possible and a numerical evaluation of the equations is required. Analytical solutions exist only in relatively simple cases; i.e. linear equations and some simple nonlinear equations. These include the logistic growth function, and the Lotka–Volterra model. More complex, nonlinear ecological

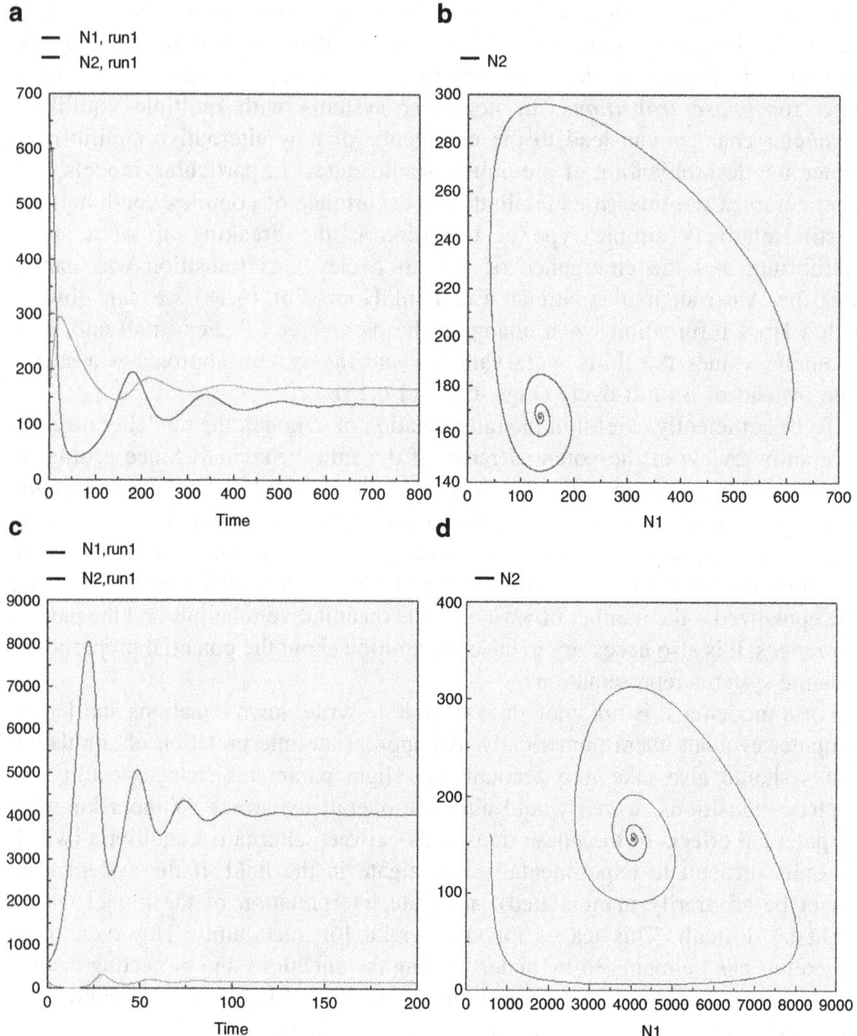

Fig. 6.15 Hopf bifurcation: The system described by (6.18) exhibits a limit cycle. If the parameter *C3* is decreased from *0.1* (original version shown in Fig. 6.11) to *0.01*, the limit cycle vanishes and a stable equilibrium emerges (*upper two figures*). The same happens in the equation if the parameter is increased to *0.4* (*lower two figures*). In between there are transition points where the phase shift between approximating a stable equilibrium and a limit cycle occur

models require a numerical approach via simulation. When assessing an ecological context with analytical approaches, the level of natural complexity has to be reduced for tractability. A negative consequence of this simplification might be the loss of the dynamical interaction structures. A prominent exception to this dilemma is represented by the trophic level analysis using steady state models as explained in Chap. 5.

For the understanding of time-dependent dynamics, numerical approximation is necessary. This, of course, poses the problems both that numerical results may be in error and that generalizations based on numerical results are difficult. Plausibility considerations are to some extent inevitable. Personal experience of the modeller who works with differential equations is not only the basis to adequately understand and formulate the code, but also generates a kind of gut feeling for solutions that are adequate to capturing the relevant ecological patterns and processes. A brief outline on some of the major points to be considered for successfully working with this model category can be found at http://www.mced-ecology.org.

With the beginning of the elementary contributions from Malthus (late eighteenth century), James Lotka and Vito Volterra (1925, 1926), and Fisher (1930–1940) the method of differential equations is well established in biology and ecology; for a detailed historical overview: see Chap 3; for further developments: see Chap. 7; example applications are given in Chaps. 17–20.

Since their introduction, differential equations have had a leading role in ecological modelling. During the last decades the field of ecological modelling has expanded considerably. Today, differential equations still contribute to scientific progress, though side by side with a wide variety of other approaches which are outlined in the following chapters.

Further Readings

Many textbooks exist on ordinary differential equations, often with a very specific focus. A list of books relating to the ecological context can be found at http://homepage.ruhr-uni-bochum.de/michael.knorrenschild/embooks.html (Knorrenschild M (2010) List of textbooks on ecological modelling). From our perspective we would select the following books and webpages that expand on the contents provided in this chapter:

Edelstein-Keshet L (2004) Mathematical models in biology, 2nd edn. SIAM, 586 p

Jeffries C (1989) A workbook in mathematical modeling for students of ecology. Springer, Heidelberg

Kot M (2001) Elements of mathematical ecology. Cambridge University Press, Cambridge, http://www.cambridge.org/us/catalogue/catalogue.asp?isbn=9780521001502

Sharov A (n.d) Quantitative population ecology. On-Line Course. http://home.comcast.net/~sharov/PopEcol/popecol.html

William SC, Gurney WSC, Nisbet RM (1989) Ecological dynamics. Oxford University Press, Oxford, New York. http://www.stams.strath.ac.uk/ecodyn/

Wiki book on differential equations. http://en.wikibooks.org/wiki/Differential_Equations

Yodzis P (1989) Introduction to theoretical ecology. Harper & Row, New York

Chapter 7
Partial Differential Equations

Michael Sieber and Horst Malchow

Abstract Spatially homogeneous processes of change are the subject of the preceding chapter. Partial differential equations are one method to model the interplay of these processes with spatial phenomena such as movement of individuals and/or a heterogeneous environment. Random motion of organisms might be described as diffusion, and directed motion as advection. The latter can be composed of locomotion and motion of the surrounding medium. The focus of this chapter is on classical systems of no more than two interacting and diffusing populations. The potential of such systems to exhibit spatiotemporal pattern formation is studied.

7.1 Introduction

Mathematical models for spatially homogeneous processes, with their potential for multiple steady states and complicated temporal dynamics, are the cornerstone of ecological modelling (see Chap. 6). Yet only by taking into account the spatial dimension of species growth, interaction, locomotion and transport is the full diversity of population dynamics realized. The possible spatiotemporal dynamics include stationary and dynamic patchy patterns, regular and irregular oscillations, propagating fronts, target patterns and spiral waves amongst others. Historically, possibly the best-known examples for spatiotemporal patterns come from physics and physical chemistry, cf. the Bénard convection cells (Bénard 1900) and the waves in the Belousov-Zhabotinskii reaction (Belousov 1959). Similar patterns have been found in biological and, in particular, population-dynamical systems, such as the bioconvection of up-swimming microorganisms (Hill and Pedley 2005; Wager 1911), travelling waves in cyclic populations (Sherratt and Smith 2008), the wavy dynamics of amoeba (Gerisch 1968) and striped vegetation patterns (White 1971), cf. Fig. 7.1. Partial differential equations (PDEs) are one way to incorporate the spatial dimension

M. Sieber (✉)
Institute of Environmental Systems Research, Department of Mathematics and Computer Science, University of Osnabrück, 49076 Osnabrück, Germany
e-mail: msieber@uni-osnabrueck.de

F. Jopp et al. (eds.), *Modelling Complex Ecological Dynamics*,
DOI 10.1007/978-3-642-05029-9_7, © Springer-Verlag Berlin Heidelberg 2011

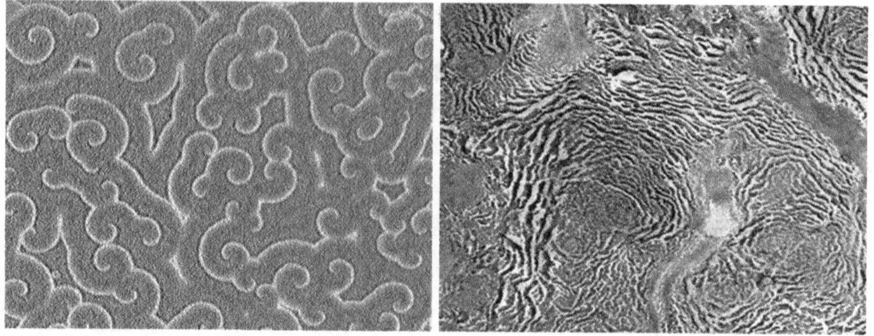

Fig. 7.1 *Left*: Spirals in an amoeba population (*Dictyostelium discoideum*). The base line of the photo is about 28.9 mm (Courtesy of Christiane Hilgardt and Stefan C. Müller, University of Magdeburg). *Right*: Satellite image of tiger bush in Niger, the darker lines of woodland are on average about 20–40 m wide and 50–100 m apart (Courtesy of the US Geological Survey)

in models of biological dynamics, and as such they are related to coupled map lattices and cellular automata. Often, PDEs take the form of reaction-diffusion equations, especially if individuals are assumed to perform a random walk, similar to small particles in a fluid whose molecules are in constant thermal motion. PDEs can also be used to model directed motion in the form of advection and even random environmental fluctuations. The next few pages give an overview of some basic PDE models and the interesting range of patterns they may generate.

7.2 Single Population Models

In the case of a single species homogeneously distributed in space, the rate of change of the species population density $Y = Y(t)$, that is, its temporal behaviour, is described by the ordinary differential equation (ODE):

$$\frac{dY}{dt} = f(Y, t, c) \tag{7.1}$$

Here, f describes all processes relevant for the species growth, i.e. reproduction, competition and predation. In general, f will depend on a set $c \in \mathbf{R}^k$ of biological parameters, like birth and mortality rates. Additionally, the growth rate parameters of the species may also explicitly depend on the time t, i.e. reflecting seasonality of reproduction or increased mortality in harsh winter conditions. In the following, we always assume that f does not explicitly depend on time, and the variables t and c are dropped from the notation. A real-valued function Y is a solution of this equation if its temporal derivative satisfies (7.1). In order to uniquely identify a particular solution, it is also necessary to specify the initial population density condition in the form $Y(0) = Y_0$.

Now, for a population that is inhomogeneously distributed across its habitat, the population density $Y = Y(t, \mathbf{x})$ of a single species changes not only over time t, but also with spatial location $\mathbf{x}$. In biological scenarios, the domain or habitat will usually be some bounded subset of three-dimensional space. In this case, $\mathbf{x}$ usually denotes the Cartesian coordinates of a point in that domain. Let us first consider only one-dimensional spatial domains, in which case the spatial location is simply denoted by the real number x.

If it is assumed that the motion of individuals on this domain can be approximated by a random walk, the rate of change is given by the reaction-diffusion equation:

$$\frac{\partial Y}{\partial t} = f(Y,x) + D\left(\frac{\partial^2 Y}{\partial x^2}\right) \tag{7.2}$$

As in the non-spatial case, the growth term f describes the species growth, which now may additionally depend on the spatial location x. To shorten notation, the variable x is also usually omitted in the following. The second term now describes the species dispersal, usually down its own spatial density gradient. The diffusion coefficient D reflects how motile the individuals of the population are. The fact that the population density now depends on two independent variables is reflected in the partial derivatives in (7.2), one with respect to time t and the other of second order with respect to the spatial variable x. A solution to this equation is a real-valued function Y, whose partial derivatives satisfy (7.2) and which has a given initial population distribution $Y(0, x) = Y_0(x)$. If the spatial domain is bounded with boundary δ, the solution additionally needs to fulfil suitable boundary conditions. An important special case is referred to as the no-flux boundary conditions, given by:

$$\frac{\partial Y(t,x)}{\partial n} = 0$$

for all points $x \in \delta$. Here, with respect to the outward pointing normal vector, the partial derivative is perpendicular to the boundary. This boundary condition simply reflects the assumption that no individual leaves or enters the domain through the boundary; i.e., because the population is physically confined to a certain habitat or the habitat is surrounded by a hostile environment. In the next sections we will see how the form of the growth term f determines which spatiotemporal patterns these solutions may exhibit. Note, that for all examples we assume no-flux boundary conditions.

7.2.1 Exponential Growth

For the Malthusian assumption (Malthus 1798) of exponential growth of a single species, the growth term takes the form:

$$f(Y) = rY \tag{7.3}$$

The intrinsic growth rate r is the only parameter. For $r > 0$, the population will grow explosively to arbitrarily high values at every location x. If the initial population distribution is a localized patch, this gives rise to a spatiotemporal wave moving outward from the initial patch. Luther (1906) proved that the speed v_F of this wave front of population spread is:

$$v_F = 2\sqrt{rD} \tag{7.4}$$

Clearly, unlimited growth, as predicted by (7.3), is never observed in nature. However, the speed of the wave front obtained from this model is seen again in the next section in a more realistic setting.

7.2.2 *Logistic Growth*

The concept of a carrying capacity K of the environment was introduced by Verhulst (1838). This yields a growth saturation for higher population densities, effectively limiting the population density to a maximal finite value. The corresponding growth term reads:

$$f(Y) = rY\left(1 - \frac{Y}{K}\right) \tag{7.5}$$

Logistic growth is a widely used standard assumption for population models. At carrying capacity the intrinsic growth rate vanishes; that is, $f(K)/Y = 0$ and this is the only stable steady state of the system.

In combination with diffusion, logistic growth was first investigated by Fisher (1937) as a model for the spread of genes in a population and simultaneously by Kolmogorov et al. (1937). Any initial, smooth density distribution from capacity K to zero will form a wave front with the speed given by (7.4), finally filling the whole space with the population at its carrying capacity. An illustration for a one-dimensional domain is shown in Fig. 7.2.

7.3 Allee Effect

Allee (1931) found that population growth is optimal and highest at medium population densities. This has been called the *Allee effect*. In its stronger formulation it implies the existence of a minimal viable population size (Courchamp et al. 2008). A population with a density below this value will die out, whereas a population with a size above this value will grow to its carrying capacity. The growth function reads:

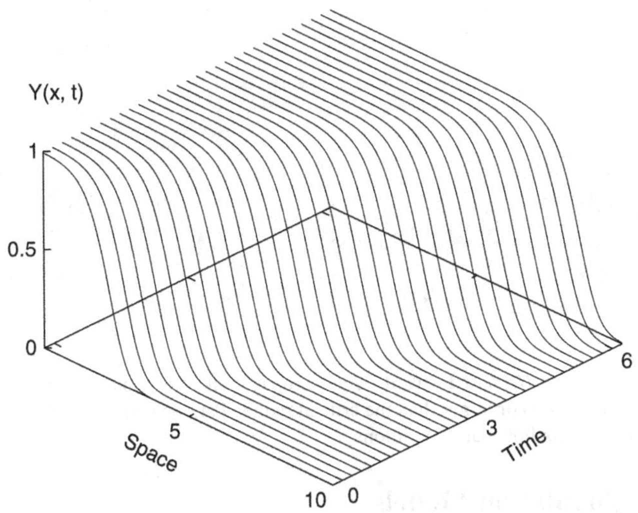

Fig. 7.2 Spatial propagation of a logistically growing population for $r = 3, K = 1, D = 10^{-1}$, all parameters given in arbitrary units (a.u.), no-flux boundary conditions

$$f(Y) = rY\left(1 - \frac{Y}{K}\right)\left(\frac{Y}{K_-} - 1\right) \tag{7.6}$$

where $K\_$ is the minimal viable population density. The system is bistable, with extinction as well as carrying capacity as stable steady states. This changes the dynamics (not only locally), such that, in the absence of noise, the initial condition determines the final steady state. With diffusion, the same initial, smooth density distribution as in Sect. 7.2.2 will not necessarily grow and propagate towards capacity but can also break down. The front moves back towards total extinction, i.e. until the population has died out everywhere. The two stable steady states introduce a critical size of the spatial extent of a population (Malchow and Schimansky-Geier 1985). Population patches greater than the critical size will survive, while the others will go extinct. In spherically symmetric coordinates, the temporal dynamics of the radius R of a population patch is:

$$\frac{dR}{dt} = 2D\left(\frac{1}{R_k} - \frac{1}{R}\right) \tag{7.7}$$

where R_k is the critical radius that has to be exceeded in order to survive, even if the local density is greater than $K\_$. This is a superposition of two critical size problems, which is demonstrated in Fig. 7.3. It is important to understand that one can find moving fronts in single population systems. In the long run however, the spatial population distribution on a finite domain will be uniform.

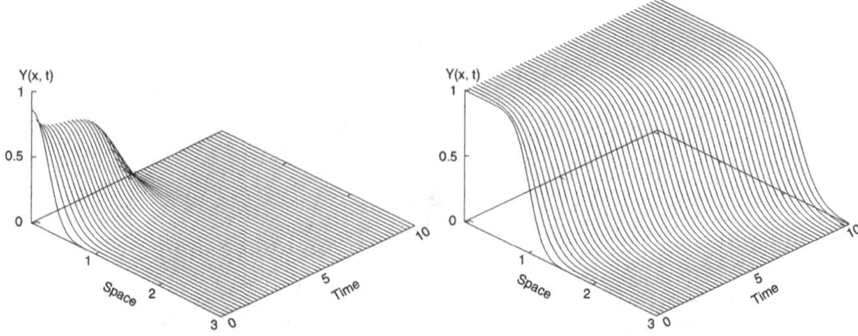

Fig. 7.3 Spatial reverse (*left*) or forward (*right*) propagation of a population with Allee effect for an initial condition less or larger than the critical radius. Parameters: $r = 3$, $K = 1$, $K_{-} = 0.4$, $D = 10^{-1}$ (a.u.), no-flux boundary conditions

7.4 Two Population Models

As we have seen, growth and dispersal of a single species in a constant and homogeneous environment does not support spatial pattern formation, it merely balances out spatial differences in population density. However, spatial patterns can appear in models with at least two interacting and moving populations. Since these patterns are more striking in two spatial dimensions, we will also move on from one spatial dimension to a two-dimensional model; that is $\mathbf{x} = (x, y)$. The interaction and dispersal of both populations is then described by the two equations:

$$\frac{\partial Y_1}{\partial_t} = f_1(\boldsymbol{Y}) + D_1 \nabla^2 Y_1 \tag{7.8}$$

$$\frac{\partial Y_2}{\partial_t} = f_2(\boldsymbol{Y}) + D_2 \nabla^2 Y_2 \tag{7.9}$$

The basic structure of each equation is the same as for the single species model of (7.2), but the respective growth terms $f_{1,2}$ now depend on the vector $\mathbf{Y} = (Y_1, Y_2)$ of both populations. Also, dispersal of the populations is now possible in two dimensions, indicated by the two-dimensional Laplacian:

$$\nabla^2 = \frac{\partial^2}{\partial x^2} + \frac{\partial^2}{\partial y^2}$$

which is simply the sum of the second order partial derivatives with respect to the spatial dimensions. The diffusivity or motility of the populations is given by D_1 and D_2, respectively. A selection of stationary and dynamic patterns will be described below. All of these patterns arise from the following, very important model of a prey species Y_1, a predator Y_2, and a constant top predator population Y_3:

$$f_1(Y) = rY_1\left(1 - \frac{Y_1}{K}\right) - \left(\frac{aY_1}{1 + bY_1}\right)Y_2 \tag{7.10}$$

$$f_2(Y) = e\left(\frac{aY_1}{1 + bY_1}\right)Y_2 - mY_2 - \frac{g^2Y_2^2}{1 + h^2Y_2^2}Y_3 \tag{7.11}$$

Here, the prey grows logistically with growth rate r and carrying capacity K. The consumption of prey by the specialist predator Y_2 is modelled with the so-called Holling-type II functional response, which assumes a linear relation between prey density and prey consumption at low prey densities, but saturates if the prey becomes abundant. This takes into account that there is maximum value of prey biomass that each predator can consume in a given time. This maximum value is given by a/b, the ratio of search rate a and prey handling time b. The parameter $e < 1$ is the predator's conversion efficiency and m its mortality. The constant top predator is assumed to be a generalist, described by a Holling-type III functional response. This functional response saturates at g^2/h^2, but assumes a lower than linear consumption at low prey densities. This reflects that the generalist predator Y_3 switches to a significant consumption of the specialist predator only when Y_2 becomes abundant. If the top predator is absent, that is $Y_3 = 0$, model (7.10 and 7.11) reduces to the classical Rosenzweig-MacArthur predator–prey model (1963). In this reduced model, the unique stationary point where both population densities are strictly positive can be stable or unstable. In the unstable case, the equilibrium is surrounded by a stable limit cycle, which corresponds to periodically varying population densities. As we will see, the form of the spatiotemporal patterns that can be observed in the full reaction-diffusion model greatly depends on whether the spatially homogeneous system given by (7.10 and 7.11) is in the parameter range of stationary or periodic dynamics.

7.4.1 Turing Patterns

Turing patterns are perhaps the most famous spatial patterns arising from reaction-diffusion systems (Turing 1952). These stationary patterns appear after diffusive instability of a stable, spatially uniform population distribution. For them to arise, the diffusion coefficients of the two species need to be sufficiently different, i.e. $D_2 \gg D_1$, and the growth terms have to obey certain conditions. For two interacting species, these conditions are called activator–inhibitor (Gierer and Meinhardt 1972) or destabilizer–stabilizer (Segel and Jackson 1972) relations. Because of their often striking polarity and symmetry, Turing had thought them as a possible mechanism of forming physiological gradients in biomorphogenesis. Applications in population dynamics soon followed (Segel and Jackson 1972). Three simulation results of (7.10 and 7.11) for different initial conditions are shown in Fig. 7.4.

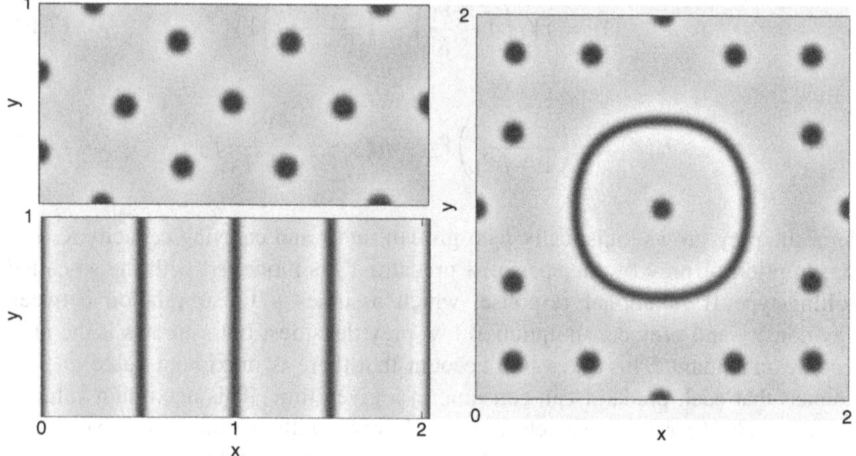

Fig. 7.4 Formation of Turing patterns for different perturbations of the spatially uniform distribution of (7.10 and 7.11). Parameters: $r = 5/14, K = 7/50, a = 2/3, b = 5/3, e = 3/5, m = 7/40, g^2 = 2/125, h^2 = 1/25, Y_3 = 1, D_1 = 10^{-3}, D_2 = 2 \times 10^{-3}$ (a.u.), no-flux boundary conditions

These stationary spots and stripes are typical representatives of Turing structures. There have been many discussions about the role of diffusive instabilities in forming population or animal coat patterns. Natural patterns never have this symmetric shape; however, the pattern can be altered by noise.

7.4.2 *Target Patterns and Spiral Waves*

In addition to stationary Turing patterns, (7.10 and 7.11) also allow for dynamic spatiotemporal patterns. Therefore, assume that our sample system (7.10 and 7.11) is in the parameter region of limit cycle oscillations. In this case, any local perturbation leads to the formation of concentric waves, the so-called target patterns. A subsequent perturbation of the circular wave fronts, like a collision with an obstacle, the domain boundary or another wave, may cause the opening of these fronts and spiral waves to appear. The two corresponding patterns are shown in Fig. 7.5.

Closely related to this phenomenon are periodic and irregular travelling waves (see Sherratt and Smith 2008 for a review of theoretical results and field studies) and the so-called 'wave of chaos' (Malchow et al. 2008), which can be seen as the spatially one-dimensional analogon to the target pattern shown in Fig. 7.5.

The effects described above are examples of irregular spatiotemporal dynamics. There are numerous other examples of temporally and spatiotemporally irregular oscillations in model systems and there is an ongoing discussion about the role and identification of deterministic chaos in ecology. It has been identified in laboratory systems (Becks et al. 2005; Cushing et al. 2003), but it will always be hard to distinguish between deterministic and stochastic effects in real data.

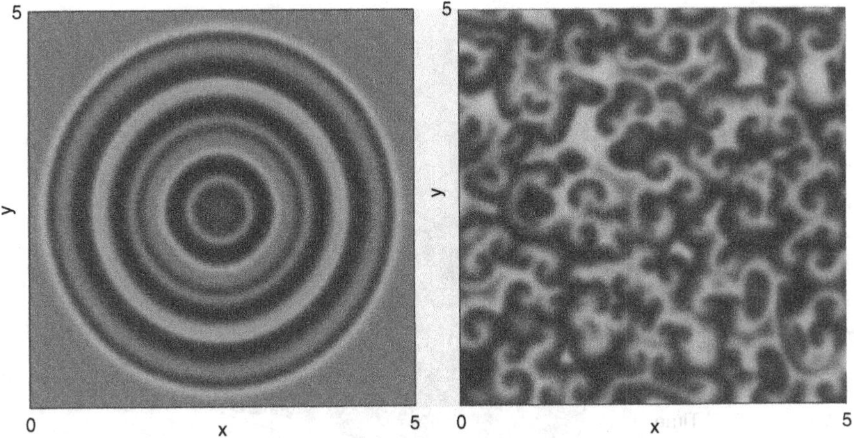

Fig. 7.5 Generation of target patterns and spiral waves of (7.10 and 7.11). Parameters: $r = K = 1$, $a = b = 10/3$, $e = 2$, $m = 4/5$, $g = h = Y_3 = 0$, $D_1 = D_2 = 1$ (a.u.), no-flux boundary conditions

7.4.3 Diffusion-Induced Chaos in Heterogeneous Environments

So far we have assumed that the environmental conditions relevant for species growth and interaction do not explicitly depend on the spatial position, i.e. the parameters are constant all over the spatial domain. However, it was already implied in (7.2), that the growth term f may explicitly depend on the spatial location x. This allows one to incorporate heterogeneous environmental conditions into the spatial model. One effect of a heterogeneous environment in a spatially one-dimensional variant of model (7.10 and 7.11) has been presented by Pascual (1993), assuming a linear increase in the prey growth rate $r(x) = r_0 + cx$. Assuming that the system is in the oscillatory regime at all spatial locations, this leads to a line of infinite diffusively coupled non-identical oscillators. Following the temporal change of population density $\mathbf{Y}(t)$ at fixed spatial locations x indicates that the local dynamics undergo a transition from regular oscillations at high prey growth rate to quasi periodic and finally chaotic oscillations at low prey growth rate. This is shown in Fig. 7.6.

7.5 Concluding Remarks

The examples given above can be generalized to N interacting species in three dimensional space, which are subject to diffusive and advective motion and environmental fluctuations. This leads to the following equation for the rate of change for the ith species:

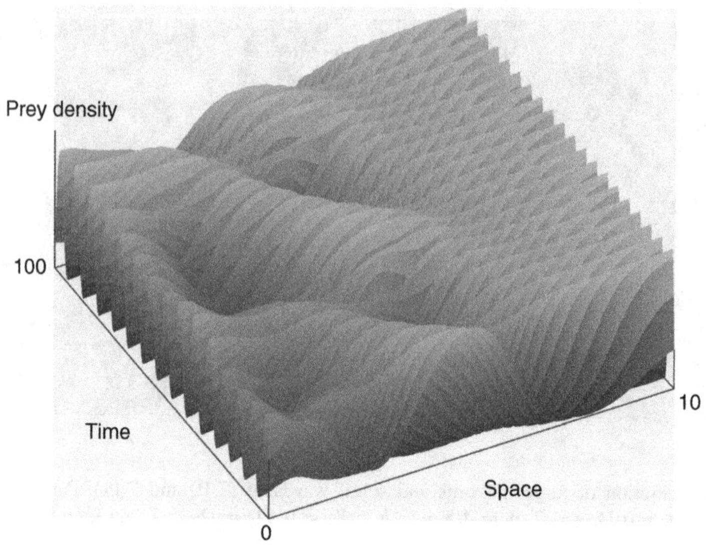

Fig. 7.6 Diffusion-induced chaos in (7.10 and 7.11) along a gradient in the prey growth rate. Parameters: $K = 1,\ a = b = 5,\ e = 1,\ m = 0.61,\ g = h = Y_3 = 0,\ D_1 = D_2 = 10^{-4}$ (a.u.), no-flux boundary conditions

$$\frac{\partial Y_i}{\partial_t} = f_i(\boldsymbol{Y}) + \sum_{j=0}^{n} D_{ij}\nabla^2 Y_j - \nabla \cdot v_i Y_i + F_i(\boldsymbol{Y}, t) \tag{7.12}$$

Here, $\mathbf{Y} = \{Y_i;\ i = 0,\ 1,\ 2,\ \ldots,\ N\}$ denotes the population densities of the N species at time t and position $\mathbf{x} = (x,\ y,\ z)$. The term f_i describes the growth, death and interactions of the ith species, which, as we have seen, may depend on the time t, spatial location $\mathbf{x}$ and a set of constant parameters. The D_{ij} are the self- and cross-diffusion coefficients. As in the previous examples, the self-diffusion coefficients D_{ij} reflect the motility of the species with respect to its own spatial gradient. Cross-diffusion is the dispersal of a species along the gradient of others, which facilitates the description of some behavioural strategies like neutrality, attraction or repulsion (Skellam 1973). The velocity vector $\mathbf{v}_i$ of the ith species gives the speed and direction for both the common passive advection, with the surrounding transport medium as water or air and the potential individual capacity of active locomotion. The nabla operator $\nabla = (\partial/\partial x,\ \partial/\partial y,\ \partial/\partial z)$ is simply the vector of the partial derivatives with respect to the spatial directions, with the dot product $\nabla 2 = \nabla \cdot \nabla$ denoting now the three-dimensional Laplacian. Environmental and/or demographic variability may be introduced into the model via a density-dependent external stochastic force F_{i} with certain noise characteristics. Note, that for $F_{\mathrm{i}} \neq 0$ (7.1) has to be interpreted as a stochastic PDE and solutions $\boldsymbol{Y}$ to (7.12) then constitute stochastic processes. This chapter has provided a very small collection and short description of selected spatiotemporal pattern forming mechanisms, focussing on a specific predator–prey model. However, general PDEs, as given by (7.12), have also been used to model bio-invasions (Shigesada and Kawasaki 1997), epidemic spread (Brauer et al. 2008; Hilker 2005) and noise-induced pattern formation and transitions between different spatiotemporal patterns (Sieber et al. 2007, 2010). Reaction-diffusion PDEs are also especially suitable for the modelling of marine plankton dynamics, where small plankton particles are subject to turbulent diffusion within the surrounding water column (Hilker et al. 2006; Malchow et al. 2002, 2004, 2005). There is a rich literature for further reading on the use of PDEs for modelling biological systems. A good overview has been provided in Holmes et al. (1994) and Allen (2003) and a very nice and already classical introduction to the role of diffusion processes in ecology has been given by Akira Okubo (1980, 2001). Another recommended reading is the book by Jim Murray (2003), and a recent overview more focused on eco-epidemiology can be found in Malchow et al. (2008).

Chapter 8
Cellular Automata in Ecological Modelling

Broder Breckling, Guy Pe'er, and Yiannis G. Matsinos

The chessboard is the world;
the pieces are the phenomena of the universe;
the rules of the games are what we call the laws of Nature.

T. H. Huxley (1870)

Abstract Cellular Automata (CA) are models that generate large-scale pattern from small-scale local processes. CA deal with spatially extended dynamics using a grid structure. Successive states of cells, which are arranged on a grid, are calculated according to a set of rules. State transitions depend on the state of the single cells and the state of the cells in the local neighbourhood. Cellular Automata are applied as a modelling approach in many scientific disciplines and are used in ecology as one of the most popular model types to study spatially extended dynamics. The chapter starts with a brief historical overview about CA. It describes how CA function, and for which types of problems they can be employed. We present simple theoretical examples, followed by a more detailed case study from plant competition and grassland community dynamics. As an outlook, we discuss major fields of application with a special focus on the ecological context. Finally, we provide a brief overview and recommendations on the use of some of the software specialized in the field of CA modelling.

8.1 Introduction and Historical Background

Cellular Automata were conceptually developed by the Austro-Hungarian mathematician John von Neumann (1903–1957) during the 1950s. He was interested in simulating self-reproducing patterns. Instead of continuous approximations, he

B. Breckling (✉)
General and Theoretical Ecology, University of Bremen, Leobener Str., 28359 Bremen, Germany
e-mail: broder@uni-bremen.de

F. Jopp et al. (eds.), *Modelling Complex Ecological Dynamics*,
DOI 10.1007/978-3-642-05029-9_8, © Springer-Verlag Berlin Heidelberg 2011

used discrete (stepwise) representations of space and time. Together with Stanislaw Ulam, the work was developed at the Los Alamos Laboratory, where both were also involved in the Manhattan Project.[1] Ulam used the idea to study crystallization processes on a two-dimensional grid (or lattice). The first CA model that made the approach widely known dated from the 1960s, when the Cambridge-based mathematician John Conway developed the "Game of Life" (Gardner 1970). This is a simple grid based process where cells can switch between two states following simple rules (Sect. 8.3). Because of its simplicity and surprisingly interesting and complex behaviour, the Game of Life created a lasting enthusiasm.

During the 1970s, a series of applications of cellular automata models were developed in physics to study gas and liquid diffusion, crystallization processes, magnetic and spin phenomena (Forrester et al. 2007). CA were further used as stepwise (discrete) approximation models for partial differential equations (see Chap. 7).

During the 1980s, following a marked increase of computer availability and computation power, the application of cellular automata has seen a significant increase, especially in mathematics and physics. Scientists started to realize that a discrete representation of systems could provide simpler and more efficient approximations of spatially complex processes compared to continuous approximations. It was then that cellular automata machines were constructed, in order to handle parallel processing more efficiently (Toffoli and Margolus 1987). An important contribution to CA was made by Wolfram (1994), who systematically explored the overall dynamics of large classes of one-dimensional cellular automata using the software "Mathematica", which he developed initially for this purpose. Wolfram showed that simple, deterministic rules can generate complex patterns in space or time that look as if they were completely random. In ecology, CA successively became one of the most frequently used approaches to model spatially extended processes. Often, they are used in combination with other techniques such as individual-based models (Chap. 12).

Due to their ease of implementation and capacity to simulate spatial patterns, CAs have been widely applied to ecological problems related to spatial processes, such as epidemic propagation (Sirakoulis et al. 2000), plant population dynamics (e.g. Iwasa et al. 1991; Pascual et al. 2002), post-disturbance resilience (Matsinos and Troumbis 2002), colonization processes (Silvertown et al. 1992; Hobbs and Hobbs 1987), land-use and land-cover change (White et al. 1997) and spatial competition of corals (Langmead and Sheppard 2004, see also Chap. 17). Rietkerk et al. (2004) used a simple cellular automaton model based on the model of Thiery et al. (1995) in order to understand how scale-dependent feedback can explain a diversity of spatial patterns in self organizing savannah ecosystems. Moustakas et al. (2006) developed a CA to analyse the interaction between fish schools and fleets of fishing vessels, in order to assess the efficacy of conservation measures.

[1]The Manhattan project covered the initial initiatives in the USA to develop nuclear weapons of mass destruction. The leading physicists worked for this project during the Second World War.

These models refer to basic biological processes: dispersal and competition. It is the variation in the strength and scale of feedbacks between cells in the automaton that influence the outcomes in terms of structure and scale of patchiness. This illustrates the general nature of scale-dependent processes underlying self-organized patchiness in ecosystems.

8.2 Cellular Automata: The Components

Though cellular automata can handle very complex spatial situations and quite difficult rule systems, the conceptual basis is quite simple, easy to understand and applicable with almost any conventional or object-oriented programming language.

A cellular automaton consists of a large number of cells, which are connected to a grid and can change their state individually. For all cells, a neighbourhood is defined that constitutes the surrounding area that influences the state transitions of each particular cell. Finally, there is a set of rules defining how each of the potential states of a cell and the states of the neighbourhood will determine the transition between cell states.

The Cells

Cellular automata models use cells as the units of operation. Cells can be considered as a storage space, with a defined number of state variables that can either be discrete or continuous. The most simplistic CAs consist of cells that can switch between two different states (binary), to be represented e.g. by black and white, on and off, dead or alive, etc. But it is also possible to have a cell's state being characterized by a larger number of variables. For example, when modelling soil processes using a CA, the cell could represent a square meter of the ground and have storage space for variables such as water content, organic material, temperature, etc.

The Grid

In a CA, each cell is surrounded by other neighbouring cells. The grid can be visualized by drawing the cells as nodes and the connection to adjacent cells as edges. A grid can be finite or infinite, (for simulation purposes only finite) and can have different topologies. For instance, cells along a line with one neighbour to the right and one to the left would represent a one-dimensional grid, cells with four neighbours (North, South, East and West) would represent a two-dimensional grid, etc. In principle, any topological structure would be possible (Fig. 8.1).

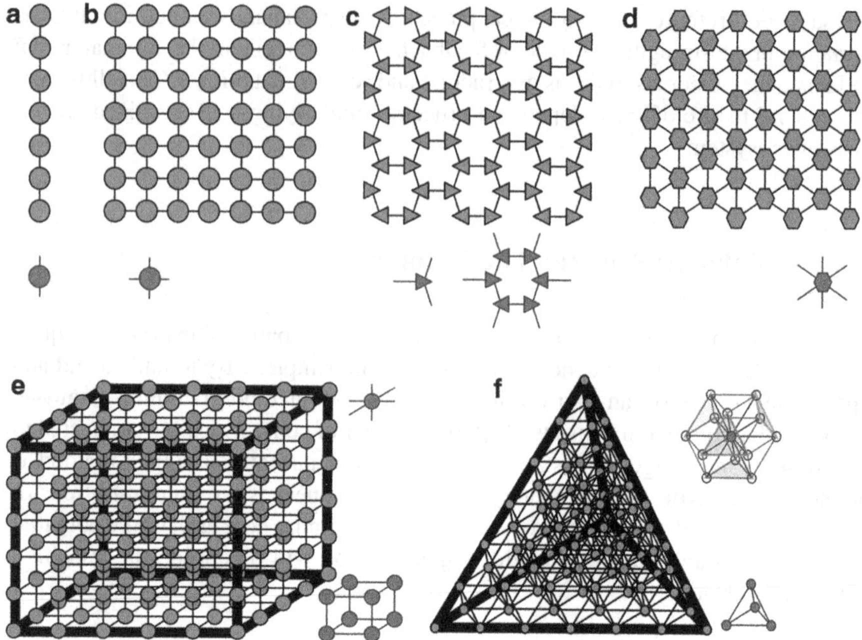

Fig. 8.1 Examples of one-, two- and three-dimensional grids of different topologies. (**a**) Linear grid: each cell has two neighbours; (**b**) 2D rectangular grid: each cell has four neighbours; (**c**) 2D triangular grid: each cell is connected to three neighbours; (**d**) 2D hexagonal grid: each cell has six neighbours; (**e**) 3D cubic grid: each cell inside the grid has six neighbours; (**f**) 3D tetrahedral grid: A cubic grid is not the only possibility to model in three dimensions. An alternative could consist of cells connected at the edges of stapled tetrahedrons. Please keep in mind that the given neighbourhood relations do not apply for margin cells

The Neighbourhood

The neighbourhood comprises the cells in the surrounding of a focal cell. The neighbourhood cells are defined as those that can influence the state of the particular focal cell. To determine which change occurs, the state of the focal cell and the states of the neighbourhood cells are evaluated. Usually, the neighbourhood consists of the directly adjacent cells, but the neighbourhood can have different extents and can vary in shape between rectangular, circular, etc. Other definitions are possible as well, e.g. that each cell selects a random number of other cells as neighbours – regardless where they are located on the grid. In case of a rectangular two-dimensional grid (Fig. 8.1b), the most commonly used neighbourhood comprises the four direct neighbours (Fig. 8.2b). In the CA terminology, this is also called von Neumann neighbourhood. If the eight directly adjacent cells are considered as neighbours, it is called Moore neighbourhood (see Fig. 8.2c), named after the US-American mathematician Edward F. Moore (1925–2003).

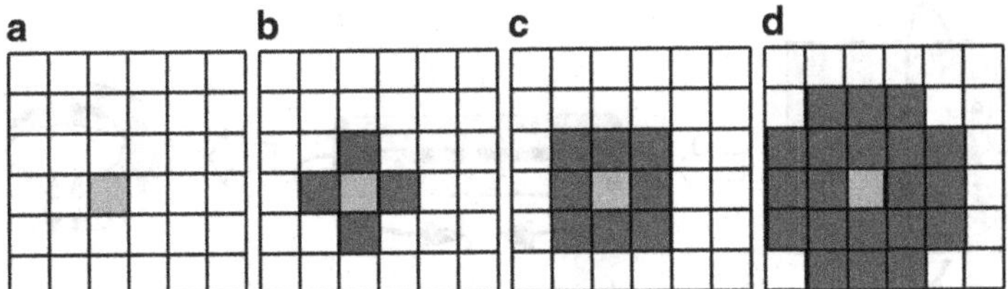

Fig. 8.2 Neighbourhoods in cellular automata: (**a**) focal cell; (**b**) von Neumann neighbourhood comprising the four adjacent cells – North, South, East, West; (**c**) Moore neighbourhood considering eight directly adjacent cells; and (**d**) a neighbourhood which consists of the nearest and second-nearest cells. Other definitions of neighbourhoods are also possible

The Rules

Rules of the CA are fundamental to specify how cells change their states. There can be an arbitrary number of rules. The rule-set for a CA applies to all cells. The current state of a particular cell and the states of the cells in the neighbourhood determine which of the rules are applied to change the cells state. The rules must consider all possible combinations of situations which can occur in the neighbourhood of a cell. A very simple example of a rule for a CA would be that the state of a cell can be any natural number, and that the subsequent state of a cell in the next step is the sum of the states of the neighbouring cells. This rule would constitute a deterministic CA. It would, however, also be possible to add stochasticity, e.g. by determining that for a given probability the state of the focal cell is zero.

Running a CA: The Iteration

A CA is processed step by step. One step (one iteration) comprises an application of the rules to all cells on the grid. To process a CA, the initial state of all cells must be set. This initial configuration is used for the first update. Cell by cell the rules are applied, taking into account the state of each cell and the state of cells in the neighbourhood. This yields the next state of all the cells. To avoid a bias of the update procedure, the new state of each cell is saved in a separate interim grid, so that the transition is applied only once after all the states (or transitions) of all cells have been calculated. Then the iteration can be repeated until a termination condition is met. The termination condition can be a maximum number of iterations, a pre-defined state of the grid, or an interruption by the user.

Boundary Conditions

Practical applications cannot work with infinite grids. The grid has to be spatially limited and a specification is required on how to process the cells at the boundary,

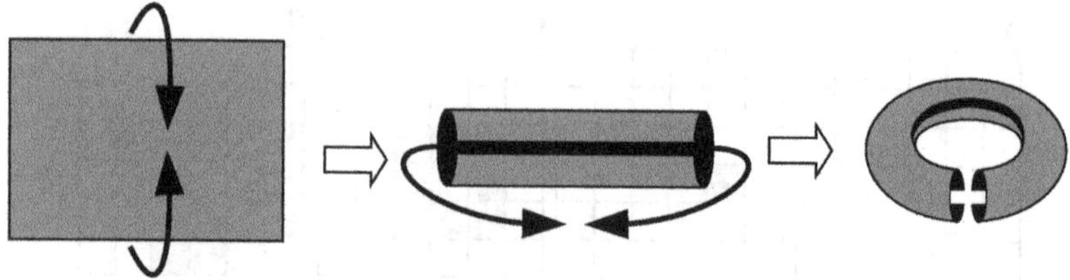

Fig. 8.3 Applying torus boundary condition in a rectangular grid by connecting opposite edges

where their neighbourhoods can be different from those situated in the inside of the grid. There are alternative ways in which boundary conditions can be specified. To this end the following solutions are frequently taken:

- Setting a different neighbourhood at the boundaries, taking into consideration that cells at the boundary have a different number of neighbours compared to the other cells and therefore require an according adaptation of the rule-set.
- The grid can be framed by a number of outer cells that maintain a particular state without being updated.
- In case of a rectangular grid, boundary cells can take the cells of the opposite boundary as their neighbours (i.e. the Eastern edge of a grid connects to the Western, the Northern connects to the Southern edge). Topologically this yields a torus (a doughnut like shape), as shown in Fig. 8.3.
- Grid extension: in the case of a homogeneous background state the grid could be dynamically extended. This solution, however, is possible only to the limits of processing capacity.

8.3 An Easy Example: Conway's Game of Life

Conway's Game of Life (Gardner 1970) is an excellent example to familiarize with the concept of CAs and with the process of updating the grid cells. Since the rules are rather simple, it is even possible to solve smaller grid iterations on paper. In more complex models, this process can of course be done only by a computer.

The Game of Life CA uses a two-dimensional rectangular grid. The cells can have two states, either "alive" (black) or "dead" (white). The state they take in a succeeding iteration (the rule set) depends on their own state and the states of their eight adjacent neighbours (Moore neighbourhood):

- A white cell becomes black (alive) if exactly three cells in its neighbourhood are black.
- A black cell remains black if two or three neighbours are black.

- A black cell turns to white, if less than two neighbourhood cells are black (it "dies of solitude") or if more than three neighbourhood cells are black (it "dies of overcrowdedness").
- The game can start with any initial configuration of black and white cells. Depending on the initial configuration, different patterns emerge. There are configurations which lead to global expansion. Others end in stationary or in repetitive pattern which re-emerge after a number of iterations. Other initial configurations can lead to "extinction", with only white cells remaining (Fig. 8.4).

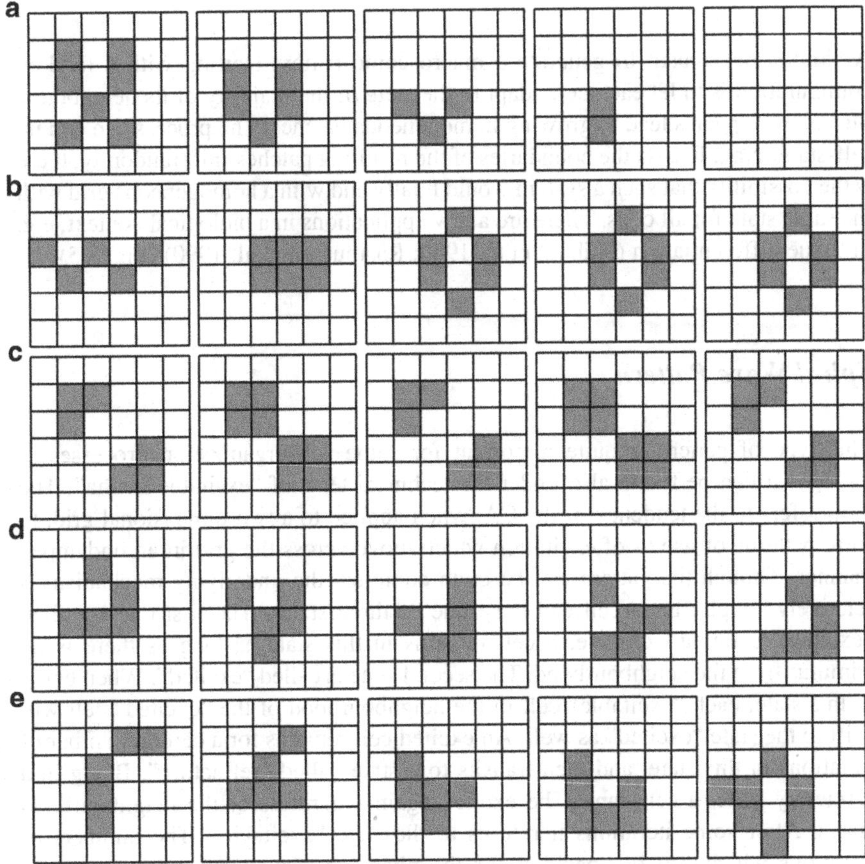

Fig. 8.4 Some pattern types which can occur in the game of life. (**a**) An initial configuration which "dies out" after four iterations; (**b**) an initial configuration which generates a stationary pattern; (**c**) an initial configuration which generates an oscillating pattern (re-emerging after a number of iterations); (**d**) a pattern that re-emerges but shifts location with time ("glider"); (**e**) an initial configuration generating a complex irregular pattern with parts that die out and others that oscillate or remain stationary. This is the so-called r-pentomino, which grows a large, irregular pattern taking more than 1,000 iterations before it becomes stationary and/or periodic

8.4 Examples of Pattern Generating Mechanisms

The Game of Life is a deterministic CA. Other applications may also include stochastic rules. There are certain types of interactions that can be found in different contexts that give rise to a specific category of pattern. We present three examples: A self-scaling random pattern, a spiral wave pattern, and a diffusion-limited aggregation.

Self-Scaling Random Pattern

A simple mechanism to generate a macroscopic pattern starting with a random configuration is to let each cell adapt to the state of the majority of its neighbours. Successive updates lead to growing homogeneous patches. The process can end up self-stabilizing. In case the boundaries of the resulting patches shift randomly, there is the possibility that such a system would finally end with a homogeneous grid with the same state for all cells. There are a few applications in a biological context, e.g. in tissue differentiation (Nijhout et al. 1986; Rasmussen et al. 1990) (Fig. 8.5).

Spiral Wave Pattern

This type of pattern is quite important for some self-organization processes in biology. It can be found also under the technical term of "excitable media". The term refers to the tendency of the CA, when applied to a two-dimensional grid, to form patterns of waves of excitation which move across the grid in an undamped manner. Though the pattern can be quite complex, the underlying mechanism is relatively simple. Each cell can have one of three states. The first one is called "excitable". Being excitable, a cell remains in this state as long as there is no stimulus from the neighbourhood. The second state is called "excited". When being in this state, each "excitable" cell in the neighbourhood of the "excited" cell will shift to the state "excited" as well. An excited cell remains for a certain number of iterations in this state, and then transits to a state called "refractory". Being in a refractory state, a cell cannot be excited again regardless of the neighbourhood states. After some iterations it returns to the state "excitable". The numbers of iterations which can be specified for the phases of "excited" and "refractory" influence the shape of the emerging macroscopic pattern.

A grid started with only "excitable" cells would remain as it is. A grid started with only "excitable" and "refractory" cells would end up in an overall state of being excitable as well. However, when there are a few excited cells, a spreading pattern can occur, which can organize the rest of the grid spatio-temporally. For random initial configurations of a sufficiently large grid, spiral wave patterns frequently occur

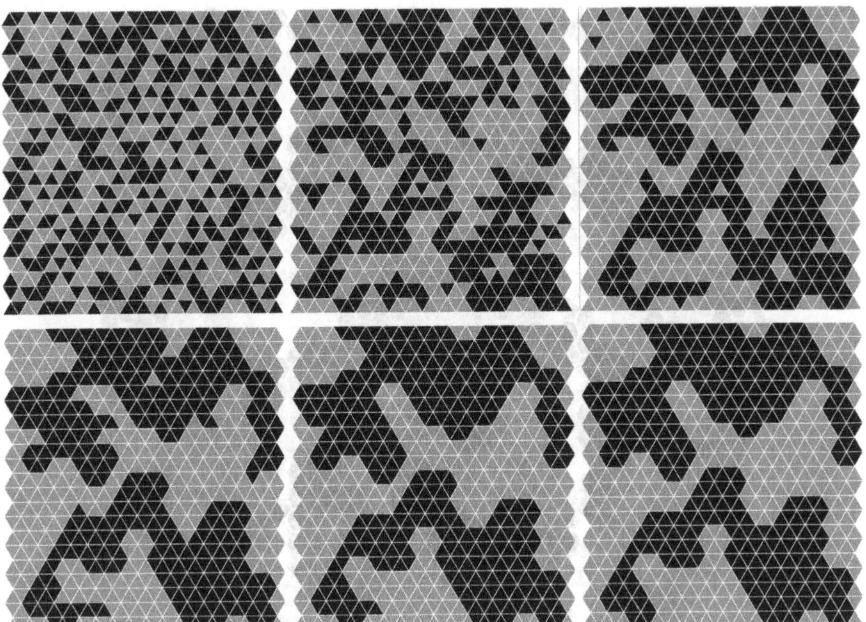

Fig. 8.5 The pattern on a triangular grid (three neighbours to each cell) was obtained by applying the following rules: The initial configuration is a random distribution. If the state of the majority of the neighbours differs from the focal cell's state, it has a 50% probability to change to the state of its neighbours. After a number of iterations, the pattern reaches a stable state, where each cell's state conforms to the majority of its neighbours. Shown are the iterations 0, 2, 5, 8, 11, and 14

(Fig. 8.6). Comins et al. (1992) used a cellular automaton employing deterministic rules to explore the spatial dynamics of a host–parasitoid interaction resulting in spiralling spatial patterns.

Diffusion-Limited Aggregation

Diffusion-limited aggregation is a process that can be observed in the successive growth of river systems, in certain forms of organic growth, involving branching, and in some inorganic immobilization processes. Again, the basic underlying rules are relatively simple. Cells can be in three types of states, which can be called "empty", "mobile" and "fixed". A "mobile" cell shifts the state of any (randomly chosen) "empty" neighbouring cell into "mobile", while turning back to an "empty" state. This simulates random movement of a particle across the grid. If a cell is in the state "mobile" and has a cell with the state "fixed" in its neighbourhood, the "mobile" cell changes its state to "fixed" and remains in this state for the rest of the simulation, regardless of the states of neighbouring cells. To obtain non-trivial results, it is required that a sufficiently large number of "mobile" cells

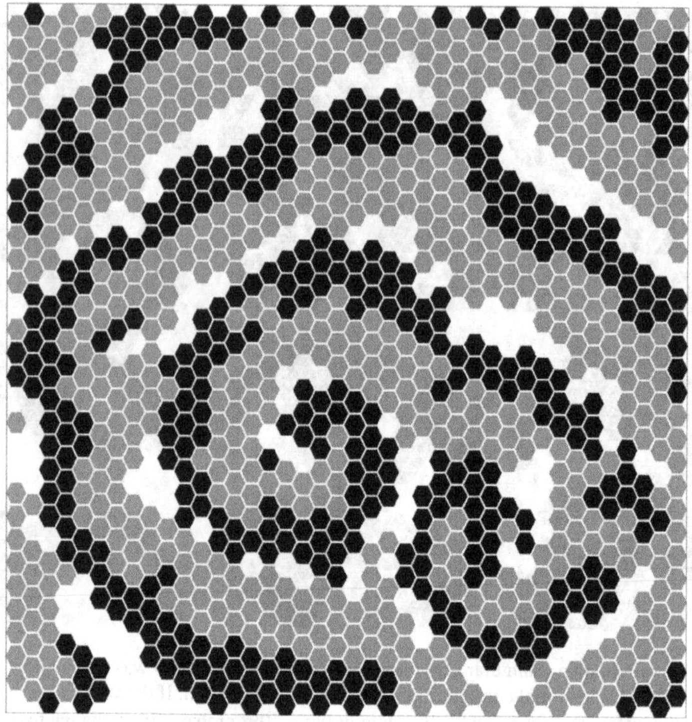

Fig. 8.6 A spiral wave pattern on a hexagonal grid; excitable cells (*white*), excited cells (*black*), refractory cells (*grey*); see text for details of state transitions

are started, and that at least one "fixed" cell exists on the grid. The emerging pattern is a random structure with a self-similar pattern (Fig. 8.7). Examples for the application of diffusion-limited aggregation models are CAs for predicting urban development (White and Engelen 1993), investigating the spatial distribution of plants and animals (which often seem to form fractal patterns; Kendal 1995), and studying pattern formation, e.g. in plant morphogenesis (Fleury 1999) (Fig. 8.7).

8.5 Case Study: Competition and Dispersal in Grassland Communities

As an example of a more complicated, recent model applied for the study of ecological questions, we elaborate on a CA model of a grassland community dynamics (Matsinos and Troumbis 2002). The model illustrates nicely how local interactions (dispersal and competition) determine overall community structure. The model focuses particularly on the effect of resilience in communities on gap-creating

Fig. 8.7 Diffusion limited aggregation, produced with the fractal generator FRACTINT. A large number of mobile cells are started in the periphery, which attach by chance to a centrally located "fixed" cell, bringing up the typical diffusion-limited aggregation-pattern

disturbances (i.e. fire), imposed at different spatial extents. Model simulations were based on data from an experimental community with five grassland species.

A lattice of 200 × 200 square cells was used, with at most one individual of each species occupying a cell at a time. Global rules applying at the local scale determine the state of the system at each time step. The degree of influence of a cell to neighbouring cells follows a negative exponential decrease. The biological processes simulated are seed dispersal and competition. The following main procedures were used:

- Every individual of each species i produces seeds at a given rate p_i. We assume that the distribution of times between individual seed production is exponential, with a mean of $1/p_i$.
- The probability that a seed disperses from one cell to another depends on the species dispersal type (local, medium or long) and the distance between cells.
- Displacement of species is modelled in occupied cells depending on the competitive advantage of the invading species. The process of seed dispersal to neighbouring empty cells is modelled using a probabilistic algorithm.

The model starts with an assignment of seeds to donor cells in a random manner but with a frequency that is inversely related to donor cell distance. If a cell receives multiple seeds from different plant species, a random variable linked to the competition coefficient of the species determines the winner at that cell. All seeds are then eligible to sprout and will germinate at the next growing season; dormancy is not considered; the model does not consider environmental variability between years.

Parameters from an experimental biodiversity study of grasslands in Lesbos, Greece were used for the model specification. The experimental study was part of the European-wide research project BIODEPTH (Hector et al. 1999), aiming to

Table 8.1 Matrix of displacement probabilities that were used in the model

Displacing species *i*	Resident species *j*				
	Phalaris	*Bituminaria*	*Hordeum*	*Hirschfeldia*	*Lagoecia*
Phalaris	–	0.2	0.6	0.15	0.3
Bituminaria	0.1	–	0.01	0.03	0.04
Hordeum	0	0	–	0.01	0.01
Hirschfeldia	0.03	0.01	0.01	–	0.02
Lagoecia	0.1	0.01	0	0.01	–

investigate diversity–productivity relationships in natural grasslands. Experimental plots consisting of 2, 4, 8, 18, and 32 native species were established and maintained since 1997 in seven European countries. For the parametrization of our model we chose the two-species configurations, extracting relative biomass changes and estimating competition strength to simulate the interactions between five plant species: *Phalaris coerulescens*, *Hordeum geniculatum*, *Hirschfeldia incana*, *Lagoecia cuminoides*, and *Bituminaria bituminosa*. Among the five species, pair experiments have shown their competitive hierarchy and biomass changes in the plots yielded information on the competitive strength. This was assumed to translate into displacement probabilities in the model (Table 8.1).

Results showed that longer distance dispersing plants have a competitive advantage in colonization success as compared to better competitors, especially in the cases of disturbance-mediated creation of gaps in coverage. An increase in species number led to more resilient communities and a higher percent cover of the landscape. A further model adaptation therefore incorporated

- A scale-related neighbourhood structure
- Asymmetrical hierarchy in competition
- Invasion processes

The neighbourhood structure in the model was based on the dispersal attributes of the different species, and showed significant change in final assemblage patterns where short-distance dispersers were found to decrease in abundance. Asymmetrical hierarchy (in terms of competition) was modelled as a stochastic process, and showed to alter the composition of steady-state communities significantly, favouring assemblages with low overall diversity. Invasion was shown to interfere and alter the overall pattern of abundance. The effect of disturbances was studied as well, examining whether the community is resilient to disturbances or tends to change subsequent to disturbances (e.g. Fig. 8.8). The approach highlighted the emergence of complex community patterns from simple local interactions. A great amount of information is necessary for the parametrization of such a model, yet the outputs of the model provide a broader understanding of patterns that are far too complex to grasp with any other tool. Therefore, despite the relative complexity of the model, it provides the means to gain understanding of complex patterns in nature, the underlying mechanisms of which are otherwise poorly understood.

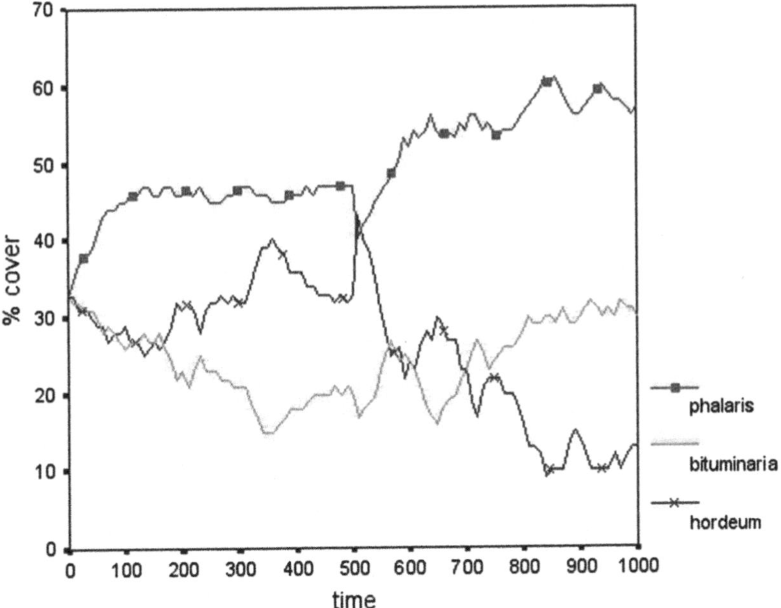

Fig. 8.8 Plot of relative abundance for 1,000 iterations for the three species: *Phalaris*, *Hordeum*, *Bituminaria* starting from same abundance (33%). Disturbance occurring on iteration 500 affects 20% of the landscape [from Matsinos and Troumbis (2002)]

8.6 Outlook and Applicability

In a unique way, Cellular Automata combine conceptual simplicity, the potential to expand simple interactions to complex structures, and an enormous range of application fields for quite demanding problems – with the potential to capture surprising self-organizing effects. This makes it worthwhile and desirable for any ecological modeller to familiarize with this approach.

It is possible to run CAs without much effort in pre-defined modelling environments, each of which specializes in a particular field of rule types. Yet, it is equally easy to escape the restrictions that customized software frequently have, and develop a unique CA according to one's specific applications, with the additional power to modify it to specific situations or explorations, e.g. by time-dependent or situation specific variations of the neighbourhood or through self-modifying rule systems. Other ecological modelling applications, and especially those that require a spatially structured input in order to provide an environment with particular statistical features, can be easily generated with a CA and used as a grid input. Clearly, Cellular Automata can contribute not only to strengthen ecological theory, but also for the development of predictive tools for ecology and conservation. In the process, one may reveal that modelling itself can be fun as well.

Chapter 9
Leslie Matrices

Dagmar Söndgerath

Abstract This chapter introduces matrix models – used to describe the dynamics of populations classified by age or other criteria like size or stage. It will be shown how characteristic values of the Leslie matrix, i.e. eigenvalues and eigenvectors, are used to determine the asymptotic behaviour of the population. Elasticity analysis deals with the effects of small parameter changes on population growth. As a result, values for the relative importance of specific life history parameters for the population dynamics are given. For example, these values can be used in conservation to identify those parts of an organism's life history where management methods should focus. Finally, an extended Leslie model for populations with both age and stage structure will be introduced and used to forecast the effects of climate change on the voltinism and range of occurrence of a dragonfly species.

9.1 Introduction

Matrix models are used to simulate structured populations. The origin of these models dates back to a paper by Leslie (1945). Originally, they were used for demographic purposes; i.e. to describe the development of human populations. For this purpose the population is divided into classes according to age. With an appropriate projection matrix composed of age-specific fertility, survival rates and a given initial distribution, it is possible to project the age distribution for every future time point. Since development, these models have been modified in multiple ways and have been applied in many fields.

For plant life cycles the vital rates cannot be regarded as functions of age because reproduction strongly depends on size and more complex life cycles have to be regarded (Caswell 1986). Therefore, stages instead of age are usually incorporated in the model to divide the population into classes. A frequently cited

D. Söndgerath
Institut für Geoökologie, Abt. Umweltsystemanalyse, Technische Universität Braunschweig, Zimmer 115, Langer Kamp 19c, 38106 Braunschweig, Germany
e-mail: d.soendgerath@tu-bs.de

F. Jopp et al. (eds.), *Modelling Complex Ecological Dynamics*,
DOI 10.1007/978-3-642-05029-9_9, © Springer-Verlag Berlin Heidelberg 2011

example (Caswell 2001; Soetaert and Herman 2009) is the life cycle of teasel (*Dipsacus sylvestris*), a European perennial weed with the following six stages: dormant seeds 1st year, dormant seeds 2nd year, small rosettes, medium rosettes, large rosettes and flowering plants. Pascarella and Horvitz (1998) also used a stage-classified matrix model to determine the importance of environmental variation to the dynamics of a tropical understory shrub caused by hurricanes.

Selhorst et al. (1991) coupled two Leslie models to describe the predator–prey relationship between two arthropods, *Scolothrips longicornis* and *Tetranychus cinnabarinus*. The difficulty in dealing with insect populations is that the development of insects strongly depends on environmental factors (e.g. temperature), so that the age of an individual is not a good indicator for its development status. Several studies dealing with insect populations therefore used an approach combining age and stage by coupling several Leslie processes (Söndgerath and Müller-Pietralla 1996; Braune et al. 2008).

Size-dependent classes are widely used when dealing with fish populations. In these cases fecundity is a function of the size of the individuals and not of their age. Jung et al. (2009) used a Leslie size-dependent approach to simulate the dynamics of the Pacific cod in order to improve the reliability of stock assessments for fisheries management. Size-dependent vital rates were also regarded by Ang and De Wreede (1990) for the simulation of algal life histories.

In order to control an African pest rodent, Stenseth et al. (2001) used a Leslie model with three stages of females and three stages of males. In this study they formulated the vital rates as functions of density and rainfall. Bieber and Ruf (2005) set up a Leslie model for the wild boar with three stages: juveniles, yearlings and adults. They found that reducing juvenile survival will have the largest effect on the population growth in good years, whereas strong hunting pressure on adult females will lead to the most effective population control in bad years. Heppel et al. (2000) did a meta-analysis of 50 mammal populations based on Leslie models. They examined the impacts of small changes in fertility, juvenile and adult survival on the population growth. On this base they were able to discriminate between carnivores, rodents, grazers, marine species and primates.

9.2 Model Description

The dynamics of age-structured populations is usually described with matrix models, which can be traced back to the papers of Leslie (1945, 1948). The main assumption of these models is that time and age are measured in the same units. For example, if the age classes consist of individuals from 0 to 1 year old, from 1 to 2 years old and so on, then the time step of the model must be one year. If the age classes denote 1-week old individuals and so on, then the time step of the simulation model has to be one week. Hence, during one time step, individuals from one age class pass into the next class according to the age-specific survival rate. The first age class consists of the offspring from all reproductive age classes.

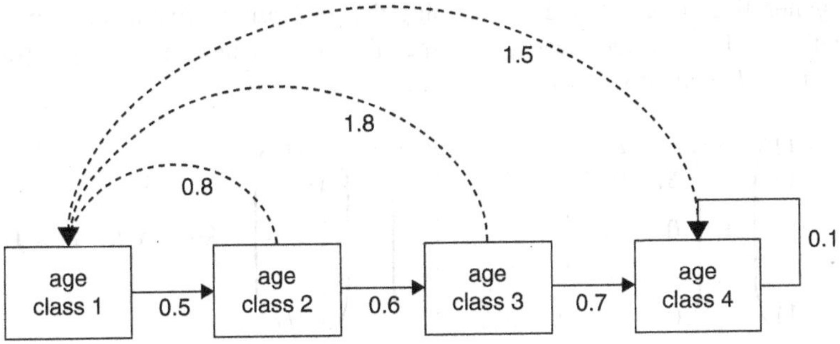

Fig. 9.1 Illustration of a life cycle with four age classes. The *solid line arrows* indicate survival rates and the *dotted line arrows* indicate fertilities

The life cycle of a population with four age classes is given in Fig. 9.1. The solid line arrows represent the survival rates to the next age class, whereas the dotted line arrows indicate the reproduction of the different age classes. The first age class is assumed to be non-reproductive, so there is no dotted line arrow starting from this class. All individuals surviving to or beyond age class 4 are aggregated in that age class. Individuals of the first age class will survive with a probability of 0.5 and will then be in age class 2. Likewise, individuals in age class 2 will survive to age class 3 with a rate of 0.6, and those of class 3 will be in age class 4 with a 70% probability. 10% of the individuals in age class 4 will survive and will remain in age class 4. Individuals of age classes 2–4 contribute to the first age class with rates of 0.8, 1.8 and 1.5, respectively. Therefore, if $x_i(t)$ denotes the number of individuals in age class i at time t, the following equations can be set up to evaluate the number of individuals at the next time step t+1:

$$\begin{aligned} x_1(t+1) &= 0.8\,x_2(t) + 1.8\,x_3(t) + 1.5\,x_4(t) \\ x_2(t+1) &= 0.5\,x_1(t) \\ x_3(t+1) &= 0.6\,x_2(t) \\ x_4(t+1) &= 0.7\,x_3(t) + 0.1\,x_4(t) \end{aligned}$$

Summarizing these equations into a matrix formulation one can write:

$$\begin{pmatrix} x_1(t+1) \\ x_2(t+1) \\ x_3(t+1) \\ x_4(t+1) \end{pmatrix} = \begin{pmatrix} 0 & 0.8 & 1.8 & 1.5 \\ 0.5 & 0 & 0 & 0 \\ 0 & 0.6 & 0 & 0 \\ 0 & 0 & 0.7 & 0.1 \end{pmatrix} \cdot \begin{pmatrix} x_1(t) \\ x_2(t) \\ x_3(t) \\ x_4(t) \end{pmatrix}$$

The vectors $\vec{\mathbf{x}}(t)$ and $\vec{\mathbf{x}}(t+1)$ consist of the number of individuals in the four age classes at time points t and t + 1, respectively. The matrix on the right hand side of the equation is the so-called projection or Leslie matrix. In the first row of this matrix the age-specific fertility rates are written and the sub-diagonal contains the respective survival rates.

To generalize, a Leslie model describing the population dynamics of a population with $i = 1, ..., n$ age classes, age-specific survival rates S_i and age-specific fertility rates F_i can be written in the form:

$$\begin{pmatrix} x_1(t+1) \\ x_2(t+1) \\ \vdots \\ \vdots \\ x_n(t+1) \end{pmatrix} = \begin{pmatrix} F_1 & F_2 & & & F_n \\ S_1 & 0 & \cdots & \cdots & 0 \\ 0 & S_2 & \ddots & & \vdots \\ \vdots & \ddots & \ddots & \ddots & \vdots \\ 0 & \cdots & 0 & S_{n-1} & S_n \end{pmatrix} \cdot \begin{pmatrix} x_1(t) \\ x_2(t) \\ \vdots \\ \vdots \\ x_n(t) \end{pmatrix} \Leftrightarrow \vec{\mathbf{x}}(t+1) = \mathbf{L}\,\vec{\mathbf{x}}(t)$$

The Leslie matrix **L** contains the life-history parameters of the population, i.e. survival and reproduction rates. Obviously some of the F_i-values in the Leslie matrix are usually zero, because in natural populations only certain age classes will be reproductive ones. Starting with a given initial population, this equation can be used to update the population vector for future points in time.

Other criteria such as size classes or different stages of the life cycle can also be considered, without changing the structure or behaviour of the model.

Long-Term Behaviour

Several characteristics concerning the long-term behaviour of the population can be derived from the projection matrix by means of some well-known results of matrix algebra (see e.g. Kaw 2008; Meyer 2000; Searle 2006). First the model equation can be written as:

$$\vec{\mathbf{x}}(t+1) = \mathbf{L}\,\vec{\mathbf{x}}(t) = \mathbf{L}^2\,\vec{\mathbf{x}}(t-1) = \ldots = \mathbf{L}^{t+1}\,\vec{\mathbf{x}}(0)$$

Second, it is known that any Leslie matrix **L** is similar to a diagonal matrix with the eigenvalues λ_i as diagonal entries. Hence $\mathbf{L}^{t+1}$ can be transformed to $W\,\Lambda^{t+1}\,V$ with W and V being composed of the right and left eigenvectors, respectively. Hence the model can be rewritten as:

$$\vec{\mathbf{x}}(t) = W\,\Lambda^t\,V\,\vec{\mathbf{x}}(0) = \sum_{i=1}^{n} \lambda_i^t \vec{\mathbf{w}}_i \vec{\mathbf{v}}_i^T \vec{\mathbf{x}}(0) = \sum_{i=1}^{n} \lambda_i^t \vec{\mathbf{w}}_i c_i$$

Third, it is known that for each projection matrix **L** there exists one positive real eigenvalue λ_1 of modulus greater than any other. This dominant eigenvalue therefore determines the asymptotic behaviour of the population as can be seen from the above model representation. If $\lambda_1 > 1$ the population will grow exponentially, whereas in case of $\lambda_1 < 1$ the population will decrease. For this reason λ_1 is often called the growth rate of the population and is related to the intrinsic rate of increase obtained from Lotka's equation via $r = \ln(\lambda_1)$.

Table 9.1 Life table of the red deer (Lowe 1969) and the resulting projection matrix

Age (years)	Birth rate	Mortality rate	Survival rate
1	0	0.14	0.86
2	0	0.10	0.90
3	0.31	0.11	0.89
4	0.28	0.12	0.88
5	0.30	0.14	0.86
6	0.40	0.16	0.84
7	0.48	0.19	0.81
8	0.36	0.50	0.50
9	0.45	0.67	0.33
10	0.28	0.37	0.63

$$\mathbf{L} = \begin{pmatrix} 0 & 0 & 0.31 & 0.28 & 0.30 & 0.40 & 0.48 & 0.36 & 0.45 & 0.28 \\ 0.86 & & & & & & & & & \\ & 0.90 & & & & & & & & \\ & & 0.89 & & & & & & & \\ & & & 0.88 & & & & & & \\ & & & & 0.86 & & & & & \\ & & & & & 0.84 & & & & \\ & & & & & & 0.81 & & & \\ & & & & & & & 0.5 & & \\ & & & & & & & & 0.33 & 0.63 \end{pmatrix}$$

maximal eigenvalue $\lambda_1 = 1.047$

A further result is that the population will finally reach an equilibrium state called stable age distribution, given by the right eigenvector $\vec{\mathbf{w}}_1$ of λ_1. This can be derived from the characteristic equation $\mathbf{L}\,\vec{\mathbf{w}}_1 = \lambda_1\,\vec{\mathbf{w}}_1$ by considering the second to nth row of this matrix, setting $w_{11} = 1$ and solving for successive values, finally arriving at: $\vec{\mathbf{w}}_1 = \left(\ 1,\ \ S_1\lambda_1^{-1},\ \ \cdots,\ \ S_1S_2\ldots S_{n-1}\lambda_1^{-n+1}\ \right)^T$. Appropriately scaled, this stable age distribution gives the proportion of individuals in the different age classes and does not depend on the initial distribution. Only for the sake of completeness it should be noted that the appropriate left eigenvector $\vec{\mathbf{v}}_1$(sometimes called the reproductive value vector of the population) can be interpreted as contribution values of the age classes to future generations.

As an example, consider the life table of the red deer (adopted after Lowe 1969, cited in Begon et al. 1996) (Table 9.1). This life table results in the Leslie matrix given on the lower panel of the table. Entries which are not explicitly given are zero. In this example we are dealing with a Leslie model with ten age classes; i.e., 1-year-old animals, 2-year-old animals and so on, up to animals 10 years old and older. Based on this matrix a projection for the next 30 years was done, starting with an initial population of 10 individuals, all belonging to the first age class or uniformly distributed over all 10 age classes, respectively.

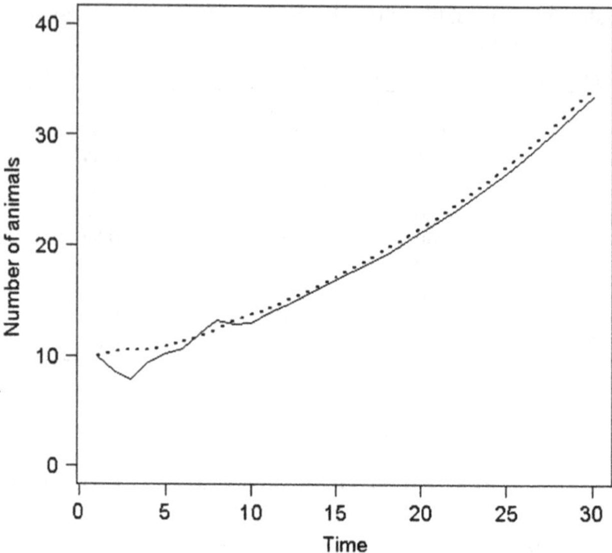

Fig. 9.2 Results of two simulation runs with the projection matrix given in Table 9.1. The initial population was ten individuals, all in age class 1 (*solid line*) and uniformly distributed over all ten age classes (*dotted line*), respectively

The sum of individuals in all age classes for every time point is shown in Fig. 9.2. Apart from small differences at the beginning the population dynamics is the same for both initial populations. Because the maximal eigenvalue of **L** is greater than one, the population will grow exponentially. The normed right eigenvector belonging λ_1 describes the stable age distribution of the deer population (i.e. the proportions of individuals belonging to the different age classes), which finally will be reached no matter what initial distribution is assumed (Fig. 9.3).

Elasticity Analysis

The impact of small changes in the life history parameters on the population dynamics is of special interest for ecologists when making recommendations for species management. For example, a specific age class should be a target for conservation or control, if a small change in survival of this age class markedly affects the population growth. Whether an increase or decrease in survival is desired depends on the population being at risk or a pest, respectively. The effects of small changes in the parameters on the growth rate can be assessed by an elasticity analysis of the projection matrix (see also sensitivity analysis Chap. 23). Elasticity is a measure of the effect of a proportional change in the life history parameters on

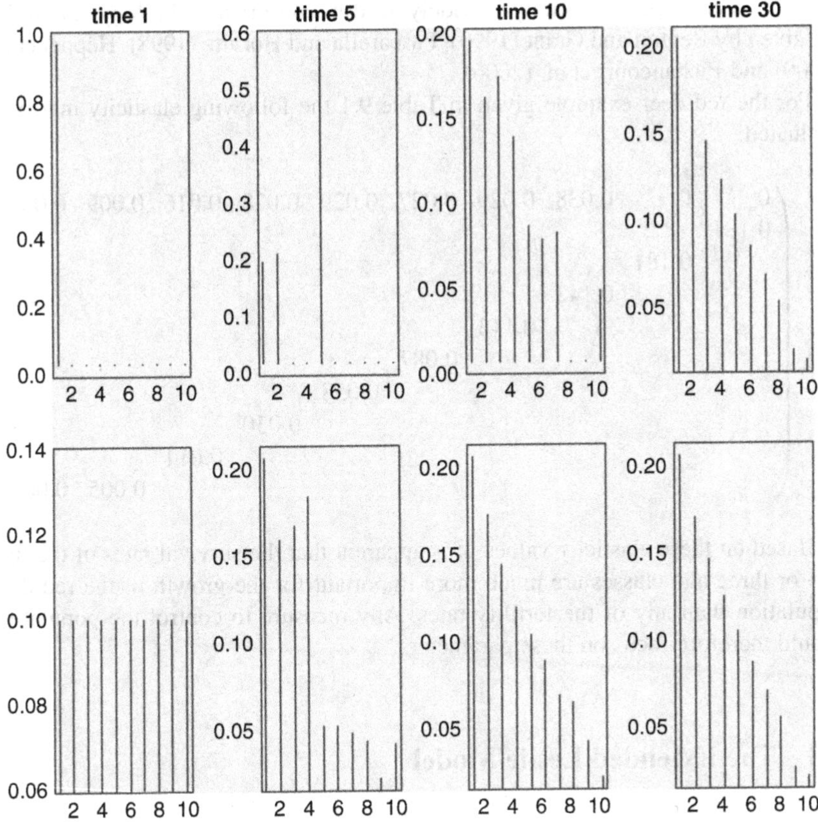

Fig. 9.3 Proportions of individuals (*y*-axis) in ten age classes (*x*-axis) at four time points for two different initial distributions of ten individuals. For the upper panel the simulation was initiated with all individuals being in the first age class, whereas in the lower panel the same number of individuals were uniformly distributed over all age classes. In both cases the same stable age distribution was reached after a few time steps

the population growth rate (Caswell 1978, 2001; de Kroon et al. 2000). According to Caswell (1978), the elasticities for each entry l_{ij} of the matrix **L** are evaluated as:

$$e_{ij} = \frac{\partial\left(\log \lambda_1\right)}{\partial\left(\log l_{ij}\right)} = \frac{l_{ij}}{\lambda_1}\frac{\partial \lambda_1}{\partial l_{ij}} = \frac{l_{ij}}{\lambda_1}\frac{v_i w_j}{\vec{v}^T \vec{w}}$$

with $\vec{v}$ and $\vec{w}$ being the left and right eigenvectors of λ_1. An important characteristic of the elasticities is that they sum up to unity (de Kroon et al. 1986). For specific life history parameters they thus can indicate the relative importance for the population dynamics. Elasticity analysis decomposes the population growth rate into contributions made by the different life history parameters; e.g. growth, survival, reproduction and, therefore, points to the parameters where management measures should focus. The concept of elasticity is widely used in conservation biology; examples are given by Benton and Grant (1999), Pascarella and Horvitz (1998), Heppel et al. (2000) and Pichancourt et al. (2006).

For the red deer example given in Table 9.1 the following elasticity matrix is evaluated:

$$\mathbf{E} = \begin{pmatrix} 0 & 0 & 0.038 & 0.029 & 0.027 & 0.029 & 0.028 & 0.016 & 0.009 & 0.005 \\ 0.181 & & & & & & & & & \\ & 0.181 & & & & & & & & \\ & & 0.143 & & & & & & & \\ & & & 0.113 & & & & & & \\ & & & & 0.087 & & & & & \\ & & & & & 0.058 & & & & \\ & & & & & & 0.030 & & & \\ & & & & & & & 0.014 & & \\ & & & & & & & & 0.005 & 0.007 \end{pmatrix}$$

Based on these elasticity values, it is apparent that the survival rates of the first two or three age classes are much more important for the growth of the red deer population than any of the fertility rates. Any measure to control the population should therefore focus on these parameters.

9.3 The Extended Leslie Model

In the original Leslie model there are several restrictions that should be mentioned: First, up to now we have dealt with a constant environment. To consider fluctuating environmental conditions (e.g. temperature) that play a crucial role in insect development, other concepts are needed. One possible way to do this is to translate the real time into another unit, taking into account the environmental conditions. This alternative unit is called the biological time (or biological age, as well) and it measures (to a certain extent) the state of development of individuals (Söndgerath and Müller-Pietralla 1996; Schröder and Söndgerath 1996).

The concept of biological time is as follows: consider a population whose development will last 100 days under constant optimal conditions. The development rate for this population is defined as $1/100 = 0.01$/day. Summing up the development rates after 100 days will yield a value of 1 (the time the development is complete).

Real conditions are never constant, so a formulation of the development rate dependent on environmental conditions is needed. For the most important environmental factor (the temperature), this relationship is often an optimum curve which, for example, can be described by an O'Neill function (Spain 1982) depending on parameters $r_{\max}$ (maximal development rate under optimal temperature), T_{opt}, T_{max} (optimal and maximal temperature, respectively) and the Q10-value, which describes

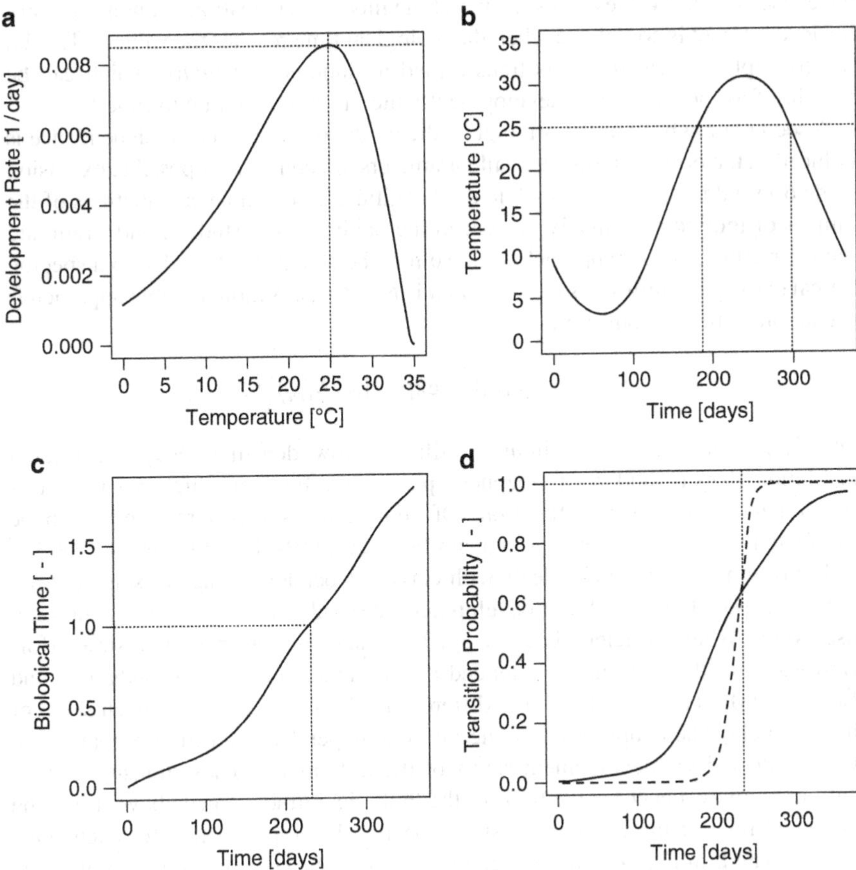

Fig. 9.4 Relation between temperature, development rate, biological time and transition probability: (**a**) Temperature-dependent development rate described by an O'Neill function with $r_{\max} = 0.0085$ and $T_{opt} = 25$. (**b**) Artificial time course of temperature described by a sine function. (**c**) Resulting biological time calculated by integrating the O'Neill function over time. (**d**) Weibull distributions with scale parameter 1 and different shape parameters (2 for the *solid* line and 12 for the *dashed* one) applied to the biological time shown in panel c lead to the transition probabilities of the extended Leslie model

the increase of the development rate as a consequence of increasing the temperature by 10°C. An example for an O'Neill function can be seen in Fig. 9.4a. Combining this function with a real temperature course yields a time dependent development rate. Finally, integrating this time-dependent rate over time will give the biological time:

$$biol(t) = \int_0^t rate(temp(\tau))\, d\tau$$

For simplicity, one can regard the biological time as the sum of the development rates. This is because an integral can be approximated by an infinite sum of piecewise constant values. Irrespective of the mode of evaluation (sum or integral), the development is completed when the biological time reaches the value of 1. With this concept, the real time t is transformed to another unit $biol(t)$ which can be thought of as the fraction of development which is completed up to time t.

A second drawback of the original Leslie model is that no density dependence is incluced in the original model and all parameters are constant. A possible extension of the original model is to formulate the fertilities, for example, as functions of the number of individuals already present in the habitat under study (Söndgerath and Schröder 2002). A first approach to do so may be a step function. Up to a specific critical density the fertility is F_1; beyond it is F_2 ($< F_1$). A more flexible approach is a function of the Weibull-type:

$$F(N) = F_{max}\,(1 - \exp(-(N/N_{crit})^{\alpha})),$$

with F_{max} denoting the maximum fertility at low densities, N_{crit}, the critical population density and α, a steepness parameter. For very high α-values this function reduces to a step function. Of course, the survival rates can also be formulated as density dependent in a similar manner (Pykh and Efremova 2000).

The original Leslie model deals with classes dependent on age or stage but not both. For populations whose development depends on the environment (e.g. insects) this is not sufficient, because age and stage are not linked in a straightforward way. For this reason, the extended Leslie model was set up (Söndgerath and Richter 1990). This model coupled different Leslie models, one for each stage of the life cycle. The coupling was done via time-dependent transition probabilities, which reflected the development status of the individuals. These transition probabilities were evaluated on the basis of the biological time defined above. First, the biological time resulting from the stage-specific development rate for each stage was evaluated. In Fig. 9.4b an artificial temperature curve is given. In Fig. 9.4c the biological time resulting from integration of the time-dependent development rates is shown. The latter can be reached by combining the O'Neill function with the time course of temperature. In the first period the biological time increases slowly because of the low temperature. Due to nearly optimal conditions, the increase is higher after approximately 150 days. Temperatures above the optimum of 25°C (between days 187 and 298) result in a flattening of the biological time curve in the specified period. As outlined above, the development of one stage is completed when the appropriate biological time reaches the value of 1. In the example shown in Fig. 9.4 this is the case after 233 days. To take biological variation into account a statistical distribution (e.g. a Weibull distribution) is applied subsequently to finally reach at the transition probabilities for the model. The Weibull distribution has two parameters: one scale parameter, which is 1 in this case, and a shape parameter, which affects the steepness of the curve. The higher the shape parameter, the less biological variation is included (see Fig. 9.4d).

The general structure of the extended Leslie model for a population with three development stages can be seen in Fig. 9.5. This kind of model was successfully used for different purposes, e.g. to forecast the dynamics of pest populations

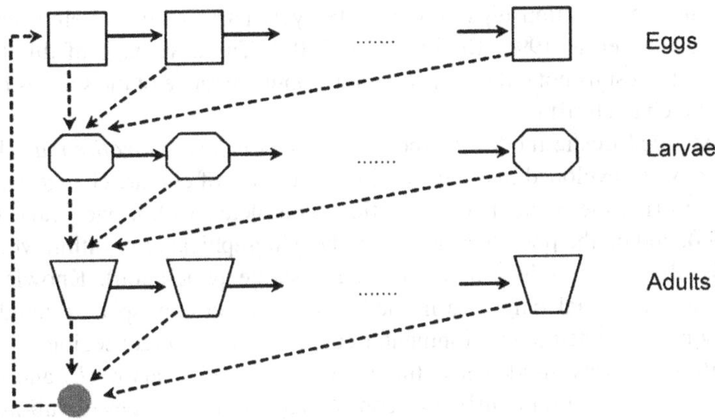

Fig. 9.5 Life cycle of an age- and stage-structured population. In each of the three development stages different age classes are passed through, for each stage a Leslie model is set up. The different stages are linked via transition probabilities depending on environmental variables. The *solid arrows* denote survival probabilities within one stage, and the *dashed* ones transition probabilities into the next stage

(Söndgerath and Müller-Pietralla 1996). In a simulation study it was used to investigate the effect of habitat fragmentation on the spread of populations (Söndgerath and Schröder 2002). Recently, the extended Leslie model was applied to describe the dynamics of the dragonfly *Gomphus vulgatissimus* along a latitudinal gradient over Europe (Braune et al. 2008). This will be described in the next section.

9.4 An Ecological Application: Effects of Global Change on the Voltinism in Dragonflies

All model simulations of the Intergovernmental Panel on Climate Change (IPCC) show a warming in the future across Europe due to climate change. Several examples for ecological consequences of recent climate change have been described; e.g. changes in populations and reproductive biology, changes in phenology, changes in geographic range and ecosystem-level changes (Hickling et al. 2005; Parmesan 2006; Hassall et al. 2007). Studies on European dragonflies show that some species already exhibit accelerated life cycles and/or their ranges have shifted northwards (Hassall and Thompson 2008).

Different approaches are in use to model species spatial distributions under different climate conditions. The species distributions can be predicted by inferring the environmental requirements of the species from their current geographical distribution (Climate Envelope Models, see e.g. Araújo et al. 2006 or Hijmans and Graham 2006). Other approaches are structured population models, which assume functional relationships between life cycle parameters and environmental factors (Caswell et al. 1996; Braune et al. 2008). The advantage of this kind of model is that it results not only in information about the range of the species but also in life cycle characteristics.

The extended Leslie model for the dragonfly *Gomphus vulgatissimus* (Braune et al. 2008) was developed in order to study the effects of climate change scenarios on voltinism (i.e. the number of generations completed within one year) (Corbet et al. 2006), and on the potential range (i.e. the geographical area within which this species is able to reproduce and establish a stable population). Knowledge of voltinism and potential range are needed to understand how species could adapt or already have adapted to environmental conditions in different regions.

The life cycle was divided into three stages: eggs (E), larvae (L) and mature adults (A). The projection matrix was constructed from two types of sub-matrices for each stage: one for survival within that stage (S) and one for transition or development into the next stage (D).

$$\mathbf{L} = \begin{pmatrix} S_E & 0 & D_A \\ D_E & S_L & 0 \\ 0 & D_L & S_A \end{pmatrix}$$

The temporal pattern of the life cycle was determined by the probability distribution function of the random variable development time, which itself depends on the time course of environmental covariates (Söndgerath and Richter 1990). In the case of *G. vulgatissimus* the major controlling environmental variables are temperature and day-length (Corbet 1999). The dependency on these variables was modelled by the accelerated life model (Cox and Oakes 1984) with a multiplicative approach, leading to the biological time (Schröder and Söndgerath 1996). As explained above, this item reflects the actual development status of an individual: the greater the biological time, the more advanced is the development. The development is completed when the biological time equals unity. Biological variation was taken into account by subsequently applying a statistical distribution function to evaluate the transition probabilities into the next stage. For further details of the model see Braune et al. (2008). With this model, simulations were done along a latitudinal gradient from southern (42°N) to northern (62°N) Europe. This latitudinal gradient describes the major distribution limits of the species. Evaluations were performed for present-day conditions, as well as for three future time points (2020, 2050 and 2080). For the latter, temperature rises according to the scenarios given by the Intergovernmental Panel of Climate Change (IPCC) were incorporated. The initial population for each simulation was 5,000 eggs.

The results of two of the four simulations can be seen in Fig. 9.6. For the present-day scenario, *G. vulgatissimus* showed a 2-year life cycle, up to about 50°N in southern Europe. Between 50°N and 52°N, both 2- and 3-year developments are shown, suggesting cohort splitting. For latitudes from 52°N to 54.5°N, the larval life cycle lasted 3 years, followed by a region with 3–4-year development (54.5°N–56°N). At the northernmost range, the larvae needed 4 years for their

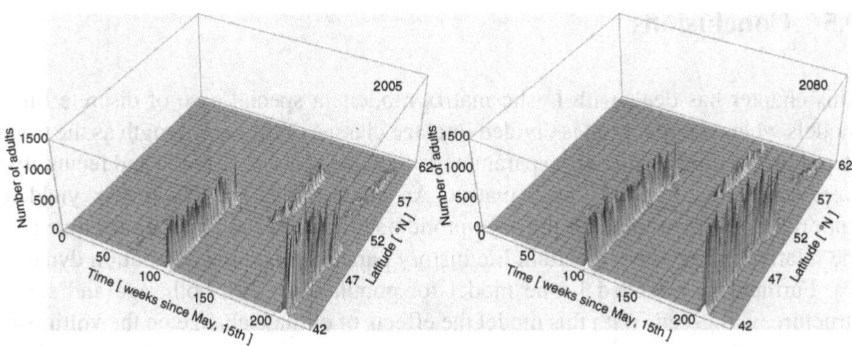

Fig. 9.6 Results of simulation runs for the dragonfly Gomphus vulgatissimus based on an extended Leslie model with temperature and day-length as environmental variables. On the *x*-axis time is shown, geographical latitude is on the y-axis and the simulated number of adult dragonflies on the *z*-axis. The results are show along the gradient from 42°N to 62°N for current climatic conditions (2005) and for the year 2080 according to the IPCC scenarios. The model shows a northward extension of the species together with a faster development (Braune et al. 2008)

development. The model predicts that from 2020 a 2-year life cycle will predominate in wide parts of the current range of the species, whereas the 3-year life cycles shift further northwards. The 4-year life cycle in the north became negligible in 2020 and disappeared even in the northernmost parts in 2050.

Common to all simulations was a shift towards a slightly later emergence from south to north induced by the decrease of day-length with more northern latitudes. Naturally, the beginning of emergence (which is reproduced by the model) ranged from April to June between southern and northern populations. In the zones of cohort overlap the model predicts a longer flight season in the future or even a bimodal phenology, with a large peak in spring and a smaller peak in late summer.

The model also predicted a slight northwards shift of the distribution range by at most 1.25° in 2080. Such northward shift for *G. vulgatissimus* has already been recorded. It adds up to 104 km from 1960–1970 to 1985–1995 (Hickling et al. 2005).

Summarizing the model revealed three main climate change effects: first, the distribution of voltinism patterns is affected with a general trend towards a spread of shorter generation times in the northward direction. Second, emergence is accelerated in southern latitudes. The pattern of earlier emergence does not shift northwards. This may be due to the additional photoperiodic control in the model. Third, the temperature scenarios provided by the IPCC led to a northward extension of the species.

However, competition with related species, prey availability, droughts and other factors may cause other reactions of the population that are not yet covered by the model. Nevertheless it appears that population models based on Leslie matrices can be powerful tools to forecast effects of climate change on voltinism patterns and distribution range of species. Combining several single-species models can help to analyze the consequences of climate change on community and ecosystem levels, e.g. a temporal decoupling ("mismatch") of up to now synchronized processes such as temporal coincidence of predators and the appropriate prey populations.

9.5 Conclusions

This chapter has dealt with Leslie matrix models, a special case of discrete time models where a population is divided into age classes of the same length as the time step. For this model population parameters, like age-specific survival and fecundity, were summarized in the Leslie matrix. An eigenanalysis of this matrix yielded important information about the asymptotic behaviour of the population as well as the relative importance of certain life history parameters on the population dynamics. Further, an extended Leslie model for populations with both, age and stage structure, was given. With this model the effects of climate change on the voltinism and range of occurrence of a dragonfly species were investigated. In this example it was shown that population models based on Leslie matrices can be powerful tools in ecological investigations, although some important factors were not yet covered by the model. Therefore, Leslie matrix models can help to analyze community and ecosystem level consequences of climate change.

Further Readings

Caswell H (2001) Matrix population models, 2nd edn. Sinauer, Sunderland

Keyfitz N, Caswell H (2005) Applied mathematical demography, 3rd edn. Springer, Heidelberg

Montshiwa MI (2007) Leslie matrix model in population dynamics. PGDipl Thesis, University of Witwatersrand. http://users.aims.ac.za/~mosimanegape/essay.pdf

Soetaert K, Herman P (2009) A practical guide to ecological modelling. Springer, Heidelberg

Chapter 10
Modelling Ecological Processes with Fuzzy Logic Approaches

Agnese Marchini

Abstract The development of an ecological model may involve problems of uncertainty. Ecologists have to deal with imprecise data, ecosystem variability, complex interactions, qualitative aspects, and expert knowledge expressed in linguistic terms. In all these cases, fuzzy logic could provide a suitable solution. Fuzzy logic allows to: use uncertain information such as individual knowledge and experience; to combine quantitative and qualitative data; to avoid artificial precision and to produce results that are found more often in the real world. Developed in the late sixties as a method to create control systems when using imprecise data, fuzzy logic has been used for a very large number of engineering applications, and more recently to develop models of air, water and soil ecosystems.The following sections of this chapter introduce the basic structure of a fuzzy model, describing the variety of options that exist at each stage. An example of fuzzy model is also outlined: the knowledge-driven development of an index of water quality having five qualitative output classes. Finally, possible future developments of fuzzy modelling in ecology are suggested.

10.1 Introduction

10.1.1 Fuzzy Logic: A Mathematical Theory for Uncertainty

Ecologists very frequently cope with imprecise and vague data. Formalizing such fragmentary information with traditional "hard computing" approaches can be difficult. Important alternative tools are provided by the so-called "soft computing" techniques. They fill a methodological gap, dealing with imprecision, uncertainty, partial truth, and approximation. Fuzzy logic, a theory developed in the mid-1960s

A. Marchini
DET – Dipartimento di Ecologia del Territorio, University of Pavia, Via S. Epifanio 14, I-27100 Pavia, Italy
e-mail: agnese.marchini@unipv.it

F. Jopp et al. (eds.), *Modelling Complex Ecological Dynamics*,
DOI 10.1007/978-3-642-05029-9_10,

by Lotfi A. Zadeh, is a soft computing approach particularly suitable for ecological models.

Fuzzy logic is an alternative method to represent complex systems, such as those encountered in biology, sociology or economy. Instead of numerical variables and mathematical formulas, fuzzy models require linguistic variables and rules. For example: the variable *temperature* might have linguistic values such as *hot* or its antonym *cold*; a linguistic rule might be: "*if* pressure is *high*, *then* volume is *low*". This approach helps ecologists who have an idea of the process under study, but have data affected by too much imprecision to be used in the development of a formal model. Fuzzy logic applies meaning to imprecise concepts and uses uncertainty as an additional source of information. The first key concept of fuzzy logic is partial truth. Rather than labeling a statement as either *true* (1) or *false* (0), as classic binary logic does, in fuzzy logic the degree of truth of a statement can assume any value between 0 and 1. The degree of truth is established by a membership function μ. Membership functions, the second key concept, represent the transition from numbers to words and allow one to "compute with words". Dealing with fuzziness and imprecision does not prevent fuzzy logic from being a mathematical formalism, and fuzzy systems being sound and not ambiguous. Bart Kosko, one of the pioneers in the development of fuzzy systems, demonstrated that "*an additive fuzzy system can uniformly approximate any real continuous function on a compact domain to any degree of accuracy*" (Kosko 1994).

Fuzzy logic has had a great impact on mathematical disciplines such as logic, algebra, topology, data analysis, etc. Technical applications include a variety of different controllers and software, for example: chips that control cameras, elevators, air conditioners, washing machines and other home appliances; automobile and other vehicle subsystems, such as automatic transmissions, ABS and cruise control; artificial intelligence in video games; edge detection in digital image processing; pattern recognition algorithms in remote sensing; and language filters for offensive text in message boards and chat rooms. The impact of fuzzy logic on ecological research has displayed an increasing trend: the first applications date from the late 1980s, when it was introduced for the ordination of vegetation data. Since then, more than two hundred publications about fuzzy applications in ecological research have appeared in scientific journals.

10.1.2 Fuzzy Logic and Ecology

Fuzzy logic has been repeatedly proposed as a powerful technique to develop models for decision support in ecosystem management (Silvert 1997, 2000; Adriaenssens et al. 2004; Fränzle 2006; Jørgensen 2008). The capability of fuzzy logic to use uncertain information that other methods cannot take into account makes this computational technique particularly important in ecology. In fact, the representation of ecosystems is affected by many sources of fuzziness:

- Uncertain input data: gaps, inaccuracy, spatial and temporal variability, heterogeneity, qualitative ecosystem attributes
- Uncertain input–output relationships: complex systems, non-linear interactions, management of multiple variables, exceptions to general rules, different importance of variables
- Uncertain output: qualitative classes, non-ordinal classes, non-sharp boundaries
- Uncertain expert knowledge, legislative requirements, opinions of administrators, end-users, etc, expressed in linguistic form

Fuzzy models of ecosystem functions have been developed in all continents, confirming that the principles of fuzzy logic theory have been established worldwide in ecological institutions. Fuzzy systems have been developed for air, water, soil ecosystems, with both biotic and abiotic ecosystem variables. Analysis and assessment of air pollution, water eutrophication, groundwater contamination, control of ozone levels, evaluation of sustainability of fishery and fish farming, analytical assessment of soil degradation, leaching or acidification risk, are just a few examples of ecological issues modelled with a fuzzy approach that can be found in the scientific literature.

In a number of cases, fuzzy logic has also been combined with GIS or with other soft computing techniques, such as artificial neural networks, self-organizing maps, and genetic algorithms in order to integrate the adaptability of fuzzy logic to human reasoning with a data-driven approach.

Most authors have acknowledged that fuzzy logic represents a suitable, feasible and effective tool to deal with ecological issues, and that it provides more reliable results when compared with other methods for environmental assessment (e.g. Altunkaynak et al. 2005). Fuzzy logic has demonstrated to be particularly suitable for the development of new indices of ecological quality. It allows one to acknowledge the subjective aspects of the index design process, and provides the tools to easily handle subjectivity, by quantifying it and manipulating it with mathematical rigour (Table 10.1).

10.2 Structure of a Fuzzy Model

A fuzzy model typically consists of three stages: fuzzification, inference and defuzzification. The latter is not always required.

Fuzzification is the process for linking a variable to its underlying characteristics by means of a membership function μ. After fuzzification, the input variable loses its numerical definition and acquires a linguistic definition: for example, it might become *small, hot, high*. Its numerical value is transformed into a fuzzy membership grade: a unit-less number in the interval [0, 1].

Once all variables are fuzzified, they can be processed through the fuzzy rules of the model. This is the inference stage. Fuzzy rules typically are in the form of logical implication (*if ... then*). Single rules are then combined to produce a fuzzy output, still in the form of fuzzy membership grade in the [0, 1] interval.

Table 10.1 The fuzzy approach can be useful in different ways to develop an ecological index. A number of issues that can be encountered in the process of index development is presented, with the respective solutions offered by the fuzzy approach

Issues of the index development process	Solutions
Boundaries	
Many "traditional" indices subdivide the range of a variable into intervals associated with different scores or quality classes. Boundaries between two intervals are often sharp. In this way, small deviations from a threshold value implies big differences in the output. This is not in agreement with the behaviour of natural systems	Fuzzy membership functions allow to represent soft boundaries and gradients. Membership functions can overlap: a variable value can display two characteristics at the same time
System complexity	
One of the most desirable features of ecological indices is the ability to combine metrics in a manner that is complex enough to capture the dynamics of essential ecological processes, but not so complex that their meaning is obscured (Borja and Dauer 2008)	Fuzzy rule-based systems are able to model non-linear, multidimensional, complex phenomena; yet, the linguistic form of the *if...then* rules make fuzzy models easily understandable
Reference conditions	
Reference conditions, required to define the class of best ecological quality, should be described by pristine, undisturbed environments. Unfortunately, such conditions may not exist anymore. Scientists are thus required to define virtual reference conditions using mathematical models and/or expert judgement	Fuzzy models are mathematical models, based on logic rules, and are also expert systems, based on expert judgement
Legislative criteria	
Classes of ecological quality (e.g. *good, moderate, poor*) do not have a clear quantitative definition. Threshold values that separate two quality classes should take into account the levels of acceptability and public concern, which are highly subjective, and not directly measurable	Fuzzy membership functions are suitable for representing purely linguistic variables, such as classes of ecological quality. They could be designed with some degree of overlap, avoiding hard thresholds and integrating different conceptions of *good* or *poor* ecological quality

Defuzzification is the process to convert the fuzzy output into a non-fuzzy value that can be used in non-fuzzy contexts.

The following subsections will present an overview of the most used techniques for fuzzification, inference and defuzzification. In fact, the three stages can be performed in many different ways (Table 10.2), depending on the type of information integrated in the model and on the required output. This plasticity is an

Table 10.2 Stages of a fuzzy model and some possible options for developing each step

Model stages	Steps requiring definition by the model developer	Possible options	Definition can be
Fuzzification	Input variables	Qualitative or quantitative	Knowledge-driven (ecological knowledge)
	Membership functions: number	Generally 2–5	Knowledge-driven or data-driven
	Membership functions: position along the x-axis	Depends on the variable range; generally in $[0, +\infty]$, or $[-\infty, +\infty]$	Knowledge-driven or data-driven
	Membership functions: shape	Linear (triangular, trapezoidal), or non-linear (gaussian, sigmoidal, others)	Knowledge-driven or data-driven
Inference	Model type	Takagi–Sugeno or Mamdani	Knowledge-driven
	if ... then rules	Depend on the problem	Knowledge-driven or data-driven
	Mathematical operators	Min and max; product, weighted sum, others	Knowledge-driven or data-driven
	Weights (optional)	Any positive real number	Knowledge-driven or data-driven
Defuzzification (optional)	Type of output	Qualitative or quantitative	Knowledge-driven
	Defuzzification method	Centroid, mean, max, mean of maximum, center of maximum, bisector, linear combination, "winner takes all", others	Knowledge-driven or data-driven

important feature of the fuzzy approach, which allows one to deal with any kind of data. On the other hand, the variety of choices that exist at each stage may be seen as a difficulty and a source of subjectivity: ecologists have no guidelines to select the most opportune techniques for a specific problem. The definition of several steps listed in Table 10.2 can be data-driven, i.e. based on statistical evaluation of data sets or machine learning techniques. Optimal parameterization of membership functions, weights, rules, etc., is recommended as it improves model performance, but it is only possible when an adequate amount of experimental data is available. Unfortunately, in ecological research it is often difficult to collect large data sets, especially when biological information (populations, communities) is involved. Therefore, the knowledge-driven approach, based on expert opinion, is the most common strategy to define membership functions and *if ... then* rules in ecological fuzzy models. Of course, it introduces more subjectivity into the model.

10.2.1 Fuzzification: Membership Functions

Membership functions are the core of a fuzzy model, and the most revolutionary concepts of fuzzy set theory. One of their advantages is that the linguistic expression of a variable is easily understandable by everyone, whereas numerical values are meaningful only for experts.

Any model variable is described by a characteristic $\boldsymbol{C}$, such as *low* or *high*. A membership function $\boldsymbol{\mu_C}$ converts the numerical value of the variable $\boldsymbol{v}$ into a membership grade to the characteristic $\boldsymbol{C}$. The membership grade ranges from 0 to 1, and can assume all values in this interval:

$\boldsymbol{\mu_C(v)} = 1$, full membership: $\boldsymbol{v}$ displays $\boldsymbol{C}$ completely
$\boldsymbol{\mu_C(v)} = 0$, null membership: $\boldsymbol{v}$ does not display $\boldsymbol{C}$
$\boldsymbol{0 < \mu_C(v)} < 1$, partial membership, $\boldsymbol{v}$ partially displays $\boldsymbol{C}$

A straightforward example: consider the variable "water temperature" and its characteristics *cold* and *hot*. A temperature 0°C could have membership 1 to the function *cold* and membership 0 to the function *hot*. Conversely, the value 100°C could have membership 0 to *cold* and membership 1 to *hot*. Membership grades between 0 and 1 represent intermediate situations. In other approaches, they would be classified as uncertain values. Fuzzy logic converts uncertainty into enhanced information. In fact, uncertain values have partial membership to more than one characteristic. This is due to an important feature of membership functions: they overlap. A variable value is allowed to have non-null membership for two functions simultaneously. For instance, a medium temperature value can be partially *cold* and partially *hot*. The amount of overlap between two functions is related to the amount of uncertainty included in the model: the more the overlap, the more the uncertainty. Classical "crisp" functions do not tolerate uncertainty, since they have null overlap (a temperature value can be either *cold* or *hot*). In other terms, whereas in classical systems there would be a steep threshold between *cold* and *hot* water temperature, the fuzzy approach offers a gradual transition, which is more similar to real world conditions.

To develop a fuzzy model, the ecologist is required to define the number of membership functions for each variable of the model, their shape and their position along the x-axis. These are parameters of the model and indicate the semantic meaning of the variable characteristics. Membership functions can be linear (triangular and trapezoidal) or nonlinear (bell-shaped, sigmoid, polynomial) (Fig. 10.1). Triangular and trapezoidal membership functions indicate that the variable characteristic changes linearly as the variable value changes. Nonlinear functions describe more complex behaviours. Triangular and trapezoidal functions are easier to define, whereas more complicated shapes are less intuitive and may require the definition of more parameters.

However, linearity is not a common behaviour of ecosystems: nonlinear functions such as Gaussian and sigmoid would probably better represent ecological data and improve model performance.

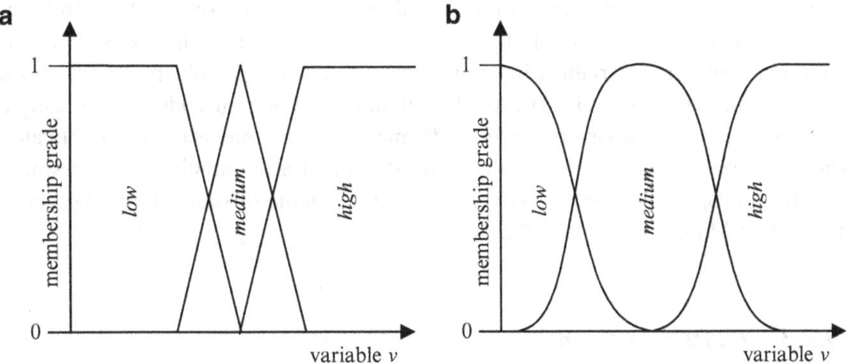

Fig. 10.1 Examples of linear (**a**) and non-linear (**b**) membership functions for the characteristics *low*, *medium* and *high* of a variable *v*

10.2.2 Model Rules

A fuzzy model is a collection of linguistic rules, thus it is transparent and easily understandable. A fuzzy rule is a logical implication, "*if* antecedent, *then* consequent". The action of the implication is to reduce the truth of the consequent according to the degree of fulfilment of the antecedent. The antecedent is a combination of several conditions connected with the logical operator of conjunction, *and*. To produce a fuzzy output, all fuzzy rules are aggregated, generally with the logical operator of disjunction, *or*. The logical operators *and*, *or*, *if*... *then* can be implemented with different mathematical operations (Table 10.2). Typically, minimum is used for *and*, maximum for *or* and product for *if*... *then*.

There are two main types of fuzzy models: Mamdani and Takagi–Sugeno. In the Mamdani type, both rule antecedent and consequent are in the form of membership functions: "*if* (variable *v* is characteristic ***C***, *and* . . ., *and* variable *z* is characteristic ***Z***), *then* output is characteristic ***O***". The model output is expressed in linguistic terms: it is a membership grade to a linguistic characteristic. In some cases, a linguistic output (for example a quality class) is desirable, thus, defuzzification can be avoided. In Mamdani models, defuzzification becomes necessary only when a numerical output is required. In Takagi–Sugeno models, the rule conclusion is already a crisp value, linear function of the inputs: "*if* (variable *v* is characteristic ***C***, *and* . . ., *and* variable *z* is characteristic ***Z***), *then* output is ***f***(*v* . . ., *z*)". Therefore, this model type does not require an explicit defuzzification procedure.

In ecological applications, the linguistic character of the Mamdani-type output is suitable to model qualitative ecosystem features, for example classes of soil or water quality (Ocampo-Duque et al. 2006; Tscherko et al. 2007; Icaga 2007), whereas the behaviour of environmental variables can be easily forecast or modelled by means of the Takagi–Sugeno approach (Jorquera et al. 1998; Ryoke et al. 2000; Altunkaynak et al. 2005; Lin and Cobourn 2007).

Fuzzy models take into account all possible combinations of membership functions (characteristics) of all variables, thus the number of rules increases exponentially with number of variables and membership functions. Simple models display a few rules, but there can be fuzzy models with hundreds or thousands of rules. Single rules are generally assessed by expert judgement, but this may become complicated when a large number of rules is involved. Methods for automatic rule assessment have been proposed, in order to overcome the problems of subjectivity (Tscherko et al. 2007; Marchini et al. 2009).

10.2.3 Defuzzification

The fuzzy output is determined by the degrees of fulfilment of several rules. Fuzzy rules in fact allow their partial and simultaneous fulfilments. Depending on the values of the input variables, some rules will be fulfilled more than others, and their own output will have more importance in the aggregation process for the computation of the final output.

Whilst in Takagi–Sugeno models the output is already in the form of scalar, the output of Mamdani models is in fuzzy form, i.e. membership grades to linguistic characteristics, and has to be defuzzified. Defuzzification is the process of combining several partial memberships to produce a crisp set, or a crisp single-valued quantity, compatible to non-fuzzy approaches. It can be performed through different techniques (Table 10.2). The selection of the most opportune technique depends on the type of information desired in the output. In ecological models the output sets are usually defuzzified by calculating their center of gravity, weighted average, and maximum; less frequently by other methods. When qualitative classes need to be evaluated, defuzzification is avoided, as it entails a loss of information.

10.2.4 How to Create and Use a Fuzzy Model

The previous sections have shown that fuzzy models can be developed using a variety of strategies. It is almost impossible to specify a detailed guideline to create a fuzzy modal that can have general validity. However, all models have essential steps to be followed, which can be implemented choosing amongst the possible options presented in Table 10.2:

1. Select input variables and their characteristics
2. Design membership functions for each variable characteristic
3. Select an output variable and its classes
4. Generate *if ... then* rules, combining all variable characteristics (antecedent of implication) and output classes (consequent of implication)
5. Define logical connectives used to manipulate rules (*and*, *if ... then*, *or*)
6. Define defuzzification strategy (optional)

It is always recommended to clearly explain the choices undertaken at each step, and the motivations that have justified them. For example: "*bell-shaped membership functions have been chosen since they better represent the Gaussian behaviour of the model variables*".

Once the model has been developed, its application with sample data involves the following steps:

1. *Fuzzify*. Calculate membership grades of the input data through membership functions
2. *Infer*. For each rule, calculate grade of antecedent fulfillment (application of connective *and*), and grade of rule fulfillment (application of connective *if ... then*). Aggregate results of all rules using connective *or*
3. *Defuzzify*. Apply defuzzification strategy (when scheduled)

10.3 An Ecological Application: Design of a Quality Index

The development of effective indices of ecological quality represents an important branch of environmental research, as the capability of measuring human disturbance on natural ecosystems is necessary for effective management (Marchini 2010). However, there is little acceptability of ecological indices by environmental managers and even by scientists. The suitability of old and new indices has been called into question during conferences and on the pages of scientific journals. Although perceived as objective procedures, ecological indices involve many steps that are based on subjective expert judgement: variables selection, data transformations, definition of thresholds, etc. (Scardi et al. 2008). Fortunately, there are advanced computational techniques such as fuzzy logic, able to handle subjectivity and effectively, by quantifying it and manipulating it with mathematical rigour (Shepard 2005).

This chapter presents the development of a multi-variable index (Marchini, unpublished data) for transitional waters (estuaries, lagoons) using a fuzzy model. The model follows the Mamdani type, which is the most commonly used for ecosystem management (Adriaenssens et al. 2004). Transitional waters are highly variable environments, therefore, distinguishing between natural and human-induced disturbance is problematic. The European Water Framework Directive (2000/60/EC) requires the inclusion of biological elements, namely phytoplankton, other aquatic flora, zoobenthos and fish fauna to measure ecological status of transitional waters. Ecological status has to be expressed by means of five quality classes: *high, good, moderate, poor* and *bad*. This requirement is extremely difficult to meet. Generally, transitional waters host low-diversity communities, with dominance of disturbance-tolerant species. For this reason, methods based on species diversity or sensitivity might be unable to identify anthropogenic impacts in these environments. Biological metrics for the definition of ecological status should be (a) ecologically relevant, i.e. their response to disturbance should be unequivocal and acknowledged by a large scientific community, and (b) easy to measure, i.e. measurements should be low-cost, technically easy to perform

Table 10.3 List of selected metrics considered in the new index, their response to disturbance (>: direct relation; <: indirect relation; ><: unimodal relation) and their importance in the definition of the index value, expressed as weight

Biological quality elements	Metrics	Response to disturbance	Weights
Phytoplankton	1. Chl *a* concentration (μg l^{-1})	>	0.14
Macroalgae	2. Percent cover of opportunistic species (%)	>	0.152
	3. Biomass (gWW m^{-2})	> <	0.168
Seagrasses	4. Presence (yes–no)	<	0.097
Benthic invertebrates	5. Shannon diversity (H')	<	0.145
Fish fauna	6. Total number of fish species	<	0.152
	7. Number of estuarine resident species	<	0.145

and unlikely to be affected by errors. Unfortunately, finding metrics with both features is a challenging task. Single metrics often display ambiguous responses to human disturbance, especially in transitional waters where the border between natural and human-induced disturbance is really fuzzy. Furthermore, different biological elements may respond in different ways to a disturbance. Therefore, an ecologically robust approach could be a combination of different biological elements and characteristics of these elements into a single classification tool, which can provide the integrated response of the community.

First, a careful survey of relevant literature on ecological indices for transitional waters was performed. The survey highlighted a set of seven metrics indicated as ecologically relevant by many authors (Table 10.3). They were used as input variables. Membership functions were drawn for all variables, trying to preserve the personal judgement of different experts. Figure 10.2 illustrates how thresholds indicated by different authors have been integrated and utilized to design membership functions for the metric "total number of fish species". For all variables, the characteristics associated with *bad* ecological status and the characteristics associated to *high* ecological status have been clearly identified. All seven metrics do not have the same importance for ecological quality assessment: they have been weighted as shown in the last column of Table 10.3.

The rule-base has been developed by setting the combination of all variables in their worst characteristics to *bad* ecological status: *if* (total number of fish species is *low, and* Chl *a* concentration is *high, and* seagrasses are *absent, and*....), *then* ecological status is *bad*. Conversely, *high* ecological status has been assigned to the combination of all the variables' best characteristics: *if* (total number of fish species is *high, and* Chl *a* concentration is *low, and* seagrasses are *present, and*....), *then* ecological status is *high*. Intermediate combinations have been associated to the other three classes of ecological status, *good, moderate, poor*. The metrics weighting and the rules assessment have been performed by means of an automatic procedure explained in Marchini et al. (2009). The following mathematical operations have been used to calculate the output: algebraic product for conjunction (*and*) and for implication (*if* ... *then*), algebraic sum for disjunction (*or*). They have been chosen as they

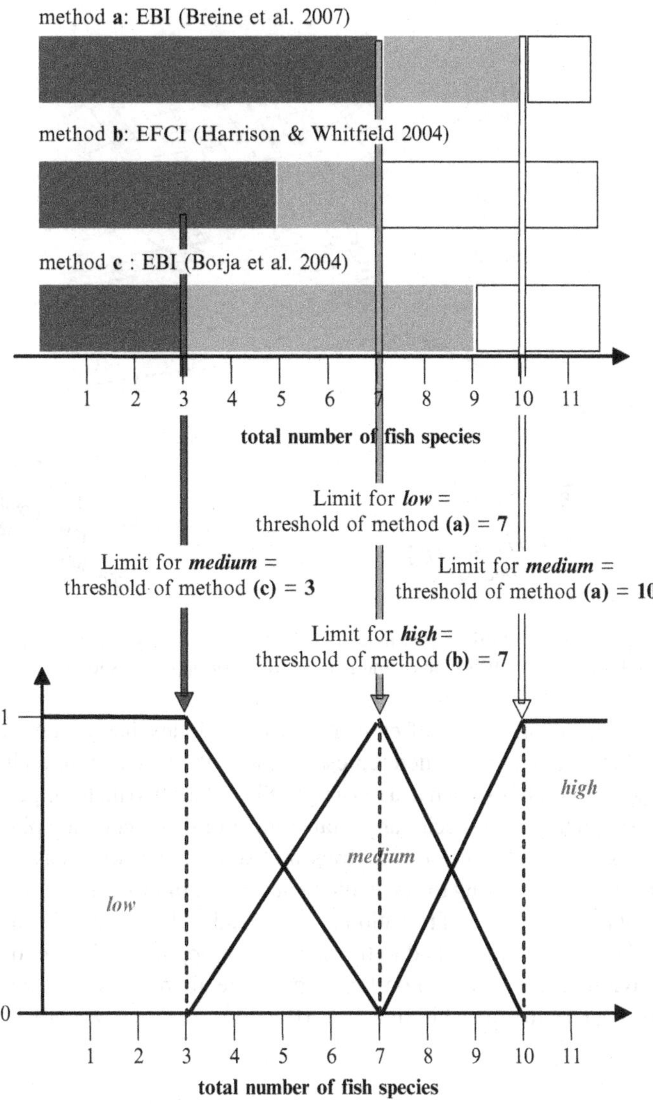

Fig. 10.2 Definition of the membership functions *low*, *medium* and *high* for the metric "total number of fish species" based on three published methods ((**a**), Breine et al. 2007; (**b**), Harrison and Whitfield 2004; (**c**), Borja et al. 2004)

guarantee a smooth variation of the index in response to input variations. Defuzzification has been performed by means of the linear combination technique: the five output classes of ecological status have been weighted to produce a final index value varying from 0 to 100, with 0 representing *bad* quality and 100 representing *high* quality status. The defuzzification weights are as follows:

$$\text{output index value} = 0\cdot\mu_{bad} + 25\cdot\mu_{poor} + 50\cdot\mu_{moderate} + 75\cdot\mu_{good} + 100\cdot\mu_{high}.$$

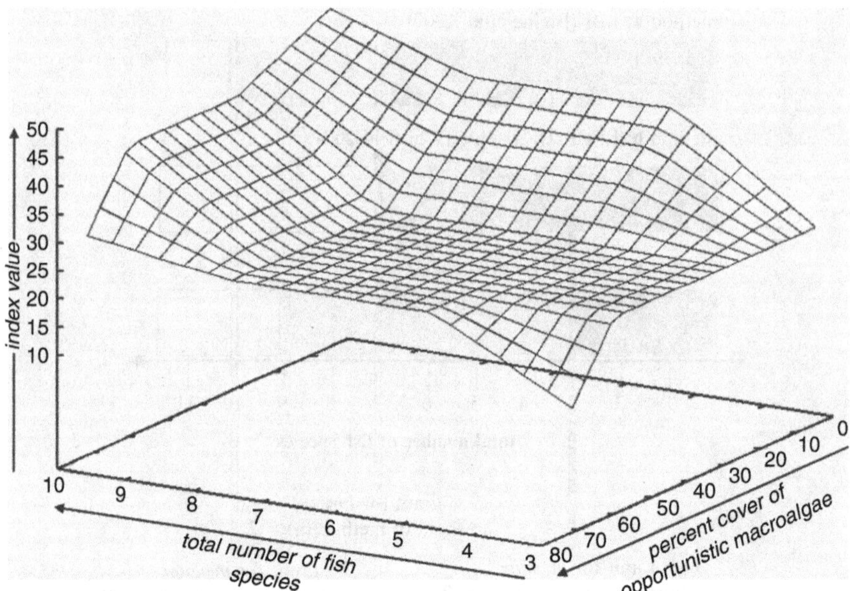

Fig. 10.3 Response of the multivariable index developed with a fuzzy system to the variation of the variables total number of fish species and percent cover of opportunistic macroalgae

Since it is based on seven different metrics, the index has a complex, multi-dimensional behaviour. A simplified representation of the index variation in response to input values variation is shown in a 3D graph (Fig. 10.3): two metrics, percent cover of opportunistic macroalgal species and total number of fish species are plotted on the horizontal axes, whereas all the other metrics are fixed in their "worst case" condition, i.e. at the values corresponding to *bad* ecological status. For this reason, the maximum index value obtained with this simulation is 50 instead of 100. The index increases in response to the increase of fish and to the decrease of opportunistic macroalgae. The general behaviour of the index is non-linear, but there are regions of linearity, due to the selected shape of membership functions (triangular and trapezoidal).

10.4 Future of Fuzzy Ecological Models

In the scientific literature, many examples of ecological fuzzy models already exist, but many more are expected to come, considering the wide range of solutions that fuzzy logic offers to ecologists and environmental scientists. Future progress of the modelling techniques is desirable, in particular relative to the fuzzification, inference and defuzzification strategies. Most of the currently published models make use of the simplest available options (linear functions, default inference operators and defuzzification methods), regardless of their semantic meaning. Furthermore, the parameterization of the model is often based on expert knowledge, without sensitivity analysis or use of optimization techniques. On the one hand, the use of personal knowledge and experience is an advantage for the fuzzy approach, as it increases the quantity of information that is possible to include in a model. On the other hand, it also includes subjectivity, which can be seen as a weakness of the model.

Such aspects need improvement to create more reliable and scientifically sound fuzzy ecological models. A possible strategy to achieve this goal is the hybridization of the human-like reasoning style of fuzzy systems with other soft-computing approaches. For example, neural networks, which are able to learn optimal membership functions and fuzzy rules from datasets. Such data-driven approaches would improve model performance by taking into account non-linear membership functions, complex model structures, alternative inference operators and defuzzification strategies.

Another aspect that can be further developed in fuzzy ecological modelling is the inclusion of biological variables (species, communities). Most of the published models have been using only abiotic variables. Physical and chemical variables are usually easier (and cheaper!) to measure and monitor. However, they might not provide sufficient information on the investigated environment. Tingey (1989) emphasized that "*there is no better indicator of the status of a species or a system than the species or system itself*". The inclusion of biotic variables, under the direction of experienced biologists would help in designing more effective fuzzy ecological models. This can only be achieved by improving the familiarity of biologists to fuzzy logic techniques.

Further Readings

Adriaenssens V, De Baets B, Goethals PLM, De Pauw N (2004) Fuzzy rule-based models for decision support in ecosystem management. Sci Total Env 319:1–12

Klir GJ, Yuan B (1995) Fuzzy sets and fuzzy logic: theory and applications. Prentice Hall, Upper Saddle River, NJ

Shepard RB (2005) Quantifying environmental impact assessments using fuzzy logic. Springer, New York

Silvert W (1997) Ecological impact classification with fuzzy sets. Ecol Model 96:1–10

Silvert W (2000) Fuzzy indices of environmental conditions. Ecol Model 130:111–119

Zadeh L (1965) Fuzzy sets. Inform Control 8(3):338–353

Zimmermann HJ (1991) Fuzzy set theory and its applications, 2nd edn. Kluwer, Dordrecht

Chapter 11
Grammar-Based Models and Fractals

Winfried Kurth and Dirk Lanwert

Abstract In ecological interactions the three-dimensional structure of organisms can play an important role. We will present an approach for modelling and simulation of the development of geometrical structures in space, which is particularly suitable for representing branching systems as they occur in plants. The related notions of self-similarity and fractality will be briefly discussed. The crucial idea for modelling is to describe the development of a modular structure by rules controlling the replacement of substructures by other substructures. Such replacement systems are also called "grammars". When the structures are encoded as strings, we speak of L-systems. A more general case are graph grammars, where the transformed structures are networks consisting of nodes and arcs. Loosely following Kurth (2007), we will first show example grammars written down in the programming language XL, which simulate the branching structures of simple plants. The final example, also implemented in XL, is about competition and resulting spatial interaction between plants. All code examples can be tested with the free software GroIMP ("Growth-grammar related Interactive Modelling Platform").

Abbreviations

FSPM	Functional-structural plant model
GroIMP	Growth-grammar related interactive modelling platform
RGG	Relational growth grammar
XL	Extended L-system language

W. Kurth (✉)
Department for Computer Science, Chair for Computer Graphics and Ecological Informatics, Georg-August University of Göttingen, Buesgenweg 4, 37077 Göttingen, Germany
e-mail: wk@informatik.uni-goettingen.de

F. Jopp et al. (eds.), *Modelling Complex Ecological Dynamics*,
DOI 10.1007/978-3-642-05029-9_11, © Springer-Verlag Berlin Heidelberg 2011

11.1 Introduction

Grammar-based or rule-based modelling tries to capture the morphological development of organisms in three-dimensional space. For example, in plants we can often observe that they are composed of structural units (internodes, root segments, buds, leaves, flowers), which are repeated in space and which grow and develop according to clear botanical rules (e.g. laws of phyllotaxis or inflorescence architecture). Thus, a seemingly complex tree crown can be described by quite a small number of geometrical units and rules. All simulations of ecological situations where a precise description of three-dimensional arrangements is needed can profit from such a rule-based description e.g. for tree crowns competing for light and space, and for root systems competing for soil resources.

The Hungarian biologist Aristid Lindenmayer (1925–1989) was the first to use a grammar-based formalism to simulate the growth of organisms, namely, of filamentous algae (Lindenmayer 1968). His rule systems were later denoted Lindenmayer systems or L-systems. For a while, the applications of this new formalism remained restricted to morphological studies of small herbaceous plants (e.g. Frijters and Lindenmayer 1974) or to the free exploration of forms which can be obtained by parameter variations in the rule systems (Hogeweg and Hesper 1974). Later, Smith (1984) and Prusinkiewicz (1987) combined L-systems with a powerful description code for static branching structures, turtle geometry, and used this approach for plant models in computer graphics. A book with numerous illustrations (Prusinkiewicz and Lindenmayer 1990) inspired further work of this sort. L-systems were also used for other organisms forming branched structures, such as fungi (Tunbridge and Jones 1995) and corals (NVIDIA 2006). However, Herman and Schiff (1975) already had a more ambitious aim: "... a general purpose simulator in which all kinds of different biological ideas can be tested with ease." To accomplish this, the purely structural representations which can be obtained with simple L-systems were not sufficient. The guiding idea was to extend the L-system approach to a formal calculus which could play a role in biology which is analogous, e.g., to that of differential equations in physics or group theory in crystallography. The first steps were already done by Prusinkiewicz and by Lindenmayer himself when they introduced numerical parameters in their L-systems. Later, a combination of the rule-based approach with process-oriented models of biological systems was realized in various forms. Room and Hanan (1995) coined the term "virtual plants" for the resulting simulated organisms, and in the preface to Sievänen et al. (1997), the notion of "functional-structural tree model" was used, which was later generalized to "functional-structural plant model" (FSPM). To provide flexible tools for creating FSPMs, several authors extended the concept of an L-system. For instance, in the language XL (which will be used below for some "virtual plant" examples), L-systems are extended to graph grammars and rule application is combined with classical imperative and object-oriented programming in order to capture the process-related aspects of plant growth in an adequate way in a model. Such an extended grammar approach is quite powerful: Recently, Cournède (2009) has also shown that the basic algorithms of the so-called Greenlab model (Guo et al. 2006) (which is the result of an independent line of FSPMs originally not employing rewriting systems) can be expressed in the form of a grammar.

11.2 Turtle Geometry

If we want to formalize rules for structural development of organisms, like those for shoot formation and branching of plants, we have to define a code. Its task will be to translate organic entities like leaves, internodes, flowers etc. into formal objects or symbols which can be referred to in precise rules and which can also readily be processed by a computer. Several codes of this sort have been designed. The most frequently used one is called "turtle geometry", a name referring to a virtual device for construction or drawing, called the "turtle" (Abelson and diSessa 1982). The concept of "turtle" was incorporated into the LOGO programming language by Seymour Papert in the late 1960s (see Harvey 1997 about LOGO). The formal symbols which will finally encode our geometrical structures are interpreted as *commands* directed one after the other to this turtle. The turtle has a memory containing information about the length s of the next geometrical entity to be constructed, its thickness d, its colour c, the turtle's current position, its current direction of movement, and more. Among the available commands are the following ones:

M0	move forward by s length units (without drawing)
F0	move forward and draw simultaneously a line of s length units
M(a)	move forward by a units (without drawing); the explicitly specified number a overrides the turtle's s
F(a)	move forward and draw simultaneously a line of a length units
L(a)	overwrite s by the value a
D(a)	overwrite d by the value a
P(a)	overwrite c by the value a (interpreted as a colour index)
RU(a)	rotate clockwise by the angle a (around the "up" axis, which is perpendicular to the plane where the turtle is moving)

In the commands M0 and F0, the zero says that there is no explicit argument; instead, the "state variable" s of the turtle is used. Strings consisting of these commands can be used to specify structures made of consecutive lines (or cylinders in three dimensions) with varying length, thickness and visibility. Each of these strings describes a fixed geometrical structure.

Loops can be used to abbreviate iterated parts of a command sequence:

for (int i:(1:n)) (X) generates n replicates of the string X (here, int i defines a counting variable for the repetitions). For example, the turtle programme:

```
L(100) for (int i:(1:60))
        (for (int j:(1:i)) (F0) RU(90))
```

generates a rectangular spiral (cf. Kurth 2007).

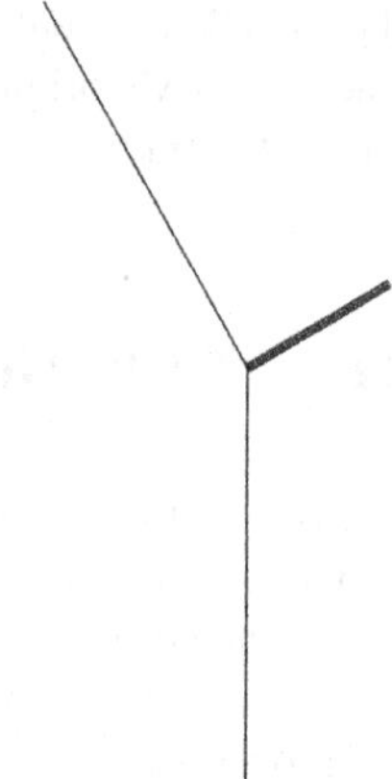

Fig. 11.1 A branched structure specified with the turtle

So far, the turtle can draw only one line after the other. To create the possibility to produce branching patterns, special commands denoted "[" and "]" are introduced: When the turtle reads "[", its current state (including the values of *s*, *d*, *c* etc.) is stored on a stack. The following string is then processed by the turtle as a branch which ends when "]" is reached: At this moment, the stored state is taken from the stack and overwrites the state, which the turtle got by finishing the branch. The turtle then "jumps back" to its old position and resumes its operation as if the construction of the branch since "[" had not taken place. Figure 11.1 shows the turtle interpretation of the string

F(50) [RU(60) D(1) F(20)] RU(-30) F(50):

After a vertical segment of length 50, specified by F(50) in the beginning, the shorter, but thicker branch to the right (with diameter 1 set by the command D(1)) is constructed. When the closed bracket is processed, the turtle resumes its old position and follows the commands RU(-30) F(50) to draw the upper-left part of the structure.

The turtle's movement can be extended to the third dimension by two further rotation commands: RL(*a*) and RH(*a*). They rotate the turtle around an axis pointing (initially) to the left, respectively around its current head direction.

11.3 L-Systems

Lindenmayer systems, also called "L-systems", are *parallel rewriting systems* which are applied to *strings*, i.e. to sequences of symbols. Mathematically, a "pure" L-system (without geometrical interpretation) has 3 components: an alphabet Σ which contains the basic symbols, a start string called "Axiom", and a finite set of rules. Each rule has the form:

symbol ==> string of symbols

where all symbols are elements of Σ. In a deterministic L-system, the left-hand side of each rule must be different from that of all other rules. An *application step* of the

L-system to a given string s consists of the simultaneous replacement of all symbols in s occurring on the left-hand side of a rule by their corresponding right-hand side. Symbols which cannot be replaced remain unchanged. By beginning with the start string and iteratively performing one application step to the result of the preceding one, we get the *developmental sequence* of strings generated by the L-system:

$$\boldsymbol{Axiom} \rightarrow s_1 \rightarrow s_2 \rightarrow s_3 \rightarrow \ldots$$

As a simple example, let us consider the L-system with $\Sigma = \{\mathrm{A; B}\}$, $Axiom = \mathrm{A}$, and with the two rules

A ==> B
B ==> AB

The resulting developmental sequence is:

A → B → AB → BAB → ABBAB → BABABBAB → ...

From Lindenmayer's original viewpoint, A and B can be interpreted as two different cell types of a filamentous organism. The rules express the facts that a cell of type A can grow into a cell of type B, and a cell of type B can divide into two cells of type A and B, respectively. The developmental sequence then reflects the growth of the filament in discrete time steps. (Note that the number of cells in this sequence increases according to the Fibonacci sequence: 1, 1, 2, 3, 5, 8, 13, ..., i.e. each number is the sum of its two predecessors.)

To let L-systems produce more interesting geometries than just linear filaments of cells, the definition of an L-system is extended by a fourth component, turtle geometry. It provides a geometrical interpretation: With each string, particularly with each S_i from the developmental sequence above, a geometrical structure S_i in two- or three-dimensional space is associated. This is realized by letting the alphabet Σ of the L-system contain the set T of all turtle commands. The strings s_i obtained from the L-system are then separately interpreted by the turtle: They are scanned from left to right, and the geometrical structure S_i is constructed by processing the occurring commands one by one. Symbols from Σ which are not in T are ignored by the turtle. Figure 11.2 summarizes the resulting scheme of interpreted L-system application.

An example (from Prusinkiewicz and Lindenmayer 1990, p. 25, adapted) demonstrates this mechanism: The rules of our L-system are:

Axiom ==> L(100) F0 and
F0 ==> F0 [RU(25.7) F0] F0 [RU(-25.7) F0] F0 .

The resulting structures S_1, S_2, S_3 and S_4 are shown in Fig. 11.3.

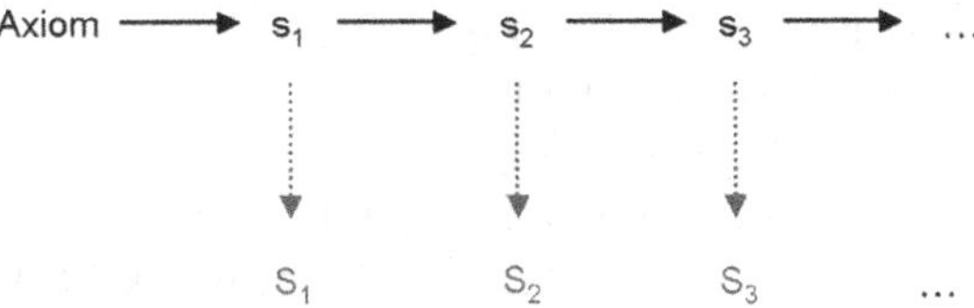

Fig. 11.2 L-system application with geometrical interpretation. The *dotted arrows* symbolize the interpretation of strings by the turtle (from Kurth 2007)

Fig. 11.3 A developmental sequence of branching structures, generated by a simple L-system (see text)

11.4 Parametric L-Systems

Some of the turtle commands (e.g. F, L, LMul, D) require parameters with numerical values. The power of L-systems to perform calculations of all kinds is considerably enhanced if we allow the use of such parameters, also in connection with other symbols. The next L-system provides an example. It produces a structure which resembles a fern leaf (Fig. 11.4a) and makes use of a parameterized symbol A standing for a bud. A has two parameters, t and k, which can take integer values. t is a time delay; it is decremented by 1 in each step until the value is 0, and thus a certain number of steps (here 6) must pass before a lateral branch starts growing. The second parameter, k, switches systematically between +1 or −1 and controls the orientation of the branch. In order to specify how many and what sorts of parameters the symbol A shall have, it is declared as a *module*:

```
module A(int t, int k);

Axiom ==> L(100) A(0, 1);
A(t, k) ==> if (t > 0) (A(t-1, k))
                    else (F(1) [RU(k*45) A(6, k)] F(1) RU(3) A(0, -k));
F(x) ==> F(1.15*x)
```

The potential of such an L-system can easily be explored by playing around with parameters: If, for example, the initial delay in the branches is reduced from 6 to 2, branches will emerge earlier and a more compact form of the structure results (Fig. 11.4b).

11.5 Fractals

Patterns like the ferns in Fig. 11.4 show a self-similar structure, i.e. affine transformations exist which map the whole pattern to some parts of it. Such *self-similarity* occurs frequently in nature, e.g., in crowns of older trees in the form of "reiterations". Self-similarity is an indication that there are some structural rules governing the pattern which can be used to specify it in a very condensed form, like our ferns were described by the three L-system rules above. Alternative methods to generate self-similar patterns also exist, e.g. the direct specification of the structure-preserving affine transformations by matrices – an approach known as "Iterative Function Systems" (IFS; see Barnsley 1988).

Self-similar structures can be characterized as *fractals*, which means, as geometrical objects which have a "broken" (or non-integer) dimension. "Dimension" in this context does not refer to the usual algebraic definition of the dimension of a manifold (as the number of coordinates which is necessary to fix positions in it), but to the degree to which the object fills the space. For instance, the ferns in Fig. 11.4 are more space-filling than a straight line (dimension 1) but less than a plane-filling object like a filled triangle (dimension 2). Hence, their fractal dimension is a broken number between 1 and 2. Exact definitions of fractal dimension are given in measurement theory, a branch of mathematical topology (see Edgar 1990). Some well-known

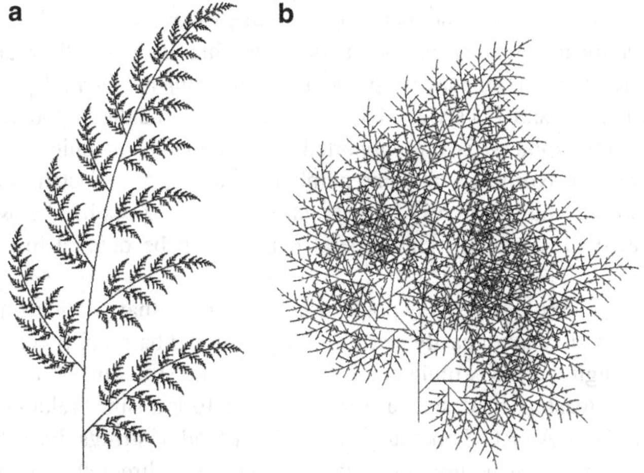

Fig. 11.4 (**a**) Fern leaf produced by a parametric L-system (see text), (**b**) variant with reduced delay parameter for branch emergence (from Kurth 2007)

fractals from mathematics are the Mandelbrot set and Julia sets (Peitgen and Richter 1986). Many natural objects can also be characterized, at least approximately, by a fractal dimension – this is true for some plants, venation patterns, coastlines, clouds, galaxies, mineral surfaces (cf. Mandelbrot 1977), or movement paths of organisms. However, fractality is more an analytic notion which is normally not directly suitable to design procedures for synthetic generation of structures. Hence we will not explore it deeper here.

11.6 Relational Growth Grammars

Despite their successful use for realistic-looking structural models of plants, L-systems still have certain limitations (even if some extensions of the original concept are included). First of all, in interpreted L-systems (with turtle geometry and with brackets for branching), only two possible relations can exist between simulated objects: A can be a *direct successor* of B or can be supported by B as a *branch*. However, many more sorts of relations between objects are possible in reality and can be worth modelling. Another problem with L-systems is that they are not an appropriate tool for the creation of truly two-dimensional or even three-dimensional arrangements, like tessellations in the plane (tilings with a collection of simple shapes without leaving any space) or cellwork systems (e.g., in tissues). There are extended formalisms like "map L-systems" and "cellwork L-systems" (see Prusinkiewicz and Lindenmayer 1990) for this purpose, but their definitions and usage are rather complicated. The core of this problem is that the classical interpretation of bracketed strings by the turtle can only yield structures with an underlying topology which is locally one-dimensional and has a structure similar to trees. Hence, *cycles* and *networks* can be created only if additional tools are allowed.

Another weakness of L-systems is apparent when they are viewed from the perspective of software engineering: As a programming language they are poor in comparison to modern languages; particularly, the *object-oriented programming* (OOP; see Chaps. 4 and 12) style, which is currently the standard approach amongst professional programmers, is not supported. The fundamental units of parametric L-systems are only symbols with some numbers attached, no objects in the sense of OOP. Furthermore, no hierarchy of object classes, where specialized classes inherit attributes and methods from more general classes, can be defined in classical L-systems.

Finally, from a biologist's viewpoint, it is a drawback that *genotype* and *phenotype* of an organism cannot easily be captured in a model based only on classical L-systems (although the DNA molecule has basically the structure of a string).

These were reasons to design a new rewriting formalism, "relational growth grammars" (RGG). An RGG operates on *graphs* instead of strings. By a "graph" we mean a mathematical structure consisting of nodes and directed arcs (also called "edges") connecting some of these nodes. "Trees" (in the abstract sense) are special graphs, but in general, cyclic substructures, which do not occur in trees, are allowed in graphs. RGG are called "relational" grammars because several types of edges (relations between nodes) are possible.

The RGG approach allows to model arbitrary relations and networks. It can also be used to represent genes, genetic processes and the development of the phenotype in one and the same formalism (Kniemeyer et al. 2004). The weakness of L-systems concerning software engineering is addressed by permitting RGG rules as construction elements in a programming language, called XL (eXtended L-system language), which at the same time is an extension of the object-oriented language Java, and by allowing Java objects to be nodes of the graphs which are rewritten. A similar approach led to the language "L+C" (Karwowski and Prusinkiewicz 2003), which is an extension of C++ by L-system rules, but this language does not include graph transformations. An exact mathematical definition of RGG and a precise description of the language XL was given by Kniemeyer (2008).

The general form of an RGG rule is as follows:

L, (* C *), (E) ==>> R {P};

where L is the left-hand side proper of the rule, i.e. a search pattern for a set of subgraphs to be replaced by the right-hand side, R, which specifies also a set of graphs. C is a set of graphs which must be present as a *context* of L in order to make the rule applicable, and E is a condition for applicability of the rule, specified by a logical *expression* which can contain parameters referring to nodes from L and C (or to their attributes). P is a piece of imperative code which is executed when the rule is applied. C, E and P are optional parts. An RGG is composed of such rules, which are usually simultaneously applied to a given graph, similar to L-system rules.

The inclusion of imperative code allows an easy link to process-oriented models, e.g., for light interception, photosynthesis and carbon allocation. As an example, Fig. 11.5 shows the simulation results of three virtual beech trees (*Fagus*

Fig. 11.5 A virtual beech tree grown under three different light conditions (incident radiation increasing from *left* to *right*). The simulation was done with GroIMP using an RGG (from Kniemeyer 2008, p. 325)

sylvatica L.) which were modelled using an RGG with process-oriented components (see Kniemeyer 2008). Another FSPM of this sort will be shown at the end of this chapter.

RGG are a special case of *graph grammars*. As for L-systems, there exists an established mathematical theory on graph grammars (Rozenberg 1997). L-Systems form a special case of RGG, because strings can be interpreted as special graphs with a linear structure, with edges of a certain, fixed type "successor" between consecutive symbols. In XL, edges are generally written down in the form "*-edgelabel->*", where "*edgelabel*" specifies the type of the edge. Since the edge type "successor" is so frequently used, a simple blank symbol is allowed instead of "-successor->". This convention allows us to write down L-system rules in XL in a familiar manner – and in fact, all L-system examples shown above were directly taken from XL programmes. In order to make them readable by an XL compiler (like that included in the software GroIMP, see http://www.grogra.de), one has only to enclose the rules (without the module declarations) in a surrounding frame of the form

public void run( ) [. . .]

This construction is necessary because RGG rules in XL can be organized in several blocks in order to enable a better control of the order of rule application, like in so-called table L-systems (Rozenberg 1973).

However, RGG have a capacity going far beyond that of L-systems. As an illustration, Kniemeyer et al. (2004) show a graph transformation rule which simulates the genetical process of "crossing over". It cannot be expressed as a simple L-system rule, but as an RGG rule.

Additional flexibility comes from the possibility offered by XL to derive new relations between graph nodes from the given ones. For example, the search pattern a -x-> -y-> b matches all pairs of nodes (a, b) in the graph (standing, e.g. for plant organs) which are connected by two consecutive edges of type x and y, respectively, whereas the pattern:

a (-x->)* y

matches all pairs which are connected by an arbitrarily long path consisting of edges of type x (in mathematics, this kind of relation is called the "reflexive-transitive hull" of the relation x). Search patterns of this sort can be used in so-called "graph queries" which, when evoked, give back all subgraphs of the current structure which are consistent to the search pattern. Queries enable to carry out calculations on all the results. Queries in XL are enclosed in starred parentheses (* . . . *) – in fact, the optional "context" part of an RGG rule is also a graph query. For example, the following expression in XL calculates the total area of all leaves which are attached to the branching system emerging from a specific node n in a tree crown:

sum((* n (– –>)* Leaf *)[area])

Here, "(– –>)*" denotes the reflexive-transitive hull of the relation "successor" or "branch", i.e. all nodes of type "Leaf" which can be reached via a directed path emerging from n are taken into account. Similar queries can be used to calculate the biomass or the stored carbon in compartments of plants or in whole individuals. An ecosystem simulation at a higher level could use these data to assess the carbon flow through the system.

11.7 An Eco-Physiological Model of a Coniferous Tree Stand

Formalisms like L-systems or RGG can be used for functional-structural plant models (FSPM). Such models combine structural and functional features of plants or plant populations in one coherent formal framework. As an example we will sketch a model of trees competing for light and a small part of its implementation in the XL language. This eco-physiological growth model is taken from Lanwert (2007). For the modelled competing conifers, only the above-ground part is represented in the model. For reasons of efficiency, the structure of the individual trees is simplified.

The model uses a spatial approach based on biomass and photosynthetic capacity which are assumed to be the main factors which control growth (Pfreundt 1988; Pfreundt and Sloboda 1996). The individual trees are modelled as one-dimensional entities, i.e. no branching is represented. The segments of the trunk axis carry all relevant information, including certain properties of the crown. The calculation of the basic processes such as photosynthesis, respiration, allocation and of growth is carried out in annual steps. First the biomass of the needles is allocated vertically along the trunk axis using a beta distribution.

Photosynthetic performance is calculated in relation to the maximum photosynthetic capacity in unshaded conditions at the tree top and depends on the corresponding weighted needle mass above the calculation point as an input quantity of a Beer-Lambert function (Monsi and Saeki 1953). Respiration is calculated separately for the five tree compartments: "needles", "branches", "trunk", "coarse roots" and "fine roots". After subtracting the maintenance respiration from gross assimilation, the remaining net assimilation pool forms the basis for growth.

After considering the mortality rates for roots, branches and needles and subtracting the proportional cost rates for needle, branch, root and height growth, the remaining pool of assimilates determines the secondary growth of the trunk. Height growth of the individual tree is calculated using a function relating height to age with a stochastic and rank preserving correction component.

After completion of growth, the new needle densities are calculated for the next year. This is carried out separately for each needle-age cohort. The new tree height, the death rates of the old needles, the newly formed needle mass as well as an upshift of the crown base is taken into account.

The implementation of the model follows an object-oriented approach (cf. Chaps. 4 and 12). A tree consists of segments whose properties, such as the needle masses, are assigned as initial parameters or calculated during simulation. All calculations, like estimation of shadow and photosynthesis, are based on these objects. The results are stored in the object properties. All objects are organized into a large graph which does not contain any cyclic path of edges. The object structure of the model is designed in analogy to a botanical object hierarchy (forest stand – tree – annual shoot). The calculation steps are carried out by applying a relational growth grammar on the graph of the model. This keeps the code comprehensible and close to the botanical structure.

The main item of every growth model based on photosynthesis is the spatial structure of the biomass which defines the light distribution and thus photosynthetic performance (see Sloboda and Pfreundt 1989). This statement emphasizes the significance of the sub-model used for photosynthesis. Its basic concept shall be explained in simplified terms below.

The starting point is the assumption of a tree-specific function F which is applied to the weighted biomass $B(\vec{x})$ above the point $\vec{x}$ and gives back a dimensionless relationship $q(\vec{x})$ between the maximum and the actual photosynthetic performance at $\vec{x}$, taking a cone into account where potential shadow-casting objects above $\vec{x}$ are located. We receive the photosynthetic performance p at $\vec{x}$ by multiplying $q(\vec{x})$ with the maximum photosynthetic rate p_0 (kgC kgC^{-1} y^{-1}):

$$q(\vec{x}) = F(B(\vec{x})) \quad \text{and} \quad p(\vec{x}) = p_0 \cdot q(\vec{x})$$

In order to calculate the influence of shadow of an object M on $\vec{x}$, contributing to B($\vec{x}$), we imagine a cone opened into the sky with its apex in $\vec{x}$ and with opening angle α. Depending on the distance of M the shadowing effect on $\vec{x}$ is varying. It is thus weighted by the reciprocal square of the distance between M and $\vec{x}$.

Taking into account the object-oriented approach of the model (see Fig. 11.6), every trunk segment *seg* of a tree represents the given needle mass $M_{total}(seg)$ (kgC) located in its height. For this height, the needle density $\rho(seg)$ corresponding to *seg* is separately calculated for each needle-age cohort using a beta distribution multiplied by the length of the segment and summed up to the overall needle mass $M_{total}(seg)$ (kgC).

Figure 11.6 shows two arbitrary, adjacent trees. For each segment seg_0 which is part of the trunk of $Tree_0$, the shadowing needle mass of all segments of the neighbouring Tree i inside a cone with the given opening angle α is calculated.

Realization in the Extended L-System Language

The object-oriented approach gives one the ability to sum up the weighted shadowing biomasses of all segments within the cone as follows:

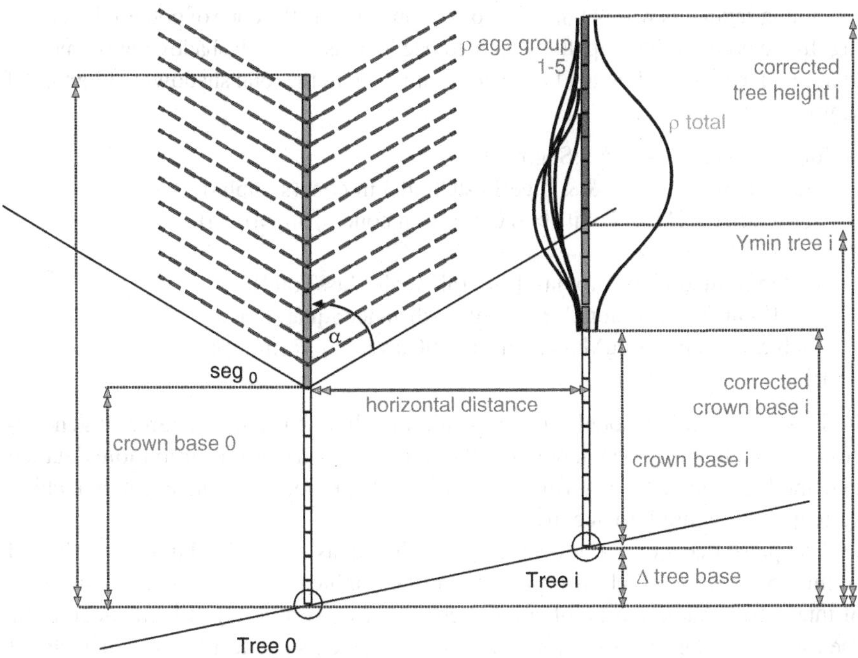

Fig. 11.6 Shadowing segments (*dark*) of a neighboring tree *i* within a light cone above the base point z of segment seg_0 (adapted from Lanwert 2007)

$$B(seg_0) = \sum_{\substack{i=1 \\ seg_i \in Cone(\alpha)}}^{n} \frac{M(seg_i) \cdot TM_{Tree(i)}}{Dist(\overrightarrow{segbase_0}, \overrightarrow{segbase_i})^2 + c_i} \quad \text{with } c_i = \frac{R^2{}_{Tree(i)}}{cf_{Tree(i)}} \qquad (11.1)$$

and with $M(seg_i)$ representing the needle biomass of shadowing segment *i* and *Dist* $(\overrightarrow{segbase_0}, \overrightarrow{segbase_i})$ giving the distance between the two segment base points. $TM_{Tree(i)}$ is the light transmission coefficient of the tree to which the segment object seg_i belongs. The quotient R^2/cf represents a correction factor taking into account the distribution of the needle mass over the crown radius *R* of the tree the segment seg_i belongs to.

The following code sample (taken from Lanwert 2007, adapted) shows a "for"-loop which is executed in a rule applied to an arbitrary segment s and which makes use of a query, enclosed by (* ... *) and defining a set of subgraphs, using a search pattern and four conditions. The search pattern looks for all segments (named a) connected by a daughter relation (directly or indirectly) with a tree element (named b). The first two conditions exclude all matches where the tree element (b) is equal to the parent tree element of the main segment s and where the top segment of b is outside the cone. Thus most trees in far distance are eliminated. The second two conditions exclude all remaining segments without needle biomass and outside the cone of potential shadowing. In the body of the loop, the weighted needle mass of each shadowing segment is then calculated according to (11.1) and summed up to the total shadowing biomass of segment s.

```
for ((* b:Tree (-->)* a:Segment,
  (s.getParent( ) != b && b.getLastChild() in cone(s, alpha)
  && a[needleMassTotal] > 0.0 && a in cone(s, alpha)) *))
{
s[externalShadowingBiomass] += (a[needleMassTotal]
  * b[lightTransmissionCoefficient] / (distanceSquared(a, s)
  + b.calculateNeedleMassDistributionCorrectingTerm( )));
};
```

This approach for modelling competition for light has a high degree of genericness, i.e. it can easily be adapted to other biomass distributions or to more detailed tree models which take asymmetric crown shapes or even an exact branching system architecture into account.

The presented conifer model was developed as part of a larger, distributed simulation system which also provides a 3D graphical interface with possibilities of interactive manipulation of single trees (see Lanwert 2007). It is intended to be used in e-learning scenarios where a group of students can test forest management practices by cutting virtual trees and evaluating the consequences on future growth of the remaining stand. Since the growth model is based on local light conditions and thus on the 3D canopy structure in the vicinity of each tree, it allows to assess different strategies of logging.

11.8 Outlook

Rule-based functional-structural modelling is suitable in situations when the three-dimensional structure of an organism or of a community has an important influence on its development. This is not only the case for tree crowns competing for light, but also, e.g. for some plant-animal interactions. There have been first attempts to model the movement and foraging behaviour of animals with extended L-systems (Kurth and Sloboda 2001). Probably, a significant number of agent-based and individual-based models, as covered by Chap. 12, can also be embedded in a grammar-based framework in the future.

Another active field of research in functional-structural plant models addresses the genetic basis of plant metabolism and regulatory networks. For example, a "barley breeder" model, implemented in the language XL, enables virtual breeding in a genetically heterogeneous population and simulates the hormonal control of internode elongation in barley via a metabolic network (Buck-Sorlin et al. 2007). On the technical side, a recent effort (Hemmerling et al. 2010) aims at providing user-friendly constructions in XL to evoke numerically suitable solvers for sub-models based on ordinary differential equations (cf. Chap. 6).

Further Readings

For further reading on the topic, the book "The Algorithmic Beauty of Plants" (Prusinkiewicz and Lindenmayer 1990) is still an excellent introduction. The original source regarding the recently developed language XL is Kniemeyer (2008), which also contains numerous examples. The use of functional-structural plant models in applications can best be traced by looking into the proceedings of the FSPM conferences and workshops: Bouchon et al. (1997), Sievänen et al. (1997), Kastner-Maresch et al. (1998), Andrieu (1999), LeRoux and Sinoquet (2000), Hu and Jaeger (2003), Godin and Sinoquet (2005), Vos et al. (2007), Fourcaud et al. (2008), Hanan and Prusinkiewicz (2008).

Chapter 12
Individual-Based Models

Hauke Reuter, Broder Breckling, and Fred Jopp

Abstract This chapter will describe the category of models that represent the behaviour and interaction of distinct individuals with specific properties. Models of this type can become very complex, but have the advantage that model structures operate on a low level of abstraction and represent ecological relations in a form similar to empirical assessment. Individual-based models facilitate studies of emergent properties, where characteristics of higher level entities like populations or communities can be generated on the basis of single actions of particular individuals. They allow to simultaneously investigate energetic and physiological aspects, behaviour, and relations to other organisms and heterogeneous environmental structures. As a technical background, object-oriented programming is frequently used for this model approach. This chapter introduces the conceptual background and describes two case studies, one that investigates spatial aspects of a predator–prey interaction, and a second one which depicts community interactions of Northern Scandinavian small mammals with oscillating population dynamics.

12.1 Introduction

Individual-Based Models (IBM) represent single organisms and their environment. They allow studying the implications of physiological processes, behavioural traits, and environmental interactions synchronously. This offers a structurally unique option for ecological modelling, because of the potential to join structural, functional, quantitative and qualitative aspects in a way that closely conforms to observation data and conceptual knowledge representation. In the context of ecological modelling applications, we use the term agent-based models synonymously.

H. Reuter (✉)
Leibniz Center for Tropical Marine Ecology GmbH (ZMT), Fahrenheitstraße 6, 28359 Bremen, Germany
e-mail: hauke.reuter@zmt-bremen.de

F. Jopp et al. (eds.), *Modelling Complex Ecological Dynamics*,
DOI 10.1007/978-3-642-05029-9_12,

Background and Development of the Approach

Early applications of this approach go back to the 1970s. They were a response to the requirements to include more biological realism and explicit spatial representations into ecological models (Łomnicki 1988). First models were introduced by Kaiser (1976; territoriality of dragonflies), Hogeweg and Hesper (1983; social interaction of bumblebees), Seitz (1984; life stage sequences in a Daphnia populations), and DeAngelis et al. (1979; development of cohort structure in small-mouth bass populations) and modelling forest stand dynamics (Botkin et al. 1972; Shugart and West 1977).

The minimum requirement for an individual-based model is the separate representation of individual entities, which can be distinguished in one or more characteristics. These characteristics of the individuals must be separately accessible and tracking of individual state changes must be possible during simulation. In most of the application cases, the number of states, and the repertoire to modify the states depends on the internal conditions. In relevant application cases, individual-based models attempt to provide a coherent picture of how particular organisms would act in a particular condition.

To use the full potential of the approach it is also possible to include different levels of entities. Complex modular organisms like trees can be represented as a set of individual branches, roots, leaves, fruits, etc. The individual organism is then a compound instance of sub-units. On the other hand, it is also possible – and sometimes useful – to operate other compound entities. For example, a representation of an environment can comprise a spatially differentiated structure where physical and chemical parameter differ locally and give rise to specific local responses of the organism's activity. Furthermore, abstract entities like populations can be represented, either as units with specific parameters like age distribution, biomass spectrum, which change during simulation, or as an aggregate that integrates over the individuals included in the model. Such an extended specification of an IBM may thus comprise configurations in which the components are not basic units but particular components of ecological systems. In principle, these may range from (sub-)cellular units, plant modules (Breckling 1996; Eschenbach 2005) to aggregations such as cohorts, social animals (nests, hives), populations, functional types, or spatial or temporal units of higher order (Köhler et al. 2003; Middelhoff et al. in print).

With such an extended understanding of how the IBM approach can be used, one can see that it is in fact structurally identical, *sensu strictu*, to agent-based models (ABMs). The term ABM originally emerged in a computational context with applications in physics as well as applications in social sciences and economics. ABM often describe robotic aggregates responding to a variable environment, or they simulate complex behaviour of humans in social networks. From an ecological perspective, it appears reasonable to use both terms (IBM and ABM) synonymously. In a similar way, the term multi-agent system or multi-agent simulations (MAS, e.g. Ferber 1999) bases on the same concept, however emphasizes the interactions of a larger number of autonomously acting software agents and is even more common in technical applications.

The approach becomes operational only if the computational challenges of such a model can be met. This was made possible by developments in computer programming and programme processing. The representation of a larger number of individuals with an interaction potential is feasible only with larger processing and data management capacities. The advancement in hardware and software development allowed more resource-demanding applications like Object-Oriented Programming (OOP). In the late 1960s, the programming language SIMULA (Dahl et al. 1968) provided the ground for the virtual representation of active agents, which was later adapted to various programming languages (Smalltalk, Delphi, C++, Java, and others).

Early IBMs often had a narrow focus and concentrated on single species investigations (e.g. Kaiser 1976; DeAngelis et al. 1979; Seitz 1984). These early models applied quasi-automatic transition between the single model-states (e.g. age, biomass, location). However, they could already illustrate the great potential of the IBM-approach. It thus added a new perspective to modelling in a close relation to the specific characteristics of ecological systems (see Chap. 4 on systems analysis), compared to the homogeneity requirements of variables as they were used in the classical systems dynamic approaches (Forrester 1968).

Further developments made IBMs applicable for investigations of behavioural decisions and interactions in social groups. Paulien Hogeweg and co-workers pioneered this field with their model on social interaction in bumblebee colonies (Hogeweg and Hesper 1983). A first paradigmatic overview was presented by Huston et al. (1988).

Facilitating a representation of variable environments, structured populations and behavioural traits, the modelling of complex life histories emerged (e.g. Wolff 1994; Colasanti and Hunt 1997). For instance, it became possible to simulate highly resolved time-energy-budgets as a basis for behavioural decision processes. The model on the reproduction phase of a robin population is such an example (Reuter and Breckling 1999) and allowed to investigate reproduction success under different environmental settings. A further development in IBM-methodology involved the number of considered and interacting species. In this context the inclusion of interaction rules plays a major role. These rules often refer to trophic relations (e.g. Kaitala et al. 2001), spatial competition or even to succession processes (Breckling 1990).

An increasing number of models combine sophisticated internal resolution of organismic processes with the representation of several species and their interactions to analyze e.g. food webs and community dynamics. Examples for this type are the simulation of plant competition including different herbivores by Parrot and Kok (2002) and the analysis of regular population cycles of small mammals (Reuter 2005, see Sect. 12.4). Often the simulated entities are designed to operate in heterogeneous environments including a spatially explicit habitat structure, seasonality and varying climate data. The models simulate explicitly designed scenarios directly, using empirical data involving e.g. GIS-derived maps and assumptions on local temporal and spatial developments. In the marine context we find successful attempts extending simulation models to include all relevant trophic levels (so-called end-to-end models, e.g. Travers et al. 2007). These approaches in marine ecology propagate the coupling of different model types in which individual-based models are thought to play an integrating role (Cury et al. 2008) because of their flexible structure, which allows to combine knowledge and data from different sub-disciplines, that can be used to analyze findings on heterogeneous ecosystem levels and to understand the corresponding complex interaction and network structure.

In particular, since radio-tracking became possible, the modelling of the behavioural repertoire within a spatial context, and especially models that explicitly investigate animal movement and dispersal, have become an important domain of IBM applications (Gustafson and Gardner 1996; Broekhuizen et al. 2003; Jopp 2003; Pe'er and Kramer-Schadt 2008). IBMs have greatly contributed to the study of population dynamics in complex landscapes (Lima and Zollner 1996; Nathan et al. 2008; Revilla and Wiegand 2008). Population development may be simulated in dependence of complex behavioural modes or context dependent changes in reactions (Shin and Cury 2001) also including the field of population viability analysis (PVA, e.g. LePage and Cury 1997; Mazaris et al. 2005). IBM of invasion processes allow combining dispersal processes with specifies properties, individual behaviour and the properties of the invaded community (Higgins et al. 2001; Nehrbass et al. 2007).

Since their beginning, IBMs have undergone a rapid development and have been applied to almost all ecological topics and a large number of research questions (DeAngelis and Gross 1992; DeAngelis and Mooiji 2005; Grimm and Railsback 2005). In the following, we illustrate the basic structure of IBMs and demonstrate their functionality on the basis of simple model applications.

12.2 The Structure of Individual-Based Models

In order to give an overview on basic formal elements of individual-based models we begin with a short introduction into the programming background and an outline of the concept of object-oriented programming (OOP, e.g. Rumbaugh et al. 1991; Silvert 1993; Hill 1996). Then we look at the application of the OOP to conveniently describe structure and interaction of individual actors.

In OOP, the source code is organized in blocks which are delimited from each other. There are different types of blocks with the so-called CLASS as the most prominent one. A CLASS is a specific programme unit which consists of storage reservations for variables and may additionally contain code (statements) how to change the values of its variables. In such a CLASS it is possible to specify further sub-classes which allows to implement a hierarchic programme structure. During programme execution, a CLASS can be copied multiple times and these class instances may be kept available in the computer storage. This is the decisive feature in object-oriented programming. The copies of a class are called OBJECTS, thus being eponymous for the whole approach. Each of these objects, which are available during runtime of the programme, consists of the same code as the class from which it is derived, but may contain specific values stored in its variables. This allows objects to differ from each other with respect to the role they play in model execution. The variations in variable values may trigger different parts of the internal code to be executed and thus may lead to a different behaviour and development of the respective object.

In fully featured OOP-languages, the command to instantiate an object may be triggered from any part of the programme. To access objects, a special kind of variable is needed. These so-called REFERENCE VARIABLES or POINTERS directly refer to a specific object and thus facilitate uni-lateral or mutual interactions between objects. OOP has revolutionized computer programming due to its more flexible design structure and clear organization of programme code. Moreover, the features of object orientation make even complex models easier to maintain and helps in tracing errors ("bug tracking"). Due to the flexible structure during programme run time (instantiation and deletion of objects, switching of pointers from one object to another) OOP easily allows to handle the structurally complex interaction networks required for advanced ecological applications (Reuter et al. 2008). This allows simulating a large variety of phenomena, in particular self-organizing spatiotemporal structures on different levels of organization (Reuter et al. 2005).

Most individual-based programmes contain relatively similar essential parts and processes which are common for this model type:

1. The representation of an individual entity as a class
2. The layout of a structured representation of the environment
3. The organization of the temporal model execution and interaction between the entities

These parts will be explained in the following.

12.2.1 Representation of Individual Entities

A class can be conveniently utilized to describe the life-history and interaction of an individual organism. Usually, it consists of three main parts (Fig. 12.1): (a) state variables which describe individual properties and attributes, (b) statements and code blocks which are used to update these variables, and (c) a scheduling mechanism to update the properties of the individuals.

a) Variables Describing Individual Properties

Which kind and how many variables are necessary to describe the properties of an individual, depends on the research question and the complexity of the individual life-history and activity repertoire in focus. In the simplest case, one property/ variable is enough to be able to distinguish the individuals. For example, describing movement or dispersal patterns would necessarily require variables to store the

```
Class Organism
    Declarations:
        Variables characterising individual state
        (e.g. location, biomass, age, alive)

        Activity Procedures („methods“) to update states
        (e.g. Move, Growth, Reproduction)

    Life Loop:
        While alive do ...
            Apply (call) Activity Procedures
            to update variables
            Hold [Delta t ] (i.e. detach this organism
                                temporarily to update others)
        End life loop
End organism
```

Fig. 12.1 Basic scheme of a class to represent an individual (Breckling et al. 2006)

coordinates related to an individual's location (Jopp and Reuter 2005). In most cases, the number of variables describing individual properties is considerably higher. Besides location, they often comprise biomass, age, sex and a relational context. For instance, when the object refers to plant modules, information on neighbouring modules is decisive to determine transport processes. For many higher animals e.g. the information on a home range (or territory) or on eventual offspring that have to be fed, may be necessary.

b) Code to Update Individual Properties

It is useful to organize the updating of an individual's state in terms of a set of "activity procedures" (changing states relevant with regard to the environment) and "physiological processes" (changing states referring to the internal condition of the individual). Common examples for activity procedures are movement, reproduction, and feeding. Physiological processes represented in a model can be ageing and energy metabolism. The according procedures access and update the involved variables. For example, an activity "movement" should change the variables storing the location information. If energetic processes are included, "movement" may also change the energetic state to include the cost of the particular activity. The description of activities can be accomplished with very simple rules and can also integrate other mathematical approaches like fuzzy logic (see Chap. 10) or differential equations (see Chap. 6). The decision which details should be included in an activity procedure depends on the focus of the model, the available knowledge and information status of the ecologist. As the ecological quality of the input information basically decides the character of the model as a whole, we advocate that ecologists, with the necessary knowledge at hand, should be intensively involved in the programming process or better yet, learn to program their own models instead of relying on specialized programmers or pre-defined software tools which usually restrict the optimal adaptation to what the specific situation requires.

c) Process Scheduling: How to Organize the Regular Update of Individual Processes and Properties

The above part focussed largely on structural descriptions. Essential for the life-history model of an organism is how exactly the update process of all individual state variables is organized. This constitutes the dynamic part and can be referred to as "process scheduling". To coordinate the concurrent execution of a larger number of entities in a model requires a loop control structure within the program code of the particular class (respectively the instantiated objects), which is iterated as long as the object has the internal status as a living entity during the simulation run. To distinguish active and no longer active objects requires the introduction of a Boolean variable, e.g. with the name "ALIVE" as one of the object's states. A Boolean variable can store the values of either TRUE or FALSE. The control mechanism that uses the distinction, is referred to as "life loop".

The "life loop" of each currently active individual entity is repeated (iterated) as long as this variable has the Boolean status TRUE. Otherwise, the object will be terminated, deleted from the storage and the storage space released as being freely available. From a top-down point of view, the execution of any specific activity can be made dependent on distinctive conditions that relate to the internal state of the individual entity (like the energetic state, age, reproductive stage) and with the external situation (e.g. availability of food, presence of predators, daytime or time of the year). It is thus necessary to implement an algorithm to determine which of the possible activities is to be executed for a given situation.

These scheduling algorithms may range from very simply to very complex. A simple scheme would e.g. execute all activity and physiology procedures in the same order during each passing of the life loop as a "cyclic activity control". This activity scheduling mechanism is adequate for configurations where it is not necessary to evaluate complex behavioural alternatives that may require changes of the sequence of activities (e.g. models concentrating on movement behaviour, see Sect. 12.3, the IPP example).

To model complex behavioural patterns, a further elaborated decision algorithm is required. For instance, with a "priority driven activity control" it is possible to assign a variable corresponding to each activity which indicates its execution priority. Consequently, during each life loop sequence the activity with the highest priority value gets executed. In the course of execution, the priority of the particular activity is reduced, while the idle activity alternatives may accumulate successively higher priority values. Thus complex behavioural decisions (including time-energy budgets, analysis of behavioural trade-off's in life history) can be represented by considering external and internal states in relation to the supposed execution time for each activity (Breckling et al. 1997).

In an individual-based model, the ecologically relevant state of any simulated entity results from all performed activities in which the relevant inner states of the individual have been evaluated in feedback processes with all relevant external states (e.g. "environment"). The activities of an individual thus can alter its own states (e.g. if hungry through food search) but can also influence the environment, and the state of other individuals (e.g. if predated then a predator will influence the "ALIVE" variable of a caught prey object).

Because the interaction structure becomes successively more complex with an increasing number of concurrent objects, it is usually not reasonable or possible to specify it directly. Instead, a self-scheduling mechanism is required. How to set-up such a mechanism is described in Sect. 12.2.3.

12.2.2 Representation of the Environment

One of the important potentials of IBMs is to facilitate an easy way for simulating spatial (and temporal) variability and heterogeneity. In the following, we focus on how to simulate heterogeneous environmental states. When starting on the organismic level, a heterogeneous spatial organization will, to some extent, already be reached by specifying location variables for the represented organisms and installing an activity procedure "movement" to change these variables adequately (see Sect. 12.3., the IPP example). When organisms interact (e.g. as predators or prey, or as schooling organisms), the presence/absence of other individuals structures an otherwise homogeneous environment. In addition, various other data structures can be used to represent spatial heterogeneities. Frequently, grid-based representations are used. Grid maps with the relevant information attached to each cell of the grid can store e.g. (water) depth and currents in aquatic surroundings, or altitude and habitat types in terrestrial environments. It is possible to include spatially heterogeneous resource levels, physical structures, local light intensity, eventually in relation to slope orientation. Often, this information is read into the programme from external sources at the beginning of a simulation run. Also, it can be generated and modified in the programme itself with particular updating routines. In more complex computer models, the environmental information is frequently generated by external modules or by other programmes. In these cases, the simulation requires the coordinated employment of an overall system of coupled models. For simulations in marine environments, an IBM can be coupled with a regional oceanography model (ROM), which provides regularly updated information on currents and physical water conditions (Penven et al. 2006). In a similar manner, it is possible to read-in weather data in order to specify seasonal changes. Sometimes, Cellular Automata (see Chap. 8) are employed to generate environmental structures (Breckling et al. 2006). The resource density and the way it is influenced by the organisms is a frequent topic considered in IBM (Reuter 2005; Charnell 2008).

12.2.3 Scheduling Programme Processes

After discussing the update of individuals, and aspects of the data management to provide dynamically changing environmental structures we now consider the coordination of the different entities. With many concurrently active units, this is an important task. A coherent solution is required – and crucial for model execution.

Individual-based programmes usually employ a *discrete event scheduling* mechanism: Any calculation result is accessible at a specific point in simulation time. It is not reasonable to attempt a simulation of a continuous approximation, in particular, if many qualitative decisions are to be taken (an organism is alive or not, etc.). In OOP simulations, very large numbers of objects can be active at the same simulation time. In principle, real parallel processing is physically not possible if the number of processors is smaller than the number of objects. Therefore, an explicit handling of execution order is necessary. This task is a general one and does not need to be newly developed for each model. It can be generally solved on the level of the simulation environment and is employed for the specific scheduling requirements.

Often, programme environments have a central instance which activates the updating of all processes. This is acceptable if a fixed scheduling scheme can be described in advance and updating is quite regular (e.g. as in the case of cellular automata). However, we advocate a more flexible solution in which updating requests are controlled by the objects themselves and allow changes depending on the objects internal state. The objects already contain the code to calculate which activities are performed under the specific conditions. As one additional task it is possible to let them calculate a time interval of how long it would take in simulation time to have the particular activities done.

This time interval is returned to a central time management and can be used to establish a so-called "event queue", which shifts programme control always to the first one in the queue and eliminates it after execution. Thus, the programme level of event control gets the function of organizing the updating process that originates from the requirements of the objects. SIMULA (Dahl et al. 1968), the first object-oriented programming language, provides a very efficient solution in form of a system class SIMULATION. The event organization is handled automatically in the background, while each object sends a message how long it is put on "HOLD" (being busy with the current activity). This event scheduling concept was revolutionary, and was the blueprint for many other programming languages, which at this time had a different concept of event handling, as the programmer explicitly had to take care of it. The different frameworks for IBM development provide specific inherent solutions for scheduling of active entities.

12.3 Application Example I: An Individual-Based Predator–Prey Model

The structure of an individual-based model will now be illustrated in detail with the Individual-based Predator–Prey Model (IPP). This model was developed as a spatially explicit, individual-based implementation of a Lotka-Volterra predator–prey interaction with one prey and one predator species (Breckling et al. 2000). The Lotka-Volterra model is a frequently considered topic in differential equation based population modelling (see Chaps. 6 and 7). It is used here to demonstrate the change of perspective, when an individual-based point of view is adopted. Both simulated species have a limited activity repertoire consisting of feeding, growing, reproduction and movement. All activities are kept at a low level of complexity, thus serving as prototypes, however, illustrating the potential for further development.

The total numbers of both prey and predators are limited to a few thousand objects, which corresponds to the processing capacity of the environment to facilitate reasonably fast computation. The potential "movement" algorithms are the same for both species. They consist either of a Brownian (random) movement (for each movement step the new direction is chosen stochastically) or a directed walk (for the next step, the old direction and speed is maintained and a small random component is added). This leads to a higher autocorrelation of the overall movement direction. The choice of the movement algorithm (random vs. directed) has to be made for all organisms of one class by setting a specific value for a switch in a parameter file ("InFile"). It then applies to the whole simulation run. The length of each movement step is calculated in relation to the biomass of the respective organism.

Feeding and *energy physiology* differ between the simulated prey and predators. The prey grow independently from any external influence. This simulates unlimited resources. However, the predators search a specified radius around their current location during each time step and feed on the prey which they find within this radius. Then, the biomass of the identified accessible prey (multiplied with a conversion factor between 0 and 1) is added to the current biomass of the predator individual. The predator loses biomass according to a biomass dependent respiration function which leads to starvation below a certain threshold, if no prey individuals are met.

Reproduction for both, prey and predators, is biomass dependent and implemented as a fission-like process. If an individual reaches a specified biomass threshold, new objects of the same species are instantiated at the same location, and adult biomass is then distributed between juveniles and the adult.

The environment for the simulation run is established as a homogeneous area, with each edge being connected to the opposite site and thus leading to torus boundary conditions (see Fig. 8.3 in Chap. 8). These conditions minimize boundary side effects (Jopp 2003).

In the beginning of each simulation run, a specified number of prey and predator individuals are distributed randomly across the simulation area. Generally, all relevant parameters describing the behaviour of prey and predator as well as other parameters of the specific simulation (e.g. duration, size of area, etc.) are set in a parameter file.

Though individual actions do not include any directional preference, and only random components change the movement paths, we obtain emerging large-scale spatial structures as a result of the interactions of the model components. The type of pattern depends largely on the combination of the involved movement processes. In the following, we will elucidate some of the characteristic emerging patterns.

In the first example prey and predator individuals both move according to a random walk. As a consequence, after the initial random distribution prey and predators exhibit a spatial segregation. The population of the predators can only grow in the proximity of the area which is dominated by the prey. As the predator aggregations successively shift towards the prey areas, spatial dynamics result in a kind of travelling wave pattern (see also Sect. 8.4) that involves both prey and predator individuals (Fig. 12.2).

The second example (Fig. 12.3) illustrates the results when the prey exhibits a Brownian (random) movement, whereas the predators move according to a correlated random walk. This scenario leads to a remarkable aggregation of the prey. We find temporarily stable prey clusters with roaming predators which rarely meet a cluster while roaming the overall area. The predators can feed during a few time steps when passing a prey cluster, however will leave it again because they maintain the momentum of their movement.

When all other factors remain unchanged (i.e. ceteris paribus condition), the degree of autocorrelation, which is represented in the value for directedness of the predator movement, is the key factor that enables transferring one class of spatial pattern (travelling wave phenomenon) into another (random distribution, see Turchin 1998). Also, further variations of different movement factors that can be specified over the parameter file between prey and predators lead to different spatial distribution patterns.

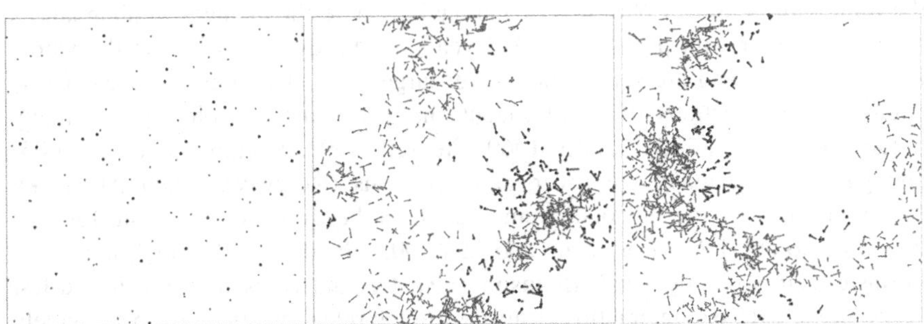

Fig. 12.2 Simulation results of the IPP model simulating a simple individual-based predator–prey interaction. *Points* indicate the current positions of prey and predators, the *line* shows the movement from the previous position. *Lighter shades* and *smaller points* indicate the prey, *darker* and *larger points* indicate the predator. *Left*: The initial distribution is random. *Center*: After 200 time steps – if both types of organisms move randomly, according to a Brownian movement, a characteristic spatial self-organization occurs. After the initial phase a dynamic change of border structures occurs where predator and prey interaction is highest. *Right*: After 400 time steps

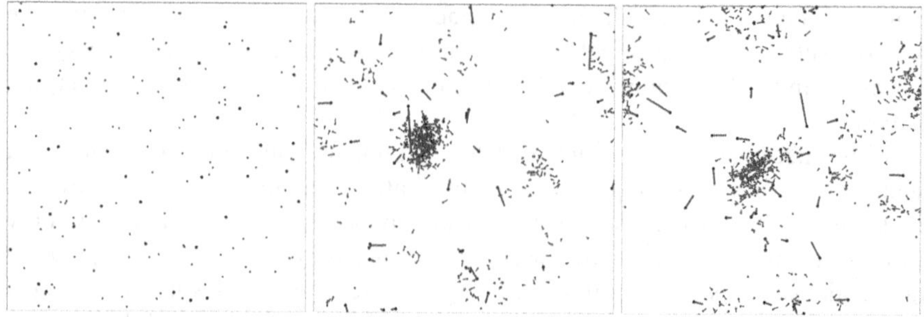

Fig. 12.3 Simulation results of the IPP model simulating a simple individual-based predator–prey interaction. For a description of symbols see Fig. 12.2. *Left*: The initial distribution is random. *Center*: After 100 time steps – if the prey moves randomly, and the predators exhibit a directed movement with a partial auto-correlation, a pronounced cluster structure of the prey occurs. *Right*: after 400 time steps

12.4 Application Example II: Cyclic Rodent Communities in Northern Scandinavia

In order to illustrate the potential of the individual based-modelling approach we outline a more complex example relating to population cycles in Northern Scandinavian rodents. In community ecology, cyclic population dynamics constitute an interesting example for complex dynamics often involving a multitude of species-intrinsic factors and environmental influences (Myers 1988; Bascompte et al. 1997; Sanderson et al. 1999; Haydon et al. 2002; Bauer et al. 2002).

Rodents, mostly in the Northern Hemisphere, often exhibit drastic changes in population size with numbers at peak times reaching up to 500-fold of numbers in the minimum phase. In Northern Scandinavia, where the changes in abundances are most regular with peaks every 3–5 years, cyclic dynamics impact the whole local biocenosis and are synchronous over large areas (Huitu et al. 2003). These community interactions of small rodents have fascinated ecologists for many decades (e.g. Elton 1927; Chitty 1960; Stenseth 1999; Korpimaeki et al. 2005) and gave rise to many controversial discussions (Rosenzweig and Abramsky 1980) on the driving factors. Numerous hypotheses have been put forward including abiotic influences (Aars and Ims 2002; Sundell et al. 2004) and biotic intrinsic factors (Chitty 1967; Boonstra 1994; Oli and Dobson 2001). In the last years, biotic extrinsic interactions (mostly trophic relationships) have been widely recognized as the most important processes. However, it is still discussed, whether rodent population dynamics are controlled by bottom-up causation (e.g. Jedrzejewski and Jedrzejewska 1996; Selas 1997) or are top-down limited (e.g. Norrdahl 1995; Klemola et al. 2003). Further more it is not clear how important the role of pathogens is (e.g. Hoernfeldt 1978; Cavanagh et al. 2004) and if driving factors change with cycle phase. Despite the long lasting controversy and the numerous field investigations, the causalities for population cycles are not yet entirely known. Especially restricting in this context are the inherent limitations of field work in relation to the temporal and spatial extent of the investigated phenomenon.

In order to analyze large-scale effects that result from complex interactions in variable cause–effect networks, an individual-based model was developed (Reuter 2005). The model allowed integrating the most essential components and their interaction structure on different integration levels. It represents small mammals' communities as a food web which is composed of three trophic levels: (1) rodent food, (2) rodents and (3) predators (Fig. 12.4). Rodents and predators were described as individual organisms with a detailed life history including an activity repertoire and physiological processes. The modelled organisms interact in an environment with a spatial arrangement of habitats under seasonally changing conditions. This concept extended previous differential equation based modelling approaches (Hanski and Korpimäki 1995; Turchin and Hanski 2001) by integrating most aspects from the ongoing debate that are relevant for rodent population dynamics.

Simulations with the model allow covering different scenarios with respect to the environment and the parameterized species. The investigated scenario included the parameterization for two rodent species, field vole (*Microtus agrestis*) and bank vole (*Clethrionomys glareolus*), and two predator species, the least weasel and the

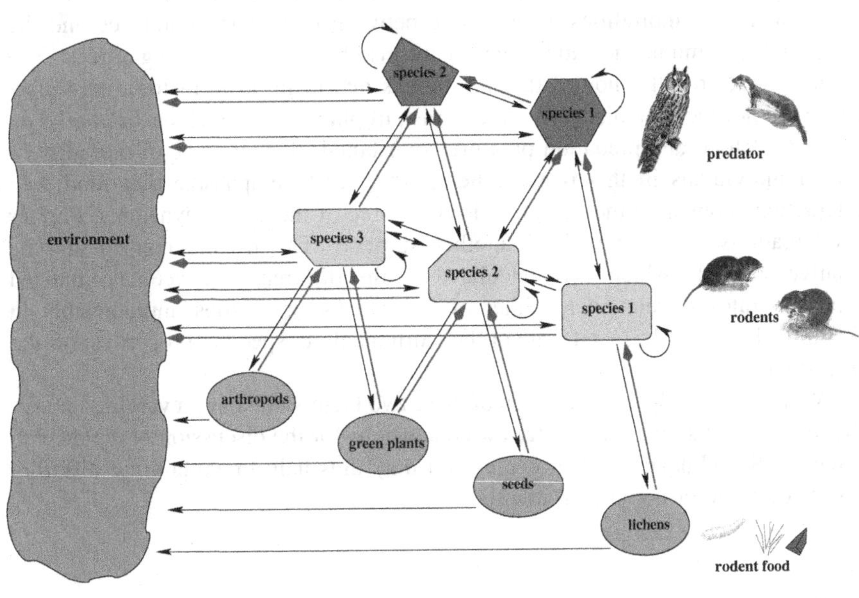

Fig. 12.4 The components and actors of the rodent cycle model: Predators (least weasel, *Mustela nivalis* and long-eared owl, *Asio otus*) and rodents (field vole, *Microtus agrestis* and bank vole, *Clethrionomys glareolus*) are represented as individual objects. The environment is represented as a grid-map which also contains the food resources for the rodents (adapted from Reuter 2005)

long-eared owl (*Mustela nivalis, Asio otus*), with respect to their different ecologically relevant properties (e.g. territoriality, food specialization, migration behaviour). Simulations take place on a grid map, with an extent of 150 ha and a spatial resolution of 30 × 30 m. The resources for the rodents are calculated for each grid cell and exhibit seasonal dynamics and allow for a feedback process with exploitation by rodents.

The model represented dynamics and interactions on different integration levels including individual life history traits, population development and community interactions. The individual level included e.g. ontology, reproduction sequences, weight development, habitat use and interaction with other rodents and predators. As a result of the individual interactions cyclic population dynamics emerged, as they are typical for Scandinavian rodent communities with an average cycle length slightly below 4 years. These cyclic population level dynamics were not implemented in the programme specification but are emergent properties produced by the model components during execution (Breckling et al. 2005; Reuter et al. 2005, 2008).

The model also allows analyzing the population structure with respect to age structure, reproduction rates and mortalities for different phases of the cycle. Specialized predators shared the cycles frequency with the rodents but the phase lags behind. The factors that cause this sudden decline of rodents are believed to be of crucial importance for the whole system dynamics (Batzli 1996).

By analyzing mortalities in the phase between maximum abundance and the following minimum, the model gave new insights into the driving forces: The overall model results showed that mortalities due to intrinsic factors like senescence did not increase distinctly. In contrast, trophic interactions that are based on the lack of food and predation pressure contributed to about 90% of mortality of rodent individuals in this phase. The results clearly emphasize that food web interactions constitute the essential driving force of the cyclic dynamics. Further investigations however yielded another surprise: The trophic control and its relative strength, which was calculated as the difference between normalized mortality rates in relation to either of the two factors, varies unpredictably in the model (Fig. 12.5) and cannot be correlated to specific properties of the respective cycle.

With respect to the identification of these two factors that have a varying impact, the model results give important new conclusions for the discussion of the driving forces in Scandinavian rodent cycles and may thus help to explain the differing results of numerous empirical investigations.

12.5 Conclusions

The relation of modelling on the one side and empirical information and biological knowledge on the other is different for IBM than it is for other modelling approaches. While the modelling approaches usually employ a specific abstraction

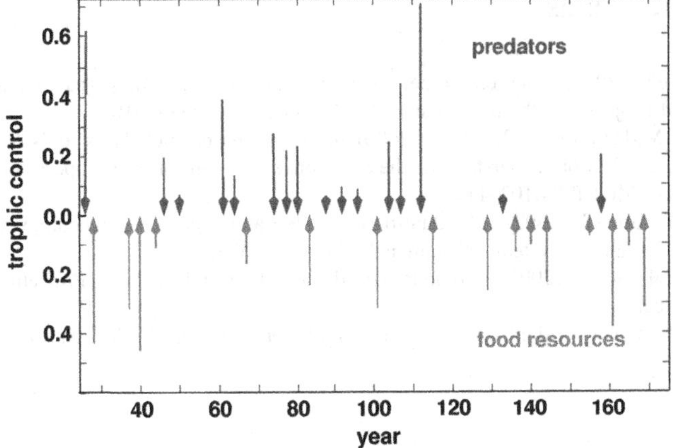

Fig. 12.5 In the rodent population oscillation model it turns out that top-down and bottom-up control of the rodent cycles change in an unpredictable way. This yields an explanation for why empirical investigations led to contradictory results (from Reuter 2005)

pattern that captures ecological relation only according to a particular scheme, IBM has the advantage to represent observation and knowledge in a form that is highly congruent with how we understand existing interaction. The level of abstraction is low, and the model represents elementary interactions that aggregate in the course of model execution in the same form as observable phenomena in empirical investigations. The iterative character of the models allows to "sample" during simulation. In that respect IBM is a crucial tool in testing consistency of ecological knowledge. Due to their potential to represent detailed biological knowledge and small-scale mechanisms, IBMs tend to have a complex model structure. This requires a particular attention to model documentation and evaluation (see Chap. 23).

The generic applicability of the structure of IBMs allows simulating a wide range of issues in terrestrial as well as in aquatic ecology. The illustrated scheme of programme organization is sufficiently flexible to capture organismic development and behaviour, environmental conditions, and the interaction of both. It is suitable to specify e.g. predator–prey interactions (Charnell 2008), schooling (Reuter and Breckling 1994), and behavioural shifts under varying conditions (Peacor et al. 2007), the formation of colonies and the description of structural–functional development of modular organisms (Eschenbach 2005).

IBMs allow to represent interaction of structural and functional features across different scales. Thus situations which do not only involve quantitative transitions but in parallel also qualitative or structural changes can be studied. Simulation results on higher organization levels emerge from the self-organizing interactions of basic units.

Further Readings

Breckling B (ed.) (2005) Emergent properties in individual-based models – case studies from the Bornhöved Project (Northern Germany). Ecol Model 186(4):375–510

Breckling B, Middelhoff U, Reuter H (2006) Individual-based models as tools for ecological theory and application: understanding the emergence of organisational properties in ecological system. Ecol Modell 94:102–113

DeAngelis DL, Gross L (1992) Individual-based models and approaches in ecology: populations, Communities and ecosystems. Chapman & Hall, New York

Grimm V, Railsback S (2005) Individual-based modelling and ecology. Princeton University Press, Princeton

Hill DRC (1996) Object oriented analysis and simulation. Addison Wesley, Harlow

Chapter 13
Modelling Species' Distributions

Carsten F. Dormann

Abstract Species distribution models have become a commonplace exercise over the last 10 years, however, analyses vary due to different traditions, aims of applications and statistical backgrounds. In this chapter, I lay out what I consider to be the most crucial steps in a species distribution analysis: data pre-processing and visualisation, dimensional reduction (including collinearity), model formulation, model simplification, model type, assessment of model performance (incl. spatial autocorrelation) and model interpretation. For each step, the most relevant considerations are discussed, mainly illustrated with Generalised Linear Models and Boosted Regression Trees as the two most contrasting methods. In the second section, I draw attention to the three most challenging problems in species distribution modelling: identifying (and incorporating into the model) the factors that limit a species range; separating the fundamental, realised and potential niche; and niche evolution.

13.1 Introduction

As more species types undergo rapid human-induced extinction, understanding why species occur where they do is becoming a highly relevant, pressing and potentially life-saving topic. Conservation actions, such as establishing protected site networks, adapting land use, providing stepping-stone habitats all require an idea of how the target species will respond. Furthermore, using organisms as a "bioassay technique", i.e. indicators of environmental trends (such as climate change, air pollution, over-fishing), demands an intimate knowledge of the organism's niche. Species distribution modelling (SDM) attempts to identify the probable causes of species whereabouts. We seek to delineate the realized niche of an organism based on its current distribution with respect to the environment (for definitions and concepts see Guisan and Thuiller 2005; Kearney 2006; Soberón 2007; Elith and Leathwick 2009b).

C.F. Dormann
Department of Computational Landscape Ecology, Helmholtz Centre for Environmental Research – UFZ, Permoserstr.15, 04318 Leipzig, Germany
e-mail: carsten.dormann@ufz.de

F. Jopp et al. (eds.), *Modelling Complex Ecological Dynamics*,
DOI 10.1007/978-3-642-05029-9_13, © Springer-Verlag Berlin Heidelberg 2011

Why Species Distribution Modelling?

There are several fundamental challenges to this approach (e.g. first and foremost that it is correlative; see Vaughan and Ormerod 2005 and Dormann 2007b for a recent critique), and before jumping into the analysis, it is worth considering whether SDMs are actually useful and fit for the purpose of our specific problem. For example, at very small spatial scales, differences in environmental conditions may be too small to be of predictive value and biotic interactions (competition, predation) may be of crucial importance. In contrast, at the global scale, data become so coarse that we "only" model the climate niche and specific habitat requirements cannot be detected.

On the other hand, SDMs try to extract ecological information from a species occurrence pattern when and where it matters. Expert knowledge usually cannot inform us which trait or limitation will be relevant for our problem at hand. We may know that a palm tree does not survive sub-zero temperatures, but the observed distribution will tell you that even 10 × 10 km grid squares with minimum temperatures well below 0°C harbour this species because of microclimatically suitable places. Thus, at the spatial resolution under investigation, the physiological threshold can be misleading even though it may be true. Overall, SDMs are useful for complementing existing approaches in at least these five areas of research:

1. Small-extent, decision-support for conservation biology (such as Biological Action Plans: Zabel et al. 2003, and numerous others)
2. Testing specific hypotheses, e.g. on the spatial scale of habitat selection (Graf et al. 2005; Mackey and Lindenmayer 2001), the species-energy hypothesis (Lennon et al. 2000) or range-size effects on diversity pattern (Jetz and Rahbek 2002)
3. Generating hypotheses, e.g. on correlation of species traits with environmental variables (Kühn et al. 2006), which can then be tested experimentally
4. Identifying hierarchies of environmental drivers (Bjorholm et al. 2005; Borcard and Legendre 2002; Pearson et al. 2004)
5. Prospective design of surveys, e.g. optimizing sampling schemes for rare species (Guisan et al. 2006)

Now, we shall focus on the technical side and assume that you know what you are doing, ecologically speaking.

Analysing the geographic distribution of species' occurrence, abundance or diversity is, essentially, a statistical task. As such, the fundamental ideas and principles of good statistics apply (and can be found in the excellent but advanced book of Hastie et al. 2009). There are at least three reasons why methods for describing or modelling these patterns have reached a higher level of sophistication than many other fields in ecology. Firstly, biogeographical data sets are nowadays large (both in terms of number of data points and potentially explanatory variables), necessitating the use of new statistical strategies. Secondly, species distribution data typically carry a largish bunch of common intrinsic statistical problems and accordingly several solutions have been tailored to these problems (presence-only data, low information content of binary data, spatial autocorrelation, multi-collinearity, model unidentifiability). Thirdly, species distribution modelling (SDM) is "sexy". As habitat of many species is continually lost, as climate changes and as environmental management becomes a matter of human survival, scientists, decision makers and the general public look for information and predictions of possible future scenarios. Consequently, substantial funding (at least for ecological topics) over the last decade has enabled talented scientists to make a career from SDMs.

Aims of This Chapter

Recent developments have made the field of SDM somewhat complex, diverse and confusing for the newcomer. The aim of this chapter is thus to (1) provide a recipe for SDM; (2) briefly discuss a few selected "hot" topics; and (3) give an overview of challenges of a more ecological modelling type (dispersal, occupancy, biotic interactions, functional variables, evolution, changing limiting resources). I shall restrict citations to fundamental or specific methodological papers and will therefore have to ignore the vast amount of good ecological papers that "only" did it right. On the other hand, I am not aware of any paper on species distribution modelling that could tick all elements of the recipe below.

13.2 A Species Distribution Modelling Recipe

A good cook needs no recipe. Alas, we are trained more in ecology than statistics. Moreover, without the right ingredients (a.k.a. data) and tools (software), no dish will be tasty. Also, I should mention other recipes along this line: see Harrell (2001) for a generic statistical recipe, and Pearson (2007) and Elith and Leathwick (2009a, b) for a specific one on SDMs. As for "cooking tools", I highly recommend using code-based software so that each step of the analysis is documented and easily reproducible. The functions mentioned in this chapter are all from the free R environment for statistical programming (R Development Core Team 2008).

The recipe falls into three sections: pre-processing, modelling and model interpretation (Fig. 13.1). These sections are somewhat arbitrary, but are useful to structure the whole endeavour. We shall assume that you have your ingredients well prepared: The observed data are as good as we need them, the explanatory variables are ecologically relevant and at the same resolution and your statistical tools are laid out in front of you. A worked example is available at http://www.mced-ecology.org ("Where's the sperm whale?"), which follows the recipe and provides example data and R-code.

Phase	SDM task		typical example
pre-processing	response	transformation	log, Box-Cox
	predictor	transformation	square-root
		imputation	multiple imputation
		standardisation	centering, scaling
	ecol. dimensional reduction		resource/direct/proxy
	stat. dimensional reduction		PCA, univariate scans
	collinearity		$\|r\| < 0.7$, sequential regression
modelling	model	formulation	splines, interactions
		simplification	BIC, pooling of levels
		type	GLM, GAM, BRT
	spatial autocorrelation		GLMM, SEVM
	diagnostics		Influential points, dispersion
	model performance		AUC, cross-validation
interpretation	functional relationship plots		shape, error margins
	Importances, significances		partial R^2, p-values
	ecological interpretation		plausibility, cf. literature

Fig. 13.1 Overview of the species distribution modelling workflow. The three phases contain various tasks, for which typical examples are given in the *right* column

13.2.1 Pre-processing and Visualization

The Response Variable

When the data are presence–absence (i.e. binary) no further preparation is needed. When data are counts or continuous, we have to make sure that assumptions of the modelling approach are met. For parametric modelling approaches (regressions by means of GLM or GAM), count data are usually assumed to be Poisson distributed but all too often are not. Continuous responses are generally assumed to be normally distributed. These assumptions can be checked only after modelling, because we need to look at the residuals or compare log-likelihoods of different distributions. Generally, if too many zeros have been observed, the data are over-dispersed and we have to resort to one of three alternative approaches: a quasi-Poisson distribution (where over-dispersion is explicitly modelled); a negative binomial distribution (where a clumping parameter is fitted); or a separate analysis of zeros and non-zeros (as in zero-inflated or other mixed distribution models:

Bolker 2008). Sometimes people log-transform count data (more precisely: `y' = log(y+1)`), and find the new `y'` to be normally distributed.

Normally (Gaussian) distributed data show a normal distribution in the model residuals and a straight 1:1 relationship in a QQ-plot of these residuals. Deviations need to be accounted for, e.g. by transforming the data (any good introductory textbook, such as Quinn and Keough (2002), will feature a section on transformations, including useful ones such as the Box-Cox[1] transformation).

When we have presence-only data (i.e. only locations where a species occurs but no information where it does not), two alternative approaches are available. We could use purpose-built presence-only methods, or we could use all locations without a presence and call them absences (pseudo-absences). Both approaches have their difficulties (Brotons et al. 2004; Pearce and Boyce 2006). The first suffers from a lack of sound methods (in fact, following Tsoar et al. (2007) and Elith and Graham (2009), I would currently only recommend MaxEnt[2] in this direction and hope for the approach of Ward et al. (2009) to become publicly available). The second approach lacks simulation tests on how to select pseudo-absences and how to weight them (see Phillips et al. 2009 for the cutting edge in this field), although it has been argued that the pseudo-absence approach can be as good or better than the purpose-built presence-only methods (Zuo et al. 2008). In what follows, I only consider presence-(pseudo)absence data.

The Explanatory Variables

Explanatory variables may also require transforming! Consider a relevant explanatory variable which is highly skewed (e.g. log-normally distributed), as is commonly the case for land-use proportions. Few high-value data points may completely dominate the regression fitted. To give a more balanced influence to all data points, we want the values of the predictors to be uniformly distributed over their range. This will rarely be achievable, and researchers mostly settle for a more or less symmetric distribution of the predictor. Note, however, that ideally we want most data points where they help most. For a linear regression, the mean is always best described, so we would want most data points at the lowest and highest end of the range. For a non-linear function, for example a Michaelis–Menton-like saturation curve, we want most data points in the steep increase, while there is little gained from many points at the high end, once the maximum is reached. As a rule of thumb we need many data points where a curve is changing its slope.

Transformation of explanatory variables is particularly needed for regression-type modelling approaches such as GLM and GAM (see below for explanation). Regression trees (used, e.g. in Boosted Regression Trees, BRT, or randomForest) are far less sensitive, if at all (Hastie et al. 2009). It is a good custom to make

[1]`boxcox` in **MASS** (`typewriter` and **bold** are used to refer to a `function` and its **R-package**)

[2]Phillips et al. (2006b): http://www.cs.princeton.edu/~schapire/maxent/

a histogram of each explanatory variable before entering it into an analysis! Transformation options are the same as for the response.

Missing data are a (very) special case of transformation. Although generally disliked by many analysts, imputation (replacement of missing data) is often a good idea (see Harrell 2001), particularly if missing data are scattered through the data set (i.e. across several variables!) and we would loose many data points if we simply omitted every data point with missing values. Standard imputation uses the other explanatory variables to interpolate a likely value for the missing one.[3] Replacement by the mean is not an option!

"Outliers" are (in general) a red herring: If there is no methodological reason why a data point is extremely high (e.g. one data set being recorded in winter, while all other data points are from the summer), then this datum should also be included in the analysis. Otherwise the data set may be poorly sampled, but the "outlier" would still represent a (potentially) valuable datum. It would be good practice to omit it later on and see if the results are robust to this omission. Furthermore, in multi-dimensional data sets (i.e. those with several explanatory variables), a datum might be an "outlier" in one dimension, but an ordinary data point in all others: why delete it?

Finally, all continuous variables should be standardized before the analysis.[4] This reduces collinearity, particularly with interactions (Quinn and Keough 2002). As a convenient side effect, regression coefficients are now directly comparable: the larger their absolute value, the more important this term is in the model (they become standardized regression coefficients).

Collinearity

Collinearity refers to the existence of correlated explanatory variables. Some predictors are only proxies for an underlying, latent variable. For example, consider temperature and rainfall, which are largely governed by distance to ocean (oceanity), altitude and regional terrain. Collinear predictors can lead to biased models due to inflated variances (Quinn and Keough 2002). There are many cures to this ailment, but no remedy. Logically speaking, if two predictors are tightly linked to an underlying (but elusive) causal variable, there is no way to find out which is the "correct" predictor for our analysis. We may choose precipitation over temperature when modelling plants (or the other way around for insects), but there is no guarantee that this choice allows us a sensible extrapolation of our model. Furthermore, using Principal Component Analysis (PCA) or any of the other tailor-made methods for collinearity (Partial Least Squares, penalized regression, latent root regression, sequential regression, and many others) will not solve the ecological problem, only the statistical. These methods will produce either a new data set of

[3]`transcan` and `aregImpute` in **Hmisc**

[4]`scale`

uncorrelated variables, or "consider" the correlation when estimating model parameters.

So where is the problem? Imagine an organism whose distribution is governed entirely by its sensitivity to frost. When we combine our climate variables into one or more principal components, model the species' distribution, and then predict to a climate change scenario, the fact that both rainfall and mean summer temperature are correlated with number of frost days will dilute its impact in the model. The total effect of "frost" is distributed over all correlated variables. As a consequence, any climate prediction will underestimate the effect of frost and hence yield a "wrong" expected future distribution. If we don't know the true underlying causal mechanism, no statistics can help us here (or at least very little). Any correct ecological knowledge used in variable pre-selection, however, will lead to a smaller bias in scenario projections!

Dimensional Reduction

Often we may have dozens or even hundreds of potential explanatory variables (e.g. from multispectral remote sensing or landscape metrics). We should try to reduce this set to as few as possible for two reasons: (1) The more variables we have, the more they will be correlated. (2) The more variables we have, the more likely one of them will spuriously contribute to our model (type I error). For SDMs, Austin (2002) and Guisan and Thuiller (2005) argue that we should choose "resource" over "direct" and "direct" over "indirect" variables. For example, the abundance of prey (hardly ever available) or nesting opportunities will be a resource variable when analysing the distribution pattern of a bird of prey. Temperature or human disturbance could be direct variables, impacting on the bird without moderation by other variables. Indirect variables would be altitude or length of road in a grid cell, which are substitutes, surrogates or proxies for other, more directly acting variables. These indirect variables are often not immediately perceivable by the organism (such as altitude by a plant or length of road verges by a rodent). So if we have two (correlated) variables, we should discard the one "further away" from the species' ecology.

If we are unable to reduce the data set sufficiently (i.e. **k**≫**N**), we should use dimensional reduction techniques, such as Principal Component Analysis[5] or its more sophisticated variants that also allow categorical variables (nMDS[6]). The scores for the most important axes in this new parameter hyperspace can be used as explanatory variables. Note that interpretation is often extremely impaired by automatic dimensional reductions. It is thus always advisable to use ecological understanding rather than statistical functions at this step!

[5]`prcomp`

[6]`isoMDS` in MASS or, more conveniently, `metaMDS` in vegan

An alternative is to "filter" the data by importance. We can use a robust and able technique to tell us which variables are important. Next, we use only those 5 or 12 variables filtered from the initial pool of variables, and continue. If variables are uncorrelated, regression-tree based methods are very useful for this, and I recommend randomForest and Boosted Regression Trees. If you plan to model your data with BRT anyway, there is little point in reducing the number of explanatory variables before.

Finally, be aware that any model can only find correlations with the variables provided. Of course, we know that our hypothetical bird of prey depends on specific prey. Without this information, we may actually be modelling the niche of the prey, not of the predator!

Exploratory Data Plotting

Can we finally start? No! It is both good practice and highly advisable to look at the data by plotting them in any reasonable combination conceivable (see, e.g., Bolker 2008). Plot thematically related explanatory variables as scatterplot[7] to detect collinearity. Plot each explanatory variable against the response (henceforth called **X** and **y**, respectively) and look for nonlinear effects. Plot data in parameter space, e.g. as function of two Xs (Fig. 13.2) and a hull-polygon around the data to see that 40% or so of the parameter space is not in your data set. This is the area outside the convex hull in Fig. 13.2. The more variables (and hence dimensions) your data set has, the more severe this problem becomes. It is so prominent among statisticians (though not among ecologists) that it is referred to as the "curse of dimensionality" (Bellman 1957; Hastie et al. 2009). Repeat this plotting for any number of variables. Getting a feeling for the data is crucial, and many later errors can be avoided. Every minute invested at this stage saves hours later on.

13.2.2 Modelling

Here, we arbitrarily divide the process of deriving a "usable" model into two steps. The first, model building, selects the variables to be included, the type of non-linearity and order of interactions considered, and the criteria for selecting the final complexity of a model. The second step, model parameterisation, performs the final step of using the data to calculate the best estimates for variable effects. It is this model that we want to use for interpolation, hypothesis testing or extrapolation. Note that in some methods these two steps are implicitly taken care of and that there is no two-step process (mainly machine learning, where model selection is done internally through cross-validation in order to prevent models from being "unreasonably"

[7] `pairs`

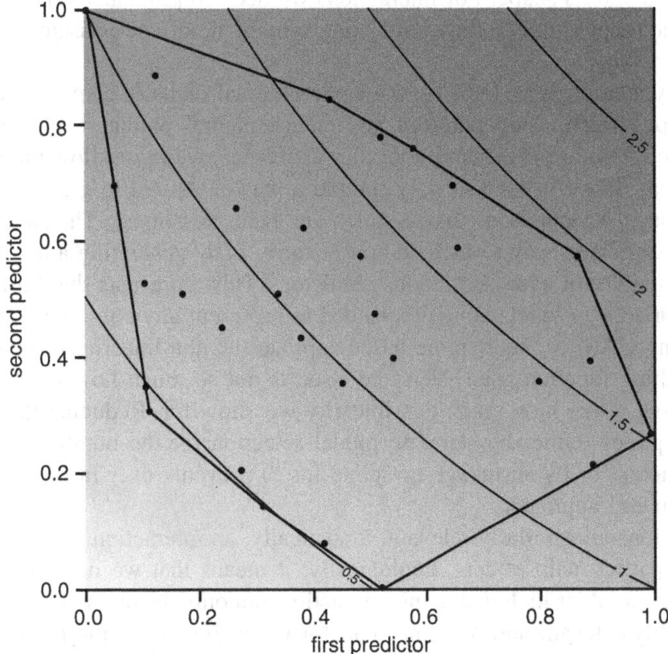

Fig. 13.2 Visualizing the parameter space supported by data. In this case, the *top-right* and *bottom-left* corner of the parameter space of the two predictors has not actually been sampled by data (despite a low correlation of r = −0.26). The parameter space actually sampled is indicated by the convex hull, covering 57% of the area, and it declines dramatically with the number of dimensions ("curse of dimensionality"). In other words: we have few data points to look at interactions of higher order

large: e.g. Hastie et al. 2009). For more traditional approaches (and here I am thinking of GLMs), we may want to have these steps functionally separated.

Model Formulation

We have reduced our data set to a moderate number of predictors in the step "Dimensional reduction" above. Now we still need to specify in which functional form the predictors are allowed to correlate with the response. In early years, both non-linear and interactive model terms were neglected, making many of their findings less trustworthy. Modern methods (such as BRT) will automatically have non-linearity and interactions build-in. It is still important to understand the relevance of non-linearity and interactions, even when using the tree-based methods, because we still have to be able to interpret the results. The information on the importance of a variable often returned by machine-learning algorithms does not allow us to see *how* the variables act. As shown in the case study at http://www.mced-ecology.org (Where's the sperm whale?), the functional relationship must be plotted to gauge its shape. For interactions we need to plot each variable at each level of the other variable, thus visualizing synergistic or compensatory effects of the two variables.

The key idea behind SDM, i.e. the environmental niche of a species, implies a hump-shaped relationship between any environmental predictor and a species' occurrence: there are lower and upper limits. Hence, we *must* allow the model to be nonlinear. If we happen to only sample a part of the entire gradient, we also need to consider saturation curves, which are again non-linear. The simplest, and generally sufficient, way to include non-linearity is by generating a new, squared dummy variable for each continuous predictor.[8] This represents the third element of a Taylor series (which can be expanded to represent any continuous function). When using GAM or other spline-based approaches, non-linearity is governed by the smoothing function used. Here the issue is not so much how to model non-linearity, but rather how much non-linearity we allow for. Reducing the "wiggliness" of splines (either by stepwise model selection for the number of knots in each predictor[9] or by shrinkage of spline fits[10]) prevents over-fitting and should be the standard approach.

Interactions are similarly relevant. Statistically, an interaction is the product of the participating main effects. Ecologically, it means that we need to know the value of all variables included in the interaction, not only the main effects. Because this is highly relevant and often difficult for the beginner, let me briefly give an example. Assume that global patterns of plant diversity are well-predicted by the predictors "annual precipitation" and "mean annual temperature" – and their interaction. For the main effects, wet or hot means more species, but not necessarily. When a site is hot, it needs to *also* be wet to have high species richness; otherwise it may well be a barren desert. But when cold, a site will never support many plant species, independent of precipitation. In this example, neither temperature nor rainfall alone is sufficient to predict species richness at any site, but we need to interpret them in concert.

Classification and regression trees (CARTs) embrace non-linearity and interactions in an elegant and natural way. Their boosted (BRT) or bagging (randomForest) extensions hence do not require specification of non-linearity and interactions.

Model Simplification

One of the fundamental problems in building statistical models is the trade-off between the variance explained by the model, and the bias it produces when

---

[8]This can be done either manually (`X1.2 <- X1^2`) or as part of the model formula (`y ~ X1 + I(X1^2)`); higher-order polynomials should be specified using poly (`y ~ poly(X1, degree=3)`), which calculates orthogonal polynomials.

[9]As proposed for the function **gam** in package **gam**: see `?gam::step.gam`.

[10]As proposed for the function **gam** in package **mgcv**: see `?mgcv::step.gam`.

validating it on a new, independent data set (variance-bias-trade-off: Hastie et al. 2009). Smaller models are more robust, i.e. less biased, at the expense of being not very good in explaining variance. The way to derive the "optimal" model size is through cross-validation (CV). For some modelling approaches this is automatically implemented, but the majority of model types require the user to carry out this step. *N*-fold cross-validation encompasses a random assignment of data points to the *N* subset, with *N* usually between 3 and 10. Care should be taken to have equal prevalence in all subsets, e.g. by randomizing 0s and 1s separately (stratified randomization). The model is then fitted to $N-1$ of the *N* subsets and evaluated on (by predicting to) the remaining subset. This is repeated for all *N* subsets and evaluations are averaged. Based on these values, we can select the best modelling strategy (both model complexity and model type). An alternative approach is to bootstrap the entire model building process and use bootstrapped measures of model performance. Since a bootstrap requires several thousand runs, and a CV only a few, CV is far more common.

Information theoretical approaches are based on analytical methods to describe this CV. Hence Akaike's Information Criterion (AIC) or Schwartz'/Bayesian Information Criterion (BIC) are implicitly also based on cross-validation. While it is clear that too large a model will be over-fitting, and that too small a model will not capture as much of the variation as it should in the data, the "true" model will always remain elusive, and our "optimal" model will only be a caricature of the truth. However, here is much to be learned from this caricature!

Model Type

At this point we have to choose one (or more) method(s) to do our analysis with. The good "traditional" approaches comprise Generalised Linear Models (GLM) and Generalised Additive Models (Guisan and Zimmermann 2000). Discriminant Analysis has been given up on, as have been Neural Networks and CARTs (Guisan and Thuiller 2005). "Modern" approaches are often based on either multidimensional extensions of GAMs (such as MARS and SVM) or machine-learning variations of CART (such as BRT and randomForest: Hastie et al. 2009). Anyone using a machine-learning method should familiarize himself with this method. The majority of them are performed on real data sets, where the truth is unknown and the performance of a method was hence assessed by cross-validation. These comparisons show, broadly speaking, that model types sometimes differ dramatically in performance, that each model type can be misused and that both GLM and BRT are reliable methods when used properly.

This is not the place to explain the differences between all of them (see Hastie et al. 2009 for a recent and comprehensive description or Elith and Leathwick 2009a). It has to suffice to make clear the main difference in the machine-learning approach to "traditional" statistical models. In traditional models (e.g. GLM), we specify the functional relationship between the response and its predictors. For example, we decide to include precipitation as a non-linear predictor for plant species richness. This model proposal is then fitted to the data. In machine learning, we propose only the set of predictors, but not the model structure. Here, an algorithm builds a model proposal, fits it to a part of the data set and evaluates its performance on the other part of the data. It then proposes a modification of the original model and so forth. Machine-learning algorithms[11] differ in scope, origin, complexity, and speed, but they all share this validation step which is used to steer the algorithm towards a better model formulation. There are plenty of studies comparing different modelling approaches (Guisan et al. 2007; Meynard and Quinn 2007; Pearson et al. 2006; Segurado and Araújo 2004). Rather, we shall continue using GLM and BRT as representatives for the two most common good approaches.

The choice of model type has much to do with availability of software, current fashion and, of course, with the specific aim of the study. Further complications arise if the design of the survey may require a mixed model approach (e.g. due to repeated measurements or surveys split across observers), if spatial autocorrelation needs to be addressed, if zero-inflated distributions have to be employed, and if corrections for detection probability shall be modeled. The more additional requirements are imposed on the model, the more GLMs become the sole possible method.[12] Alternatively, you may want to go for a Bayesian SDM (see Latimer et al. 2006, for a primer).

If your data and model require an unusual combination of steps (say a combination of zero-inflated data with nested design and spatial autocorrelation, while predictors are highly correlated and many values missing), and you develop a way to cook this dish, then you should do (at least) two things: Firstly, evaluate your method for its ability to detect an effect that you *know* is there ("sensitivity"). Secondly, evaluate your method for its specificity to detect effects that you know are *not* there. Both evaluations should be amply replicated, should be based on simulated data (so that you know the truth) and should (finally) confirm that your new methods is reliable!

Spatial Autocorrelation

Spatial autocorrelation (SAC) refers to the phenomenon that data points close to each other in space are more alike than those further apart. For example, species richness in a given site is likely to be similar to a site nearby, but very different from sites far away. This is mainly due to the fact that the environment is more similar within a shorter distance. Hence, SAC in the raw data (species occurrence) is a consequence of SAC in the environment (topography, climate), something Legendre (1993) termed "spatial

[11]http://www.machinelearning.org/ is a good place to start exploring this field.

[12]Most of these "complications" can be handled by standard extensions of GLMs (see, e.g. Bolker 2008, and various dedicated R-packages). They will, however, make the model less stable, require larger run-times and still rely on getting the distribution right. There is, of course, the alternative of Bayesian implementations. Since these are also fundamentally maximum likelihood approaches, they are similar to sophisticated GLMs. In any case, there is no Bayesian Boosted Regression Tree (not to speak of a combination with spatial terms and mixed effects). It runs against the Bayesian philosophy to use boosting or bagging, and there is no efficient implementation either.

dependence". In SDMs, we do not care about SAC per se, but about SAC in the model's residuals (i.e. unexplained by the environment), because it distorts model coefficients (Bini et al. 2009; Dormann 2007a). To date it is unclear whether this residual SAC is mainly due to model misspecification (omission of non-linearity and interactions), due to variation in sampling coverage, due to omission of important predictors, or due to ecological processes (territoriality, dispersal). Only with respect to some of these problems a *statistical* solution can be found. The spatial toolbox is rich in approaches (Beale et al. 2010; Carl et al. 2008; Dormann et al. 2007; Mahecha and Schmidtlein 2008). In any case, SDM residuals should be investigated for spatial autocorrelation, and attempts should be made to correct for it. If spatial models yield similar coefficient estimates (GLM) as non-spatial models, then there seems to be little value in "going spatial": the ranges of the spatial autocorrelation may or may not be related to the ecological scale of movement or behavioral patterns (Betts et al. 2009; Dormann 2009).

Tweaking the Model

There are several ways in which the quality of the model can be increased (Maggini et al. 2006). One important start is to investigate the model residuals. They indicate whether model assumptions were violated (e.g. when residuals are highly skewed or their variance is not the same throughout the range of fitted values) or if some non-linear relationship went unnoticed (residuals may show a hump-shaped trend against fitted values).

Model diagnostics[13] will also indicate outliers, i.e. data points that have a high influence on the model coefficients. We can use weights to decrease an outlier's impact. Weights are also useful when the balance between presences and absences is very disturbed. Down-weighting the more common category so that model weights sum to the same value for 0s and 1s has been shown to increase the sensitivity of binomial models (Maggini et al. 2006). The same approach is recommended when using pseudo-absences (Elith and Leathwick 2009a).

By including data from other scales or broader geographic coverage, regional or local SDMs can also be improved. Pearson et al. (2004) used European distribution and climate data to fit a niche model for four plant species. Predicted probabilities of occurrence from this model were then used as input variable alongside land-cover variables in the second-step model for the UK. Thereby the authors avoided the problem that the climate gradient in the UK is much shorter than of the species' global distribution.

Assessing Model Performance

To quantify how well our model fits the data, we compare model predictions with field data (usually on a hold-out sample; e.g. the subset of a cross-validation).

[13]Diagnostics for GLMs fitted in R are given by plotting the model object.

Traditionally, the probability predictions from the model were converted into presences and absences and then a confusion matrix could be used to calculate various parameters of choice (e.g. commission and omission error, kappa, etc: Fielding 2002). The AUC ("area under curve") is currently the most commonly used measure of discriminatory power of a model. Its value (between 0.5 for random and 1 for perfect) quantifies the ability of the model to put the data points into the correct class (i.e. presence or absence), independent of the threshold required by the other measures mentioned. It has recently received justified criticism because its values are not comparable across different prevalences (and the criticism extends to kappa, too; see Lobo et al. 2008). Currently, misclassification rates, commission and omission errors are more *en vogue* again, because they can be intuitively interpreted. Furthermore, by assigning different weights to false negatives (omission error) and to false positives (commission error), conservation management can come to more sophisticated and balanced decisions (Rondinini et al. 2006).

Only rarely will a second set of data be available to investigate the quality of our model(s) through external validation. A different recording strategy, another time slice or data from a different geographic location represent really independent data, and could thus be considered an external validation. The internal validation (described above as cross-validation) is an optimistic assessment of model quality. When using SDMs to infer underlying mechanisms, external validation is less of an issue than when using them to extrapolate to a future climate or other sites. Because the cross-validated models are optimistic, they give narrower error bands than they should.

13.2.3 Interpretation

Once we have arrived at what we regard as a final model, we should make every effort to understand what it means. A first and most relevant step is to visualize the functional relationships within the model. The plot of how occurrence probability is related to, say, annual precipitation should be accompanied by a confidence band around this line. It may be useful to plot the data as rug (ticks on the axis representing positions of the data values) into this figure to visualise the support at each point in parameter space (Fig. 13.3).

For interactions, visualization becomes more difficult. Two-way interactions can still be plotted (e.g. as a 3-D plot or as a contour plot). No confidence bands can be included, though. Here it is again very important to indicate the position of the samples to identify regions of the parameter space that have not been sampled. For higher dimensions, or for a model that averages across many sub-models, we can do the same plots (called marginal plots for main effects because they represent the marginal changes to a predictor, averaging across all other predictors). We can also slice through higher dimensions, i.e. calculate a marginal plot for specific values of other predictors (often their median).

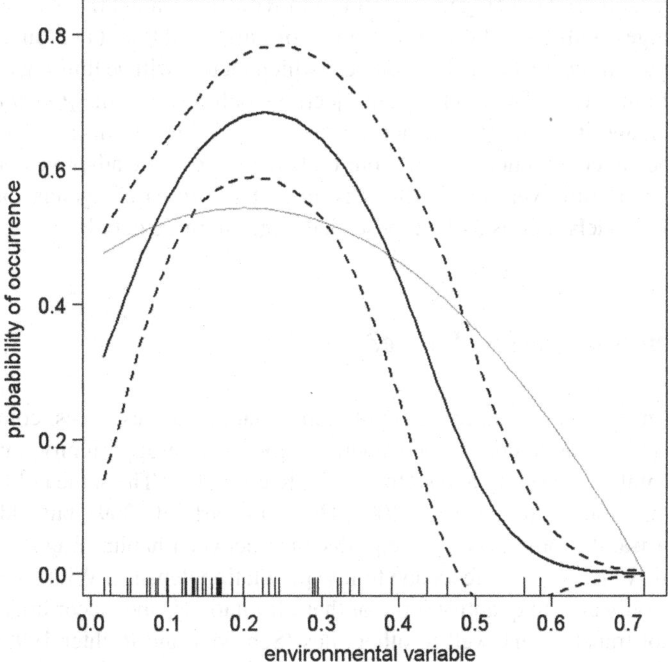

Fig. 13.3 Functional relationship between an environmental variable and a binary response. Rug (ticks on lower axis) indicate for which x-values data were available. *Lines* represent a quadratic fit (*solid*) and its standard deviation (*dashed*). Thin *grey line* is the true, underlying, data-generating function. Note the few data points upon which the declining half of the function is based (6 of a total of 50 have a value >0.35)

Spending time plotting is again well invested. We will detect errors in the model, scratch our head over inexplicable (and hence overly complex) patterns, and be forced to extract the main conclusions from it. It is this phase where the traditional GLM is superior to the BRT, because variable interpretation is easier. It is, however, also this phase where we may realize that BRTs are superior to GLMs because they can model step-changes and thresholds much better. Personally, I think we should not publish patterns we do not understand. There are, as the previous steps have shown, several decisions that could generate artifacts and their publication cannot be seen as progress.

13.3 Beyond Recipes: New Challenges for Species Distribution Models

The above recipe can be used to derive a static description of environmental correlates with distribution data. But they often leave the analyst unsatisfied. Many assumptions might be suspected to be violated (Dormann 2007b), such as stationarity, unbiased coverage, or equilibrium with environment.[14] There are three key challenges with the following problems of current SDMs: (1) A niche model describes the *current* niche, and it is unclear which factors will be limiting elsewhere or in the future. (2) Climate change projections delimit only the *potential* future distribution, and it is unclear whether the species will ever fill this new range; e.g., due to dispersal constraints. And, (3), our understanding of the adaptive potential of a species is currently very poor. This sets, in part, the research agenda for species distribution models. Let us look at these challenges in more detail.

What Limits a Species' Range?

An organism is constrained in its population dynamics by resources, competitors, predators and diseases, density dependence, reproductive opportunity, mutualists, environmental stochasticity and so forth (e.g. Krebs 2002). The same holds true for its spatial distribution (e.g. Gaston 2009; Holt and Barfield 2009), but additionally spatial constraints come into play (e.g. distance between habitat fragments, minimum territory size, Allee effects due to low population density). With an SDM, we are usually only able to quantify some of these limitations, and, accordingly, SDMs often do not transfer very well to other sites (Schröder and Richter 1999; Randin et al. 2006; Duncan et al. 2009, but see Herborg et al. 2007). In particular, biotic interactions are hardly ever quantified explicitly within SDMs (although some of them will also correlate with the environmental data used). But they matter in real data (e.g. Preston et al. 2008; Schweiger et al. 2008), and they impact model performance and predictive ability (as shown, in a simulation study, by Zurell et al. 2009). It is thus a key challenge to incorporate biotic interactions into SDMs, but to date such attempts are few and far between (e.g. Bjornstad et al. 2002). A recent review (Thuiller et al. 2008) indicates some avenues to do so, but all of them are based on the explicit modelling of populations within cells, if not individuals. It is unclear how to derive a more general dynamic SDM, without spending years per species on incorporating detailed ecological knowledge.

Fundamental, Potential and Realized Niches

The reason why many biogeographers refer to SDMs as "Species Distribution Models" and not as "niche models" is because they do not believe that we model the niche of the target species. In fact, as Jiménez-Valverde et al. (2008) argue,

[14]Actually, the term "equilibrium" is a bit misleading. What is meant is that the entire width of its niche is filled. Within this niche, there may well be unoccupied sites, e.g. due to metapopulation dynamics. A problem arises, when a species does not occupy say the dry end of its soil moisture niche for historic reasons. Then the estimate of this end of the niche will be biased.

because we do not know *why* a species is absent in some sites, we are in the dark about its niche. The discussion of what a niche is, and what we are modelling, has sparked several interesting and not always compatible publications (e.g. Kearney 2006; Soberón 2007). Hence, Araújo and Guisan (2006) have named the "clarification of the niche concept" the first of five challenges for SDMs. While we cannot resolve this issue here, it is important to realize that the "niche" based on the correlation between geographic distributions and environmental conditions is quite a bit more vague than the niche discussed in evolutionary ecology, where resources and other causal drivers are envisaged (see, e.g. Losos 2008).

More to the point in this context is the challenge to quantify how much of the fundamental niche is actually covered by the realized niche as extracted from SDMs. If, on one extreme, the realized niche is pretty much also the fundamental niche (i.e. there are no biotic interactions alike to constrain the distribution at the scale we are analyzing), then we can merrily predict future distributions of this species (e.g. under climate or land-use change). At worst, we are overestimating the future, "potential" distribution (if in the future biotic interactions may become limiting or if species do not reach the sites). At the other extreme, if the fundamental niche is considerably wider than what we model, any projection can be fundamentally flawed (Dormann et al. 2010). I am not aware of any study assessing the overlap of realized and fundamental niche for geographic distributions (see also Nogués-Bravo 2009). It could require transplant experiments into areas beyond the current range and the manipulation of biotic interactions there. The few studies going into this direction point at a large discrepancy between fundamental and realized niche. Battisti et al. (2006), for example report on a range shift after a particularly warm summer, which was not reverted afterwards, indicating that it was dispersal limitation that prevented a filling of the niche. Similarly, several studies point at the importance of dispersal limitation (Nekola 1999; Ozinga et al. 2005; Samu et al. 1999; Svenning and Skov 2004), leading to both a bias in the modeled environment-occurrence relationship as well as the width of the niche itself. There is, as yet, no standard way to wed SDMs and dispersal (for attempts see, Johst et al. 2002; King and With 2002; Lavorel et al. 2000; Lischke et al. 2006; Midgley et al. 2006; Schurr et al. 2007; Thuiller 2004).

Niche Evolution

Another important and fast developing field related to species distribution modelling is the study of niche evolution. I shall use this term very loosely, as is often done, to also include micro-evolutionary changes, genetic (and ecological) drift within species and genotypic plasticity (Pfenninger et al. 2007). Climate change projections using SDMs rely on the assumption that species are not able to adapt significantly to altering environmental conditions. This assumption is implicit in the extrapolation of the fitted niche: if a species was able to adapt rapidly, then the present niche would not be related to its future niche.

The problem is that we have considerable, if patchy, evidence that niches can rapidly evolve (reviewed in Thompson 1998), change within the fundamental niche (Dormann et al. 2010) or at least that variability within a species is large enough to allow it to shift its niche when confronted with novel environments (e.g. Ackerly et al. 2006; Broennimann et al. 2007; Hajkova et al. 2008; Holt 2003; Holt and Gaines 1992). There is, as yet, no synthesis of niche evolution nor, to my knowledge, any mechanistic approach to incorporate geno- and phenotypic plasticity into SDMs or spatial population models. There is, on the other hand, more than anecdotal evidence that microevolutionary processes are at play and matter ecologically (Hampe and Petit 2005; Phillips et al. 2006a). Hence, this field still awaits being embraced by species distribution models.

13.4 Concluding Remarks

This chapter tries to strike a balance between guidance for novices to species distribution modelling – by providing a recipe for the most crucial elements of SDMs – and an embedding of SDMs into the currently most relevant statistical challenges. Analysing a species' distribution can be a very useful starting point for further investigations or process-based modelling attempts. The correlative nature of modelling in general, and species distribution modelling specifically, should always be remembered. Tempting as it may be to incorporate a lot of ecological knowledge into mechanistic or statistical models, only little of this information will actually be relevant *at the focal scale*. The main intellectual challenges that remain are to not over-interpret one's findings and to seek independent corroboration.

Acknowledgments Over the years, many colleagues helped develop the above recipe. I am particularly grateful to Boris Schröder, Björn Reineking and Jane Elith, as well as the many participants of statistical workshops on this topic. I am also grateful to Fred Jopp, Hauke Reuter and Dietmar Kraft for improving a previous version. Funding by the Helmholtz Association is acknowledged (VH-NG-247).

Chapter 14
Decision Trees in Ecological Modelling

Marko Debeljak and Sašo Džeroski

Abstract Decision tree learning is among the most popular machine learning techniques used for ecological modelling. Decision trees can be used to predict the value of one or several target (dependent) variables. They are hierarchical structures, where each internal node contains a test on an attribute, each branch corresponding to an outcome of the test, and each leaf node giving a prediction for the value of the class variable. Depending on whether we are dealing with a classification (discrete target) or a regression problem (continuous target), the decision tree is called a classification or a regression tree, respectively. The common way to induce decision trees is the so-called Top-Down Induction of Decision Tress (TDIDT). In this chapter, we introduce different types of decision trees, present basic algorithms to learn them, and give an overview of their applications in ecological modelling. The applications include modelling population dynamics and habitat suitability for different organisms (e.g. soil fauna, red deer, brown bears, bark beetles) in different ecosystems (e.g. aquatic, arable and forest ecosystems) exposed to different environmental pressures (e.g. agriculture, forestry, pollution, global warming).

14.1 Introduction

Machine learning is one of the most essential and active research areas in the field of artificial intelligence. In short, it is the study of computer programmes that automatically improve with experience (Mitchell 1997). The most widely investigated type of machine learning is inductive machine learning, where the experience is given in the form of learning examples. Supervised inductive machine learning, sometimes also called *predictive modelling*, assumes that each learning example includes some target property, which should be predicted. The final goal is then to

M. Debeljak (✉)
Department of Knowledge Technologies, Jozef Stefan Institute, Jamova 39, 1000 Ljubljana, Slovenia
e-mail: marko.debeljak@ijs.si

F. Jopp et al. (eds.), *Modelling Complex Ecological Dynamics*,
DOI 10.1007/978-3-642-05029-9_14, © Springer-Verlag Berlin Heidelberg 2011

learn a predictive model (such as a decision tree or a set of rules) that accurately predicts this property.

Machine learning (and in particular predictive modelling) can be used to automate the construction of certain ecological models, such as models of habitat suitability and models of population dynamics from measured data. The most popular machine learning techniques used for ecological modelling include decision tree induction (Breiman et al. 1984), rule induction (Clark and Boswell 1991), and neural networks (Lek and Guegan 1999).

This chapter first introduces the task of predictive modelling. It then describes the different types of decision trees (classification, regression and multi-target trees) and presents techniques for learning them. Finally, it gives examples of the use of decision trees in ecological modelling, including examples of both population dynamics and habitat suitability modelling.

14.2 The Machine Learning Task of Predictive Modelling

The input to a machine learning algorithm is most commonly a single flat table comprising a number of fields (columns) and records (rows). In general, each row represents an object and each column represents a property (of the object). In machine learning terminology, rows are called examples and columns are called attributes (or sometimes features). Attributes that have numeric (real) values are called continuous attributes. Attributes that have nominal values are called discrete attributes.

The tasks of classification and regression are the two most commonly addressed tasks in machine learning. They deal with predicting the value of one field from the values of other fields. The target field is called the class (dependent variable in statistical terminology). The other fields are called attributes (independent variables in statistical terminology).

If the class is continuous, the task at hand is called regression. If the class is discrete (it has a finite set of nominal values), the task at hand is called classification. In both cases, a set of data (dataset) is taken as input, and a predictive model is generated. This model can then be used to predict values of the class for new data. The common term predictive modelling refers to both classification and regression.

Given a set of data (a table), only a part of it is typically used to generate (induce, learn) a predictive model. This part is referred to as the training set. The remaining (hold-out) part is reserved for evaluating the quality of the learned model and is called the testing set. The testing set is used to estimate the quality of the model when applied to unseen data, i.e. the predictive performance of the model.

More reliable estimates of performance on new data (not seen in the process of learning) are obtained by using cross-validation (Alpaydin 2010). Cross-validation partitions the entire set of data into k (with k typically set to 10) subsets of roughly equal size. Each of these subsets is in turn used as a testing set, with all of the remaining data used as a training set. The performance figures for each of the testing sets are averaged to obtain an overall estimate of the performance on unseen data.

14.3 Decision Tree Induction

14.3.1 Types of Decision Trees

Decision trees (Breiman et al. 1984) are hierarchical structures, where each internal node contains a test on an attribute, each branch corresponds to an outcome of the test, and each leaf (terminal) node gives a prediction for the value of the class variable. Depending on whether we are dealing with a classification or a regression problem, the decision tree is called a classification or a regression tree, respectively.

Classification trees predict the values of a discrete variable with a final set of nominal values. An example classification tree modelling the habitat of oilseed rape by plant abundance is given in Fig. 14.5. The tree has been derived from real-world data by using decision tree induction (Debeljak et al. 2008).

Regression tree leaves contain constant values as predictions for the class value. They thus represent piece-wise constant functions. Model trees, a type of regression tree where leaf nodes can contain linear models predicting the class value, represent piece-wise linear functions. An example model tree that predicts the abundance of anecic earthworms is given in Fig. 14.1 (Debeljak et al. 2007).

Multi-target trees (Blockeel et al. 1998), sometimes also called multi-objective trees (Struyf and Džeroski 2006) generalize decision trees to the prediction of several target attributes simultaneously. The leaves of a multi-target tree store a vector of class values, one for each target, instead of storing a single class value for one target. Each component of this vector is a prediction for one of the target attributes.

Depending on whether the targets are all discrete-valued or real-valued, we can talk about multi-target classification trees or multi-objective regression trees. An example of a multi-objective regression tree, giving predictions for three

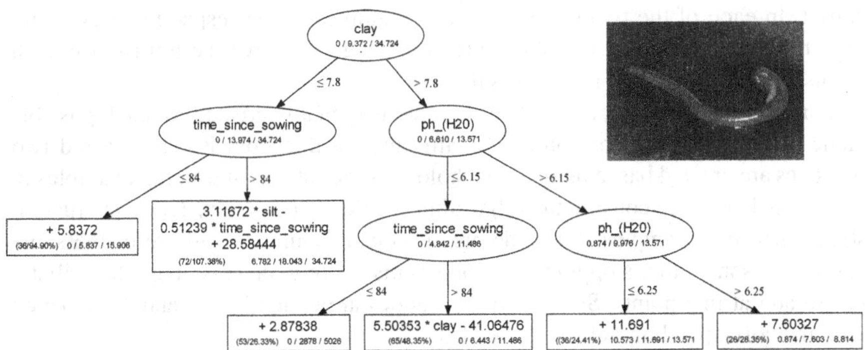

Fig. 14.1 Regression tree for predicting the abundance of anecic earthworms. The additional information given in each node is the min/mean/max of earthworm biomass. In the leaves, this information is extended with the number of examples and relative root mean square error (Debeljak et al. 2007); *upper right*: epigeic earthworm *Eisenia fetida* (Lumbricidae) (Courtesy of Paul Henning Krogh)

real-valued targets, is given in Fig. 14.2 (Demšar et al. 2006). The tree predicts three targets simultaneously: the abundance of Acari and Collembola, as well as their biodiversity in soil.

14.3.2 Learning Decision Trees

Given a set of training examples, we want to find a decision tree that fits the data well and is as small (and thus as understandable) as possible. Finding the smallest decision tree that will fit a given data set is known to be computationally expensive. Heuristic search techniques are thus employed to build decision trees, guided by measures of impurity or dispersion of the target attribute. Greedy search, considering only one test/split at a time, is typically used.

The typical way to induce decision trees is the so-called Top-Down Induction of Decision Trees (TDIDT, Quinlan 1986). Tree construction proceeds recursively starting with the entire set of training examples (entire table). At each step, the algorithm first checks if the stopping criterion is satisfied (e.g. all examples belong to the same class): If not, an attribute (test) is selected as the root of the (sub-)tree, the current training set is split into subsets according to the values of the selected attribute, and the algorithm is called recursively on each of the subsets. The attribute/test is chosen so that the resulting subsets have as homogeneous class values as possible.

Consider for example the tree in Fig. 14.1. At the root node, the algorithm addresses each of the independent variables (including silt, clay, pH and time since sowing) and selects one variable (clay)/test (clay >7.8) that splits the entire set of examples best, i.e. results in subsets with homogeneous values of the class (as compared to other attributes/tests). The examples are then split into two subsets (those with clay >7.8 go down the right branch, the others to the left), and the algorithm is started again twice, once for the left and once for the right subset. In each of the two cases, only the examples in the respective branch are used to build the respective sub-tree (examples going down the left/right branch are used to construct the left/right sub-tree).

For discrete attributes, a branch of the tree is typically created for each possible value of the attribute. For continuous attributes, a threshold is selected and two branches are created based on that threshold. For the subsets of training examples in each branch, the tree construction algorithm is called recursively. Tree construction stops when the examples in a node are sufficiently pure (i.e. all are of the same class) or if some other stopping criterion is satisfied (e.g. there is no good attribute/test to add at that point). Such (terminal) nodes are called leaves and are labelled with the corresponding values of the class.

Different measures can be used to select an attribute in the attribute selection step. Common to all of them is that they measure the homogeneity (or the opposite, dispersion) of the values of the target and its increase (decrease) after selecting the attribute/test for the current node. They differ for classification and regression trees (Breiman et al. 1984) and a number of choices exists for each case. For classification,

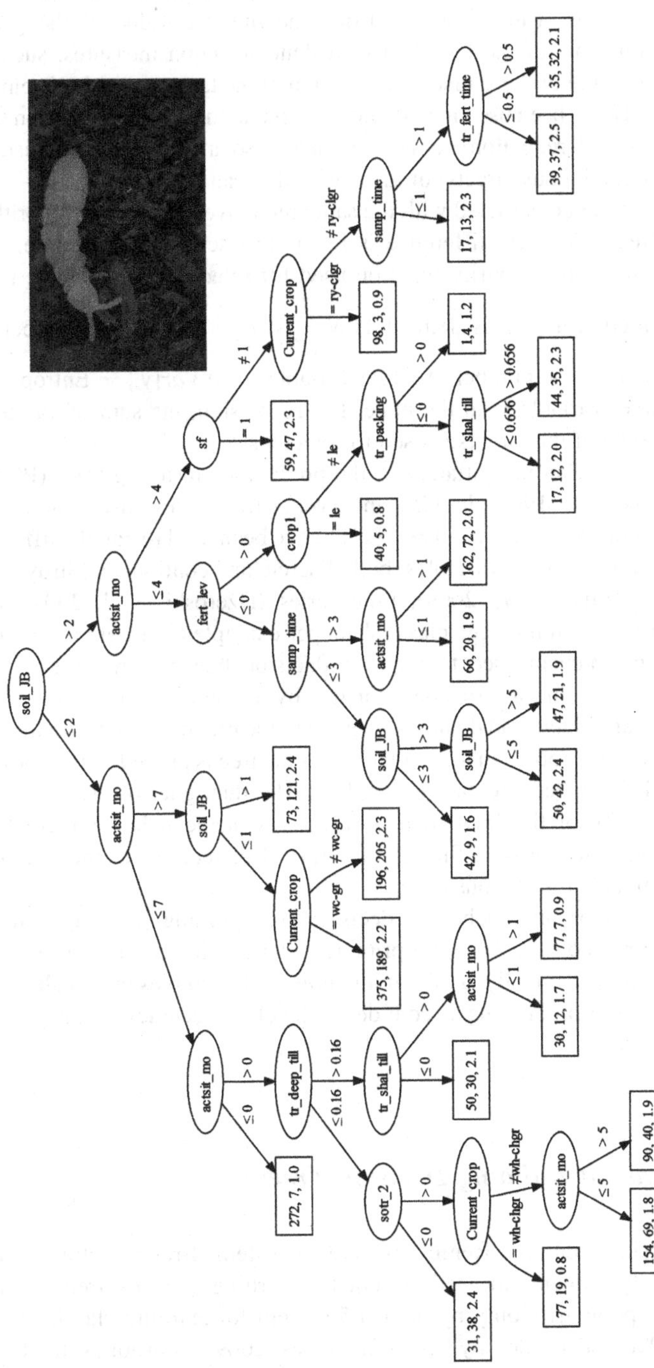

Fig. 14.2 The multi-objective regression tree modelling Acari abundance, Collembola abundance and biodiversity. The numbers in the leaves are the number of Acari individuals divided by 1,000, the number of Collembola individuals divided by 1,000 and diversity, respectively (Demšar et al. 2006); *upper right*: two Collembolan species *Protaphorura fimata* (Onychiuridae) – the largest white one and *Proisotoma minuta* (Isotomidae) – small grey ones (Courtesy of Paul Henning Krogh and Thomas Larsen)

Quinlan (1986) uses information gain, which is the expected reduction in entropy (uncertainty) of the class value resulting from knowing the value of the given attribute and the outcome of the test. Other attribute selection measures, such as the Gini index, a measure of the statistical dispersion of the target variable (Breiman et al. 1984), can and have been used in classification tree induction. In regression tree induction, the expected reduction in the variance (also a measure of statistical dispersion, but for continuous targets) of the class value can be used.

Multi-target trees are constructed with the same recursive partitioning algorithm as single-target trees. The key difference is in the test selection procedure. For classification, the heuristic impurity function used for selecting the attribute tests (that define the internal nodes) is defined as $N \sum_{t=1}^{T} Var[y_t]$ with N the number of examples in the node, T the number of target variables, and $Var[y_t] = \text{Entropy}[y_t]$ the entropy of target variable y_t in the node. For regression, the sum of variance reductions along each of the targets is used to select tests.

Multi-target trees are an instantiation of the predictive clustering trees (PCTs) framework (Blockeel et al. 1998). In this framework, a tree is viewed as a hierarchy of clusters: a node corresponds to a cluster. PCTs have been used to handle different types of targets: multiple target variables, both discrete and continuous (Struyf and Džeroski 2006; Debeljak et al. 2009), time series (Džeroski et al. 2007) and hierarchies of classes, with multiple class-labels per example (Vens et al. 2008).

An important mechanism used to improve decision tree performance is tree pruning. Pruning reduces the size of a decision tree by removing sections of the tree (sub-trees) that are unreliable and do not contribute to the predictive performance of the tree. When a sub-tree rooted in a certain node of the tree is pruned, it is removed from the tree and the node is replaced by a leaf. The dual goal of pruning is to reduce the complexity of the final tree as well as to achieve better predictive accuracy by the reduction of over-fitting and removal of sections of the tree that may be based on noisy or erroneous data.

There are two major approaches to decision tree pruning. Pruning can be employed during tree construction (pre-pruning) or after the tree has been constructed (post-pruning). Typically, a minimum number of examples in branches can be prescribed for pre-pruning and a confidence level in accuracy estimates for leaves for post-pruning.

14.3.3 Systems for Building Decision Trees

The CART (Classification And Regression Trees) system (Breiman et al. 1984) was the first widely known and used system for learning decision trees. It has been surpassed in popularity only by the C4.5 system for learning classification trees (Quinlan 1986), succeeded by C5.0 (RuleQuest 2009). Nowadays, the most

Table 14.1 An overview of decision tree types and systems for learning them, with respect to the number and type of target variables (targets)

Type of targets (decision trees)	Number of targets	
	Single-target	Multi-target
Discrete (classification trees)	C4.5, C5.0, J4.8, CART, CLUS	CLUS
Continuous (regression trees)	CART, CLUS	CLUS
Continuous (model trees)	M5, M5′, Cubist	MT-SMOTI

commonly used implementation of classification trees is likely J4.8, the Java reimplementation of C4.5 within the WEKA suite (Witten and Frank 2005).

Besides CART, the M5 system (Quinlan 1992) builds regression trees. As compared to CART, the novelty in M5 is that it can also build model trees (with linear models in the leaves). The commercial successor of M5 is Cubist (RuleQuest 2009), which transcribes the learned regression and model trees into rules (which are further post-processed/simplified). The publicly available reimplementation of M5 is called M5′ and is part of the WEKA suite (Witten and Frank 2005).

The construction of multi-target trees is implemented in the software system CLUS (Blockeel and Struyf 2002; Struyf and Džeroski 2006; Struyf et al. 2010). CLUS can build trees predicting a single target or multiple targets. It can also consider discrete and continuous targets, i.e., can build multi-target classification and regression trees. The system MT-SMOTI (Appice and Dzeroski 2007) builds multi-target model trees, whose leaves can contain multiple linear equations for predicting the values of each target.

An overview of the different systems for building different types of decision trees is given in Table 14.1.

14.4 Modelling Population Dynamics with Decision Tree Approaches

Population dynamics studies changes of the size and structure of populations over time, taking into account environmental and biological processes influencing these changes. For example, one might study the size of a brown bear population as affected by its initial size, sex and age structure, reproduction age, fertility and mortality of different age classes. The modelling formalism most often used by ecological experts is the formalism of differential equations, which describe the change of state of a dynamic system over time (see Chaps. 6, 7, 9). A typical approach to modelling population dynamics can be as follows: an ecological expert writes a set of differential equations that capture the most important relationships in the domain. These are often linear differential equations. The coefficients of these equations are then determined (calibrated) using measured data.

Relationships among attributes describing internal demographic properties of a population and the set of external environmental attributes influencing changes of population's parameters can be highly non-predictable and non-linear. This has caused a surge of interest in the use of different non-linear modelling techniques for modelling population dynamics (see e.g. Chaps. 8, 10, 12). Furthermore, these include neural networks (Lek and Guegan 1999; Recknagel et al. 1997; Schleiter et al. 1999), equation discovery (Džeroski et al. 1999; Todorovski et al. 1998) and decision trees.

Classification and regression trees can be used for modelling population dynamics as follows. The task of predictive modelling is to forecast the future state of the population or the change in the state of the population over a specified time period, given the current state of the population and the environment. For instance, Kompare and Džeroski (1995) used regression trees discovery to model the growth of the dominant species of algae (*Ulva rigida*) in the lagoon of Venice in relation to water temperature, dissolved nitrogen and phosphorus, and dissolved oxygen.

In the area of forestry, decision trees have been successfully used to model population dynamics of red deer and spruce bark beetles population dynamics in forest ecosystems. The study about the population dynamics of red deer focused on the effects of different meteorological conditions, habitat properties and hunting regimes on the population dynamic of red deer (Stankovski et al. 1998; Debeljak et al. 1999). A highlight of the results of the red deer studies was the discovery of the strong influence of meteorological parameters on the browsing intensity for new growth of woody plants (beech and maple) and consequently the body weight of 1-year-olds, 2-year-olds, and hinds (important parameters of the studied red deer population). These results challenge previous simplistic approaches, assuming simpler and more direct relationships between the density of the red deer population and its parameters and the browsing rate of new forest growth.

The study of spruce bark beetles (Ogris and Jurc 2010) focused on environmental conditions that stimulate population growth of the spruce bark beetles *Ips typographus* and *Pityogenes chalcographus*. The results show a strong correlation between the appearance of *I. typographus* at Northeast (NE) expositions, while *P. chalcographus* prefers West (W) and North (N) sites. The discovered habitat preferences of bark beetles confirm the adaptation of spruce to drought conditions at southern expositions, where its root system penetrates deeper in the soil. At N, NE and W sites, the individual trees are more sensitive to drought and mechanical destabilization due to the shallow root system and thus they are more prone to attack by bark beetles.

Decision trees are also used in agro-ecology. The population dynamics of soil organisms is affected by the changes of different biological and physicochemical environmental attributes and agricultural practices. A study about the effects of growing Bt-maize cultivation on abundances of earthworms populations (Oligochaeta) (Debeljak et al. 2007) used farming practices, soil parameters, the biological structure of soil communities, and the type and age of the crop at the time of sampling as attributes to predict the total abundance of three functional groups of earthworms (epigeic – live and feed on plant litter (Fig. 14.1); endogeic – geophagous and live in the soil; anecic – live in soil but feed on plant litter on the surface). The highly accurate ($r^2 = 0.83$) regression tree model for anecic worms

(Fig. 14.1) shows that this functional group of earthworms prefers less clay and more silt soil with medium pH. It has been shown that the seasonal effect (autumn/spring sampling) has stronger influence on anecic biomass compared to the inter-annual effect (autumn 2002/autumn 2003). Indeed, it is very well known that in temperate arable ecosystems, anecic earthworms reach their minimum in winter, due to low temperature, and their maximum in autumn, after spring and summer reproduction and development. Finally, agricultural practices, such as tillage or maize variety have no effects on anecic earthworm biomass.

Soil dwelling populations in arable ecosystems are exposed to various anthropogenic pressures. To identify attributes influencing the abundance of soil mites and springtails and the biodiversity of soil micro-arthropods, a multi objective regression tree has been induced from data collected under different crop management practices (Demšar et al. 2006). Figure 14.2 shows an example of such a decision tree predicting the target attributes abundances of Acari ($r^2 = 0.653$) and Collembola ($r^2 = 0.675$) and the diversity of Collembola ($r^2 = 0.562$). The model indicates that the most important parameters are the soil type, the time (number of months) since the establishment of the current situation, and the different forms of tillage. Hence, the model can adequately reproduce the known empirical knowledge on this phenomenon.

14.5 Habitat Modelling Using Decision Trees

Habitat modelling typically relates properties of the environment with the presence, abundance or diversity of organisms (for other detailed examples, see Chap. 13 on spatial distribution models). For example, one might study the influence of soil characteristics, such as soil temperature, water content, and proportion of mineral soil on the abundance and species richness of Collembola (springtails; the most abundant insects in soil (Kampichler et al. 2000)). Habitat modelling can also be linked with spatial information derived from geographic information systems (GIS) on the studied area (Debeljak et al. 2001; Jerina et al. 2003) (see also Chap. 22).

A number of habitat-suitability modelling applications of other machine learning methods (e.g. neural networks, genetic algorithms) were surveyed by Fielding (1999). Lek-Ang et al. (1999) used neural networks to build a number of predictive models for Collembola diversity. Bell (1999) used decision trees to describe the winter habitat of pronghorn antelope. Jeffers (1999) used a genetic algorithm to discover rules that describe habitat preferences for aquatic species in British rivers. Rule inductions were also used to relate the presence or absence of a number of species in Slovenian rivers to physical and chemical properties of river water, such as temperature, dissolved oxygen, pollutant concentrations, chemical oxygen demand, etc. (Džeroski and Grbović 1995).

Decision trees are applied widely in habitat modelling. Džeroski and Drumm (2003) have used classification tree models to predict the suitability for the sea cucumber species *Holothuria leucospilota* on Rarotonga, Cook Islands. Kobler and Adamič (1999) have used decision tree models to identify locations for construction of wildlife bridges across highways in Slovenia. Decision trees were used to model habitat suitability for red deer in Slovenian forests using GIS data, such as elevation, slope, and forest composition (Debeljak et al. 2001). Models of potential and actual habitat for brown bears have been induced from GIS data and data on brown bear sightings using decision trees (Jerina et al. 2003). Ogris and Jurc (2007) applied decision trees to identify potential habitats for different tree species under varying climate change scenarios. Decisions trees are used in habitat modelling of soil organisms that are under the influence of different soil characteristics and crop practices (Kampichler et al. 2000; Debeljak et al. 2007).

Habitat modelling is also becoming relevant in agriculture due to problems with crops, such as oilseed rape, sunflower, wheat or sorghum, which can escape from cultivation and colonise field margins as feral populations. To control the processes leading to the formation of new feral populations, habitat models enable us to identify suitable growing conditions for new potential feral populations. Such research has been conducted on a 41 km^2 production area of winter oilseed rape in Loir-et-Cher region, France (Pivard et al. 2008). Based on attributes describing locations of all cultivated oilseed rape fields and feral populations and their demographic properties, a habitat model for feral oil seed rape was developed (Fig. 14.3). The model predicts the probability of the presence of a feral population in the studied area.

Side effects of oilseed rape (OSR) cultivation include volunteer plants that emerge on the field after cultivation of OSR and may cause crop impurity or weed control problems. To understand the suitable conditions for formation of volunteer populations of OSR, a habitat model to predict presence and abundance of volunteer oilseed rape (*Brassica napus* L.) was induced from a dataset about the seedbank at 257 arable fields used for baseline sampling in the British Farm Scale Evaluations of genetically modified herbicide tolerant (GMHT) crops (Debeljak et al. 2008). Volunteer OSR was most likely present if a previous OSR crop had been grown in the same field (Fig. 14.4). However, machine learning also indicated previously unknown correlations between the abundance of volunteer oilseed rape, total seedbank and several other factors like the percent of nitrogen and carbon in the soil. Once OSR has been cultivated at a site volunteers are not excluded specifically from any part of the country or from sites having particular abiotic characters such as high pH or low % of nitrogen. Volunteers had, moreover, become present at 24% of sites where there had been no OSR crop in the last 8 years, presumably as a result of a previous crop (beyond the 8 years recorded) or imported to the site with farm machinery. Their abundance, moreover, varied systematically with factors that are generally associated with the intensity of farming, notably total seedbank abundance, species number and plant life history groups (Fig. 14.5), and most consistently with percentage of nitrogen and carbon in the soil. All these factors were linked to an extent with geographical region, being smallest in the arable south-central and south-east and largest in the north and south-west.

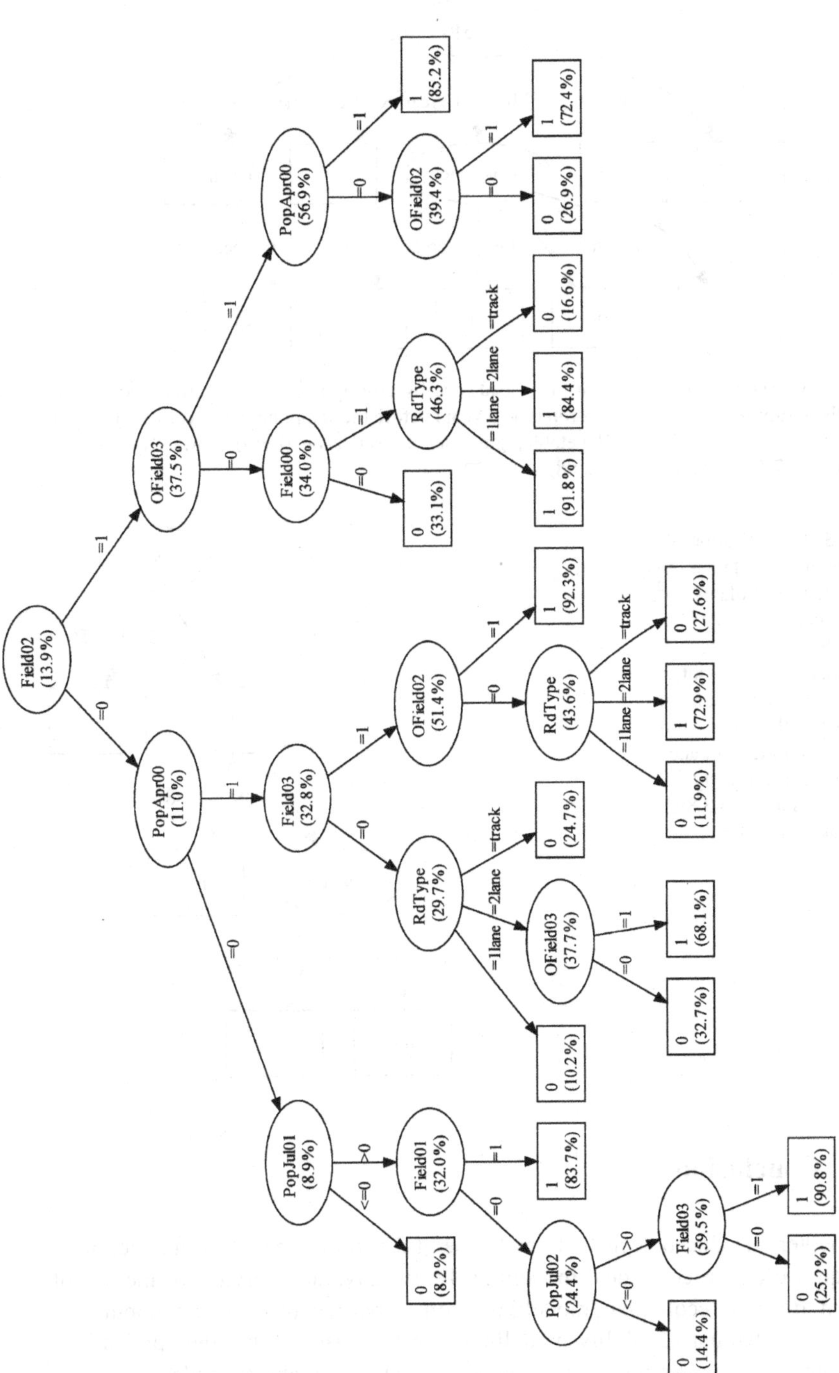

Fig. 14.3 A classification tree modelling the presence of oilseed rape feral populations. The percentages give the predicted probability of presence of a feral population in 2003 according to the situation. For instance, this probability on the whole area is 14% (at the root); in the absence of an adjacent field in 2002 (Field02 = "0"), which is the best attribute to explain the presence of a feral population in 2003, it is only 11% (*left* branch); while in the presence of such a field (Field02 = "1"), it increases to 38% (*right* branch) (Debeljak et al. 2008)

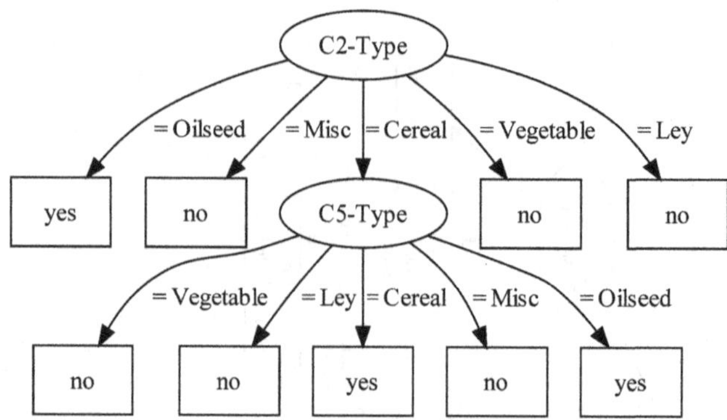

Fig. 14.4 Classification of presence of oilseed rape by crop type (C2-Type: crop type 2 years before the sampling date; C5-Type: crop type 5 years before the sampling date; types are Oilseed, Miscellaneous (Misc.), Cereal, Vegetable, grass ley or set aside (Ley)) (correctly classified instances: 60.7%) (Debeljak et al. 2008)

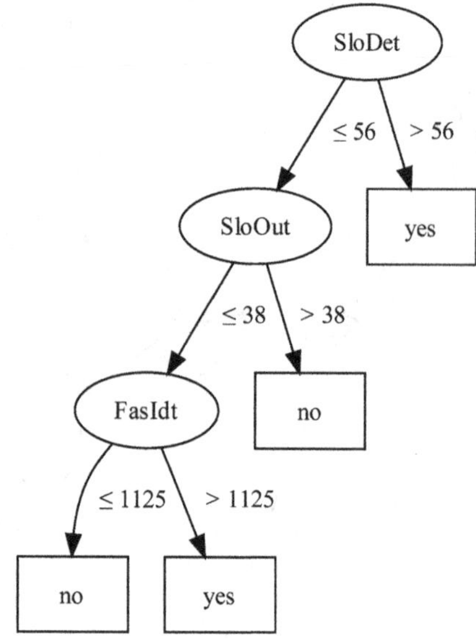

Fig. 14.5 Classification of presence of oilseed rape by the abundance of plants (m^2) of particular functional groups (SloDet – slow, determinate development; SloOut – slow development living below the crop canopy; FasIdt – fast indeterminate development) (correctly classified instances: 63.8%) (Debeljak et al. 2008)

14.6 Conclusion

This chapter introduced decision trees as one of the most popular machine learning techniques used for ecological modelling. It also gave an overview of the use of decision trees in ecological modelling with a particular focus on population dynamic and habitat suitability modelling. We have shown that the applications of machine learning to population dynamic and habitat suitability modelling can be grouped along two dimensions. One dimension is the type of environment where the studied group of organisms lives, e.g. aquatic (river or sea) or terrestrial (forest or agricultural fields). Another dimension is the type of applied machine learning technique.

The major advantages of decision tree methods include the ability to capture interactions between the variables used for modelling, the understandability of the produced models (trees) and their efficiency. Decision tree learning methods can establish models fast from large quantities of data, involving either a large number of records (example) or a large number of columns (variables) or both. Also, decision tree models make predictions very fast and can be used to classify large numbers of examples: This is important in the context of pixel-based classification in geographical information systems, where very large numbers of spatial units/ points need to be classified.

Decision tree learning is also capable of identifying the relevant variables from a large set of independent variables. The resulting trees typically use only a few of the variables available. This, however, can easily be a disadvantage in some situations: If all the variables available contribute to the classification, it is very likely that the tree will not use them all and will hence have lower performance.

Other situations where decision trees may encounter problems are domains where the variables are completely independent. In addition, small numbers of examples/records are quite problematic for decision trees. In both situations, using methods like linear or logistic regression would be more appropriate.

Decision trees are derived from data only. No domain knowledge or limited amounts thereof are used in the learning process. As such, they represent the data driven or empirical approach to ecological model construction, which is more appropriate when we have plenty of high-quality (reliable and relevant) measured data and little knowledge about the studied system. When only few or low-quality (unreliable or irrelevant) data are available, and/or there is a considerable knowledge about the studied system, the classical knowledge-based paradigm of manual model construction could be more appropriate.

Part III
Application Fields, Case Studies and Examples

Model Applications to Understand Complex Ecological Dynamics

In the previous section of this book, the most common and important modelling approaches for ecological purposes were introduced. The following section will focus on application examples from ecological systems in different regions of the world, which demonstrate how models contribute to a large variety of issues in basic and applied research in ecology. With the given examples we intend to give a motivating insight into the wide field of model applications, from small scale population processes to landscape representations involving a combination of different approaches and techniques. Thus, the examples contain both detailed research results and for specific systems, an overview of model applications. Although the examples refer to a wide set of different ecological systems, they all require an approach to deal with complex dynamic behaviour.

Hypothesis Testing on the Landscape Level

Spatial patterns in landscapes can be assessed from two different perspectives: first, as a result of interactions of the underlying biotic and abiotic processes, and second, as influenced by the constraints that direct and modify the underlying interactions. This mutual dependence is further complicated by the permanent change in land use conditions, which often leads to habitat loss and fragmentation. Here, models can be of great help to answer hypotheses on causes, effects and potential future developments. For this purpose, a neutral perspective is often adopted, meaning that the model generates characteristic spatial patterns without considering specifically targeted mechanisms referring to underlying ecological processes. In this sense, these kinds of models are used as a null hypothesis to test deviations in the empirical data from theoretical expectations.

In **Chap. 15, Gardner** outlines the historical development of this approach for Landscape Ecology. He explains the working principles and presents a case study from the North-American Piedmont Forests in Maryland.

The role of dispersal and how it can be formalised by using different dispersal kernels is the topic of **Chap. 16** by **Garnier and Lecomte**, which tests hypotheses on the conditions that support the survival of organisms in structured landscapes. They explain models that have been developed to understand the spread of oilseed rape along road verges and the invasion of grasslands by pines. The models are applicable largely for European and Northern-American habitats.

In **Chap. 17, Kubicek and Borell** investigate the importance of path dependency and ecosystem phase shifts for tropical coral reefs, which are among the most diverse ecosystems worldwide and are currently severely threatened by anthropogenic impacts. Here, standard model types are reviewed on how effectively they can be used to understand the main driving forces of tropical coral reefs, interacting sensitively between resilience and phase shifts.

Environmental Management and the Integrative Power of Modelling

Modelling plays an increasingly important role for management and ecological conservation. When phenomena are investigated that occur over a wide range of hierarchical ecosystem levels or over several spatial and temporal scales, modelling is often the only solution to analyse and understand the ongoing dynamics. Furthermore, applying empirical experiments on the landscape level is not always a possible option. In this context, ecological modelling offers many possibilities for management and planning. Moreover, ongoing research models collect data that derive from many different sources, sometimes over decades, and display different statistical properties.

In **Chap. 18, Jopp, DeAngelis and Trexler** investigate the possible consequences of changing and reducing the water table in the fluctuating marshlands of the South-Florida Everglades and how changes may affect the trophic structure.

A special contribution is **Chap. 19** by **Nielsen and Jørgenson** which presents a case study on the modelling activities on the Danish post-glacial Lake Glumsø. Here, the advancements of ecological modelling can be seen over more than three decades: moreover, this research project had profound impacts on the advancement of ecological modelling as a scientific discipline.

In **Chap. 20 Gallego** uses model-coupling to understand the complex dynamics of marine zoo- and ichthyoplankton and what importance these dynamics have for the dramatic stock collapses of cod in the Northern Atlantic.

In **Chap. 21, Jopp and DeAngelis** demonstrate the integrative power of modelling for the comprehensive Everglades restoration plan. The plan aims to support the recovery of many endangered species and biodiversity of the overall landscape of the Everglades and its neighbouring ecosystems in southern Florida. Here, ecological modelling plays a central role for science-based decision making, and hence, ecosystem restoration.

In **Chap. 22, Kraft** focuses on integrated modelling: He describes how models from different disciplines are coupled and their complex results are aggregated into a modelling framework, for which Geographical Information Systems (GIS) are commonly used. The author brings several application examples from catchment models and models that are dealing with coastal and marine landscapes.

The eight contributions in this section illustrate how to use models to understand complex problems and issues that could not easily be grasped otherwise. Although we present the chapters in a sequence following complexity considerations, you, the reader, should feel free to use any starting point that suits your preferences as the chapters represent independent cases.

Chapter 15
Neutral Models and the Analysis of Landscape Structure

Robert H. Gardner

Abstract Neutral landscape models were originally developed to test the hypothesis that human-induced fragmentation produces patterns distinctly different from those associated with random processes. Other uses for neutral models were immediately apparent, including the development and testing of metrics to characterize landscape change. Although metric development proved to be significant, the focus on metrics obscured the need for iterative hypothesis testing fundamental to the advancement of science. In this chapter, we will present an example of an alternative neutral model and hypothesis designed to relate the process of landscape change to the resulting patterns observed. The methods and programme, Qrule, are described and options for statistical testing outlined. The results show that human fragmentation of landscapes results in a non-random association of land-cover types. Options for additional landscape studies are discussed and access to Qrule described in hope that these methods may be employed to advance our understanding of the process that affect the structure and function of our landscapes.

15.1 Introduction

The accelerating rate and global extent of habitat loss and fragmentation are having a direct and significant impact on species diversity (Lindborg and Eriksson 2004; Vellend et al. 2006), the rate and extent of exotic species invasions (Hooper et al. 2005; Vitousek et al. 1997), water quality and availability (Ferrari et al. 2009; Meyer and Turner 1992) and the productivity of ecosystems that humans depend on for food and materials (Osher et al. 2003; Williams et al. 2004). Even local and regional weather patterns are recognized as being significantly affected by landscape change (Pyke 2004; Stohlgren et al. 1998). The importance and difficulty of understanding the immediate and long-term consequences of landscape change has

R.H. Gardner
Appalachian Laboratory, 301 Braddock Road, Frostburg, MD 21532, USA
e-mail: gardner@al.umces.edu

F. Jopp et al. (eds.), *Modelling Complex Ecological Dynamics*,
DOI 10.1007/978-3-642-05029-9_15,

lead the National Research Council to declare this problem an urgent scientific challenge (NRC 2003).

There are many reasons why the study of ecosystem change at landscape studies is a daunting task: Ecosystem dynamics result from complex interactions of numerous physical and biological attributes that vary in space and time. Each landscape has a unique, and often unknown history whose effects may be subtle and long lasting and experimental manipulations of landscapes are nearly impossible (the "N = 1" problem). Consequently, models of landscape dynamics have become an integral part of landscape ecology. However, the use of a landscape model presents its own special issues. Each new problem requires the selection of an appropriate (and often new) model; the parameters required by the model are often difficult to know or are estimated with high uncertainty; and the errors associated with predictions are impossible to determine. Hagen-Zankder and Lajoie (2008) and Gardner and Urban (2007) have noted an additional problem: the geographic constraints of landscapes (i.e. the boundary conditions of map extent, the presence of rivers and lakes, etc.) may restrict the dynamic range of model output, resulting in alternative model formulations producing similar sets of predictions. These constraints result in confusion with regards to the causes and effects of altered landscape dynamics. Under all of these conditions a neutral modeling approach can be of great value.

The ideal neutral landscape model is a simple construct, which generates landscape pattern without including specific physical and biological processes of interest. The magnitude of difference of landscape data, or more complex process-based models from the neutral model results is a measure of the important role that these processes may play in the development of landscape pattern (Gardner et al. 1987; Pearson and Gardner 1997). The simplest neutral model is a random map (Gardner et al. 1987) that produces patterns based on a single parameter, p_i, the fraction of the map occupied by each habitat type, *i*. Although the appearance of random maps obviously differs from actual landscapes, the lessons learned have been diverse and significant (see Gardner and Urban 2007 for a review).

The background for the development of neutral landscape models has a number of key elements. The first occurred when colleagues at Oak Ridge National Laboratory began analyzing USGS digital land use and land cover (LUDA) data (Krummel et al. 1987). The USGS LUDA database (Fegeas et al. 1983) originated from NASA U2/RB-57 high-altitude aerial photo coverage in 1973, which were subsequently hand-delineated into 1 of 37 land cover categories. For the first time, these data provided a spatial description of habitat distribution for the entire U.S. Krummel et al. (1987) selected the Natchez Quadrangle for analysis, 1:250,000 quadrangle composed of 24 separate sections. To analyze this quadrangle a special computer programme was written (remember, this was pre Arc-Info days!) to reformat the arc and node topology and remove section boundaries. It was evident that the shape of habitat boundaries changed with scale. George Sugihara suggested that fractal geometry could be used to characterize patch shape. The large size of the data set allowed a moving-window regression analysis to detect the scale at which the fractal index changed.

It was obvious that fractal geometry was both interesting and useful for landscape studies and that we had only scratched the surface of what might be done. Simulated by this and by Mandelbrot's work (e.g. Mandelbrot 1967, 1983) I attended a Gordon Conference on fractals in the summer of 1986. A broad spectrum of issues and applications were discussed, but the most interesting was the presentation by Dietrich Stauffer on percolation theory. The images he presented (which were random 2D structures) looked remarkably like landscapes. Discussions with Stauffer at the conference and subsequent study of his first book on percolation theory (Stauffer 1985) made it clear that this theoretical framework would provide a rich set of tools for landscape ecology.

Yet how does one use random maps and percolation theory in landscape ecology? Caswell's paper on neutral models and community theory (Caswell 1976) was well known and widely discussed, but had not been of particular interest to problems in landscape ecology. However, it was clear that combining Caswell's approach with Stauffer's percolation theory would provide a neutral model for landscape ecology. In the fall of 1986 we began writing the programme RULE (Gardner 1999) and running a series of simulations to describe this approach. In the spring of 1987, with the completed manuscript in hand, we presented the results during the 2nd Annual Meeting of the U.S. chapter of the International Association of Landscape Ecology at the University of Virginia. Frank Golley, who was instrumental in the development of landscape ecology in the U.S. had just established the new journal appropriately titled "Landscape Ecology". Frank requested at the Charlotte meeting that we submit this manuscript for publication in this new journal and by the end of that summer our paper (Gardner et al. 1987) had appeared in print.

The diversity of issues for which neutral models may be useful has been significant, yet the full range of applications has yet to be fully explored (Gardner and Urban 2007). Although new methods have been developed (e.g. Hagen-Zanker and Lajoie 2008; Li et al. 2009; Wang and Malanson 2007), existing methods for analysis and comparison of maps available within the original software (now called "Qrule", http://www.al.umces.edu/Qrule.htm) have rarely been fully employed. This chapter will explore these two areas by first developing a new neutral for the analysis of landscape change within the Piedmont of Maryland, a large area that is being rapidly altered by urban development. Analysis methods will then be used to test the hypothesis that the contagion processes associated with human development have been responsible for non-random changes in the patterns of forest patches. Specifically, there may be an insignificant reduction in the total amount of forest cover, but significant changes in the adjacency of land cover types and the probability that distribution of the size of forest patches has occurred.

15.2 A Programming Philosophy

Documentation and practical examples illustrating the use of Qrule are available in a number of publications (Gardner 1999; Gardner and Walters 2001; Pearson and Gardner 1997). Recently added revisions have also been described (Gardner and

Urban 2007) and the latest version of the programme is available on the web (http://www.al.umces.edu/Qrule.htm). Consequently, we will not explain the details required to perform Monte Carlo simulations with Qrule, assuming that the reader is either aware of the above resources or can refer to them directly.

RULE (now called Qrule) was originally designed for a very specific purpose: the generation and analysis of random maps and comparison of these neutral models with actual landscapes. Because other software for landscape analysis was available when RULE was written; first a programme by Monica Turner (SPAN, described in Turner and Ruscher 1988) and later FRAGSTATS (McGarigal et al. 2002), a parsimonious subset of metrics were implemented in Qrule. The choice of metrics was based on the need for a minimum summary of pattern characteristics for each land cover type (the metrics *p*, the total fraction of the map occupied; the total number of patches; and the total amount of edge were selected for this purpose) and the introduction of new metrics and principles of analysis to landscape ecology, based on percolation theory. Stauffer's book (1985) emphasized two significant problems in the analysis of random structures generated on gridded landscapes: the first problem was that the shape of the grid, whether triangular, rectangular or hexagonal, dominated the shape of smaller clusters. Thus, for many metrics there was a need to use large maps and to restrict the analysis to the largest clusters. The fractal index was sensitive to grid effects approaching the value of 2.0 on a rectangular lattice when cluster sizes where small (the rectangular lattice is typically used in landscape studies, though hexagonal lattices have been employed, e.g. Roberts 1987). The second problem was that random maps and actual landscapes often have a large number of small, isolated clusters. These small patches, which usually compose only a minor fraction of the total area occupied by that land cover type, will bias the estimates of metrics that are based on arithmetic mean values. These two problems are avoided in percolation theory and subsequently in Qrule, which adopted this approach by estimating the fractal dimension for only the largest cluster on the map and by using geometric averages for summary statistics. The discussion of the effect of map extent on the reliability of landscape metrics was extensively evaluated in Gardner et al. (1987). With the exception of the addition of lacunarity analysis to Qrule circa 1991 (Plotnick et al. 1993, 1996), changes to the programme through time have focused on fixing bugs and increasing the convenience and efficiency of programme execution.

The philosophy of a limited scope for Qrule resulted in an emphasis on the import and export of data. It is not uncommon for landscape analysis projects that use Qrule to export spatial data from ArcInfo, to rescale these maps using PDW (Gardner et al. 2008), to analyze the resulting patterns with Qrule, and display summary statistics using R (R project). To enable this flexibility, an extensive suite of data files are used in Qrule (Table 15.1): there are four different options for output of generated maps, with a 5th map type always created to visualize and analyze results in ArcInfo; and there are four data sets used to statistically summarize results (Table 15.1).

The comparison of neutral models with the observed pattern of actual landscape has always been based on traditional principles of statistical inference. If the

Table 15.1 Input and output files used or created by Qrule

Routine	Unit	Name	Purpose
main.f90	10	rulerun.log	Output file recording programme input and statistical summary of results
main.f90	11	patch_cfd.dat	Output data file of cumulative frequency of patch sizes (compressed mode)
main.f90	12	variable	User-defined name of output map of individual habitat patches
main.f90	13	variable	User-defined name of output map of rank-ordered sizes of habitat patches
main.f90	14	variable	User defined output map labeled by habitat type
genmap.f90	15	variable	User named input file of landscape map for analysis by Qrule
main.f90	16	assmat.dat	Output file of adjacency matrices
main.f90	17	stats.csv	Comma delimited output file of summary statistics (landscape metrics)
genmap.f90	19	arcgrid.map	Output file of Qrule generated map for input into ArcInfo
main.f90	20	lacun.dat	Output file of summary results from lacunarity analysis

landscape pattern lies beyond the 95% confidence region generated by a sufficiently large set of Monte Carlo iterations, then one may be confident that the observed patterns are statistically different from the random patterns at $\alpha = 0.05$ (Gardner and O'Neill 1990; Pearson and Gardner 1997). This statistical comparison is, of course, subject to Type II error (Zar 1996) when multiple metrics are employed to describe landscape patterns. One may avoid this important pitfall by first forming a specific question per the example of Krummel et al. (1987), selecting a single appropriate metric (or limited subset of metrics) and making the appropriate statistical test(s). Multivariate approaches reduce the dimensionality of the analysis (Fauth et al. 2000), providing a more succinct summary and avoiding the problem of correlated parameters (Riitters et al. 1995; Wang and Malanson 2007), but the utility of multivariate statistics still depends on the formation of a specific testable *a priori* question.

There is extensive literature in landscape ecology that has focused on the development and interpretation of landscape metrics (e.g. Gustafson 1998; Hargis et al. 1998; Li et al. 2005; Neel et al. 2004; O'Neill et al. 1987; Wickham and Riitters 1995). However, the usefulness of robust statistical testing for comparing neutral models with actual landscapes remains under-appreciated. Consider the direct effects of habitat fragmentation on the frequency distribution of patch sizes for a given cover type. The effects of a small amount of habitat loss will have dramatically different effects depending on p, the amount of habitat that exists on the landscape: when p is high, the effects of habitat loss on the frequency distribution of cluster sizes is small, however, when p is ~0.6 and near the critical threshold defined by percolation theory, small changes will have dramatic effects on the frequency distribution of cluster sizes (Gardner et al. 1987). Landscape metrics are generally poor indicators of these effects because they are usually a single-numbered, averaged value (see Li et al. 2009 for an exception). The alternative is an examination of the cumulative frequency distributions (cfd) of patch sizes where statistical differences can be examined by the familiar distributional tests (e.g. the KS test, Zar 1996). Even more powerful methods exist if the expected change in the cfd can be defined *a priori*, allowing parametric comparison of the observed with the appropriate theoretical distributions (Johnson and Kotz 1969, 1970). Because the cfd of patch sizes varies with map dimension, resolution and methods used to classify land-cover types, it is usually impossible to pre-define an expected shift in the cfd and compare this shift with an appropriate probability distribution. In these cases one must rely on neutral models to generate the "expected" cfd to compare to an observed landscape using the appropriate statistical test(s). It is these principles of analysis that are illustrated in the following example.

15.3 Methods

The neutral models available in previous versions of Qrule have all assumed that landscapes could be represented as a rectangular lattice (grid). Large, irregular landscapes were simply truncated to form a rectangular map. Although remote imagery is composed of square pixels, the boundaries of most landscapes are rarely rectangular. A new neutral model was created for the analysis presented here; one primarily designed to eliminate this truncation effect. This method, labeled "Random-with-Constraints" (RwC) first examines the actual landscape, extracts the boundaries and other embedded constraints (i.e. user-defined areas such as rivers and lakes where habitat cannot be located), constructs a mask from this information, and then randomly generates habitat within the area permitted by the mask. The number of land-cover types generated is also user controlled but must be equal to or less than the number of cover types in the original mask. During programme execution and before the original land cover map is evaluated, the user can aggregate cover types. In the current analysis, non-habitat (class "0") was combined with open water (class "1") to create the mask; classes 5, 6, and 7, which had no representation within the original map were re-classed to 0; and the remaining cover types of urban, barren lands, forest, agriculture and wetlands were sequentially renumbered to cover classes 1–5, respectively. The fraction of sites in each of these five classes, p_i, within the actual land-cover map was recorded and subsequently used to generate random cover types in the ten Monte Carlo iterations of the RwC model. The result was ten random maps with the distributional area equal to the original map (each map having five land cover types in proportion to the observed frequencies of the original map).

A second revision to Qrule was the reconfiguration of the data files used for statistical testing. The four files (Table 15.1) produced for each execution of Qrule are: *rulerun.log* which records all programme input and provides a statistical summary of results, including the land cover association matrix and indices characterizing patch attributes; *assmat.dat*, the association matrices of land cover types (one matrix for each map iteration); *stats.csv,* the output record for each iteration and habitat type in comma delimited format, including the ten indices summarized in rulerun.log; and *patch_cfd.dat,* the cumulative frequency distribution (cfd) of patch sizes for each iteration and habitat type in summary form (i.e. the cumulative frequency in each size class).

Manipulation of output files generated by Qrule by other programmes is necessary for visualization and statistical testing. To assist in this sometimes tedious process, a series of programmes have been written in R (R Team 2008) to display results and test for significant differences. The statistical results and visualization reported here were produced by these programmes.

Maryland Piedmont Maps

Data from the National Land Cover Database for the Piedmont of Maryland were downloaded and were selected for this analysis (http://www.mrlc.gov/download_data.php). The multiple categories for 1992 and 2001 were aggregated into seven classification level comparisons: open water, urban, barren land, forest, grassland, agriculture and wetlands. The 30-m resolution of the maps resulted in grids with dimensions of 4,365 rows and 5,550 columns. These two maps were first analyzed with Qrule using the next-nearest-neighbor rule for patch identification. Qrule results provide a select set of indices characterizing patterns (see Gardner 1999 for a listing and description of these indices). The indices reported here include three calculations characterizing the largest cluster: LC.sz, the size of the largest cluster in hectares; LC.ed, the amount of edge for the largest cluster in meters; and LC.frc, the fractal index of the largest cluster. Also considered are five general indices for each land cover type: S.frq, the total number of pixels for each land-cover type (pixel units); T.cltr, the total number of patches; C.len, the average correlation length of patches in meters; Sav, the area-weighted average patch size in hectares; T.eg, the total amount of edge in meters.

Many comparisons could be constructed from this information-rich data set. The analysis reported here focuses on patterns of loss of forested areas because this landscape-attribute has been threatened by population growth and urban development within this region (Lookingbill et al. 2009) and has resulted in special legislation to protect forested areas (http://www.dnr.state.md.us/forests).

15.4 Analysis and Hypothesis Testing

Forested areas occupied 28.2% of the Maryland Piedmont in 1992, declining to 27.3% in 2001 (Table 15.2), a change equivalent to the loss of nearly 6,000 ha. Urban areas showed the opposite trend, increasing from 13.5% in 1992 to 14.3% in 2001, a change that produced an additional 5,673 ha of urban development. The size distribution of forest patches also shifted although the median value of 0.9 ha

Table 15.2 Statistical description of the statistical distribution of forest and urban patch sizes (ha) within the Maryland Piedmont in 1992 and 2001

Year	Land cover type	p[1]	N	Minimum	Median	Maximum	Sav
1992	Forest	0.2820	16,145	0.09	0.9	14,072	2,742.5
1992	Urban	0.1346	33,926	0.09	0.18	26,706.7	13,753.1
2001	Forest	0.2734	14,807	0.09	0.9	13,247	2,479.4
2001	Urban	0.1429	32,816	0.09	0.18	29,308.5	15,410.9

Table 15.3 Land cover change within the Maryland Piedmont from 1992 and 2001

Land cover type	Change (1,000 ha)	Percent change
Open water	24.5	0.40
Urban	141.2	2.80
Barren lands	50.1	16.90
Forest	–516.6	–1.50
Grassland	155.1	12.50
Agriculture	244.4	1.00
Wetlands	129.6	1.20

remained constant over this 9-year period. However, the geometric average patch size (Sav) declined by 263.1 ha and the size of the largest forested area decreased by 825 ha. The size distribution of urban areas showed an 11% increase in Sav from 13,753 ha to 15,411 ha and a simultaneous increase in the largest urban area from 26,707 to 29,309 ha. The increase in urban area occurred with a declining number of urban patches from 33,926 in 1992 to 32,816 in 2001 (Table 15.2). The pattern of urban growth occurred as a result of an increase in the average patch size and the subsequent joining of adjacent urban areas. Because the median value of the size distribution for urban areas was constant (Table 15.2), the process of absorption appears to have affected all size classes of urban development.

The assessment of map accuracy for MRLC data is an important consideration, but also a complex topic (see Homer et al. 2004, 2007; Vogelmann et al. 1998; Wickham et al. 2004; Yang et al. 2001) that is beyond the scope of this analysis. If we assume that open water changed little between 1992 and 2001, then the 0.4% increase in open water over this time period (Table 15.3) provides a simple and convenient index of relative accuracy. Table 15.3 shows that the only category that lost area was forest, with a net decline of 516.6 thousand ha; All other land cover categories increased in area, with the greatest gain for the agriculture, grassland and urban categories.

Figure 15.1 illustrates the cumulative frequency distribution (cfd) of the size of forest patches in 1992 with the cfd generated by the RwC model of Qrule. The cfd for 2001 was similar in form and is not illustrated here. The random maps generated by the RwC model had a greater number of small, isolated clusters with 50% of the patches smaller than 0.18 ha, while the median patch size for the 1992 land cover map was nearly a hectare in size (0.9 ha, Table 15.2). The largest patch for the random maps was 18.3 ha while the largest for the 1992 land cover data was 14,072 ha. It is clear that the random and empirical distributions (Fig. 15.1) are different, and the KS test confirms this result ($D = 0.021$, $\alpha = 0.02$). The RwC

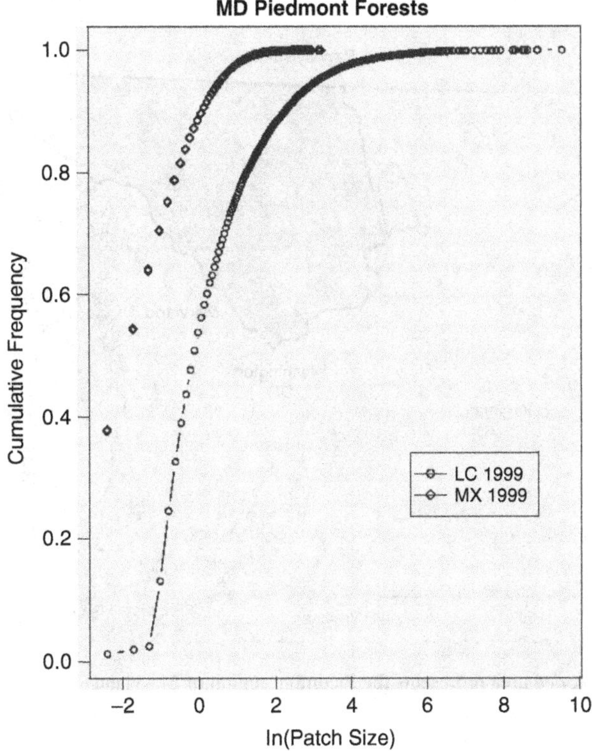

Fig. 15.1 The cumulative frequency distribution of forest patches within the Maryland Piedmont (*open circles*) in 1999 contrasted with the cumulative frequency distribution of random patches (*open diamonds*) generated by the RwC method using Qrule (see text for details)

model generated random patterns within the restricted area defined by the 1992 land cover map (i.e. its selected boundaries shown in Fig. 15.2 plus lakes and rivers). Under some circumstances the map constraints may be sufficient to cause the RwC model to produce patterns that differ substantially from a random map without constraints (see discussion of the association matrix below). Because the restrictions of the Maryland Piedmont map are not severe (Fig. 15.2), the general patterns are similar to those of a simple random map (cf. Gardner et al. 1987; Gardner and O'Neill 1990; Zar 1996) (Fig. 15.2).

The statistical differences among landscape metrics can be graphically illustrated with box and whisker diagrams. Figure 15.3 plots the distribution of Sav for forested and urban areas, contrasting the distributions of values from the RwC simulations for 1992 and 2001. The values for the actual landscapes are beyond the range of the RwC generated values (Table 15.2) and are not plotted here. Figure 15.3 shows that the range of values for Sav from the RwC simulations were very small (C.V. $<1\%$) with clearly different values for 1992 and 2001. Because patterns generated by the RwC model are due to random processes, this difference is entirely due to the shifts in p for

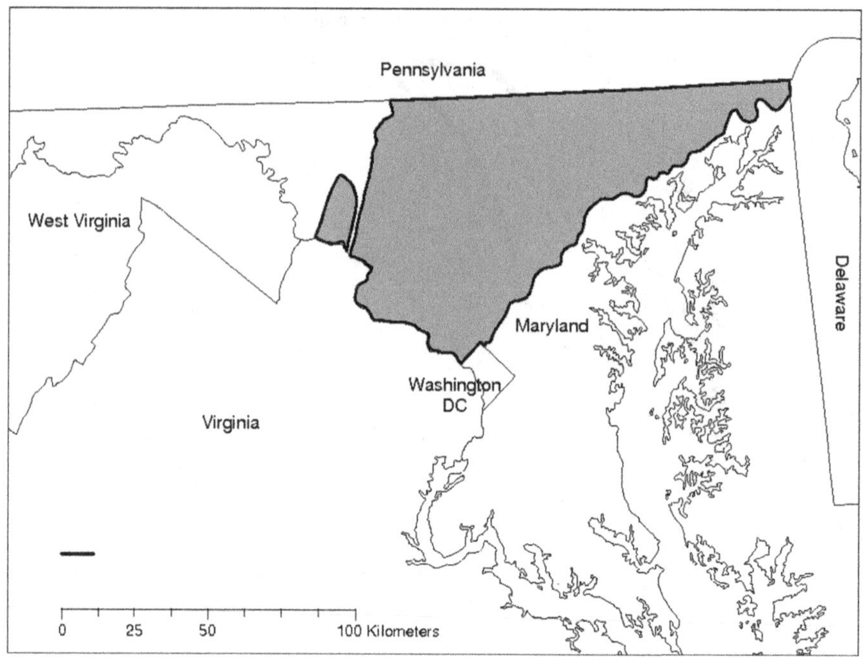

Fig. 15.2 The *shaded* area represents the Piedmont region of Maryland

forest and urban areas that occurred over this time period. The high precision in estimates of Sav (and other landscape metrics) is due, in part, to the large landscape with an equally large number of patches. If one assumes that the confidence intervals around the actual land cover data (Table 15.2) can be estimated with equal confidence, then the observed changes in the characteristic size of forest and urban patches (Sav) should be regarded as significant. Bootstrap estimates of confidence intervals could be performed to provide a quantitative verification of this assertion.

Other landscape metrics generated by Qrule (or by other similar software products) are interesting to examine and can provide insights into the change in landscape pattern with time. However, hypothesis testing using multiple metrics is dangerous because of the strong correlations that exist among metrics (Riitters et al. 1995; Wang and Malanson 2007). Figure 15.4 uses the "splom" procedure of R (see Appendix) to illustrate correlations among metrics for the RwC simulations for forested land cover in 1992. The ten Monte Carlo simulations produced 4,091,630 forest patches with the largest patch of 25.3 ha and a characteristic patch size (Sav) of 1.77 ha. Figure 15.4 illustrates three strong relationships that exist among metrics: the amount of edge on the largest cluster (LC.ed) was perfectly correlated ($r = 1.0$) with the size of the largest cluster (LC.sz); the total amount of edge (T.edg) was highly correlated ($r = 0.99$) with the total number of forested pixels (S.frq); and the correlation length of patches (C.len) was highly correlated ($r = 0.88$) with the characteristic patch size (Sav). These values were all significant at

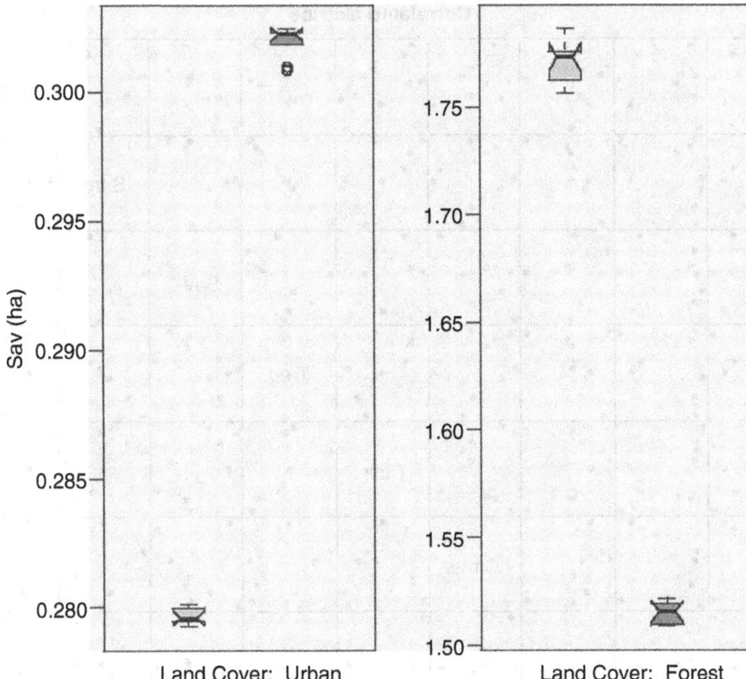

Fig. 15.3 A box and whisker plot of Sav contrasting the shifts in the size distribution of urban and forested patches generated by the RwC model using the 1992 and 2001 data for the Maryland Piedmont (see text for details)

$\alpha = 0.01$ and d.f. $= 8$. Other strong relationships exist ($r > 0.58$) among metrics for edge (LC.eg, T.eg), the total number of clusters (T.cltr), and the fractal index for the largest cluster (LC.frc). Because landscape metrics are often non-monotonic with p (Gardner and Urban 2003), the nature of the correlation structure can be expected to shift as the level of abundance of land cover type changes. Consequently, correlations among metrics should always be explored by methods like those illustrated in Fig. 15.4 before hypothesis testing begins.

Qrule produces an association matrix in the data "assmat.dat" that records the frequency of adjacency among land cover types for each simulation. The association at each time period, and the change in association over the interval 1992–2001 can be statistically tested with the usual χ^2 methods. The expected frequencies are usually estimated as a function of p, the observed frequency of each land cover type. In fact, this type of test is a neutral model before neutral landscape models were suggested! Because the simulations reported here were performed using the constraints of the landscape (the RwC method), the "expected" values for χ^2 tests the observed frequencies produced by the RwC simulations. For the large landscapes considered here, the use of the RwC association values had a minor impact on results (as noted above). We also know from the above results that the pattern of association for the actual landscapes will certainly differ from that of the RwC

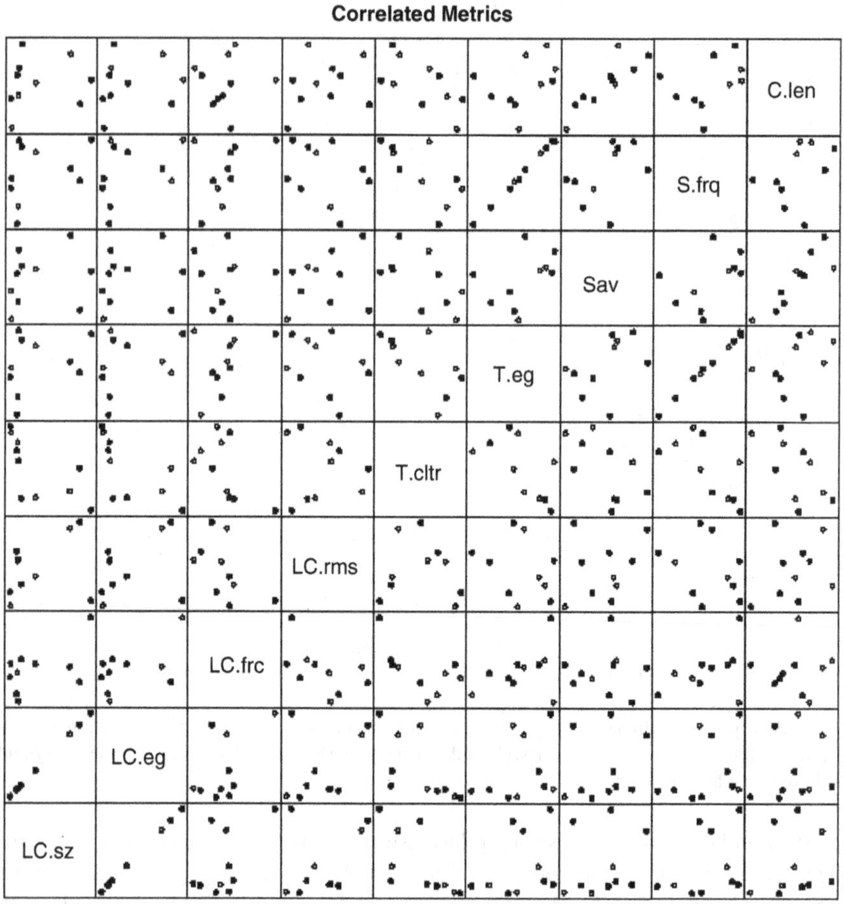

Fig. 15.4 Scatter plots of landscape metrics from ten iterations of Qrule using a neutral model with constraints (the RwC model) derived from the 1992 map of forested areas in the Maryland Piedmont. See text for a description of these metrics

simulations. Therefore, one interesting test is to examine a subset of the association matrix: the frequency of urban areas associated with forested areas from 1992 to 2001. The observed frequencies of urban:forest association for the actual landscapes were 2.14×10^{-3} and 1.82×10^{-3} for 1992 and 2001, respectively; the expected frequency for the RwC simulations were 1.2×10^{-2} and 1.23×10^{-2}, respectively. Converting to counts, we can test these differences with a 2×2 χ^2 contingency table yielding 8.9×10^9 and 7.1×10^9 for 1992 and 2001, respectively. The critical χ^2 value for $\alpha = 0.0001$with 1 d.f., as estimated by Conover (1971, p. 367), was 2.4×10^7. Thus, there is significantly less urban:forest association than expected by chance alone for both time periods. The decline in forest occurred with the simultaneous increase in urban areas, producing a loss of 235 km of forest edge over this 9-year period. Applying an χ^2 test to this change gave a value of

3.19×10^7, a difference that was also significant at $\alpha = 0.0001$. The conclusion is that the loss of forest edge was greater than expected from increases in urban areas and decreases in forest alone.

15.5 Discussion

Measuring the rate and extent of land cover change, and determining the causes of these changes continues to be a significant challenge for landscape ecology. Although extensive data sets from satellite sources are now widely available (Vogelmann et al. 1998, 2001), issues of accuracy and changing technologies continues to complicate the detection of change through time (Homer et al. 2004; Wickham et al. 2004). This chapter addresses a second issue associated with change detection; that of making scientifically rigorous comparisons from these pattern-rich data sets. The argument made here is that the appropriate approach should be parsimonious, achieved by a careful match of specific questions with a minimum set of relevant measures of landscape change. Analysis based on multiple metrics results in multiple comparisons and consequently an uncontrolled statistical error associated with inferences drawn from these comparisons. The familiar α level, or the probability of accepting a false hypothesis is usually defined for a single comparison as $\alpha = 0.05$, meaning that there will be a 1 in 20 chance of accepting a false hypothesis. With each additional comparison, C , the α-level or more appropriately, the "experiment-wise error rate", α^e (Kirk 1968) increases as $[1.0 - (1.0 - \alpha)^C]$. This means that if α for the comparison-wise error rate is fixed at 0.05 and one makes four comparisons, then $\alpha^e = 0.1855$, a nearly 1 in 5 chance of accepting a false hypothesis. A further complication is the lack of independence among metrics (as illustrated in Fig. 15.4) that complicates conclusions drawn from the examination of a suite of landscape metrics.

These issues are not new and have been discussed within the landscape literature (e.g. Dale et al. 2002; Gardner and Urban 2007; Wagner and Fortin 2005). The chief impediment for adopting a parsimonious approach that avoids this problem seems to be the three-pronged issue of determining the specific question, selecting the appropriate metric and performing the relevant statistical test. These steps may all be performed using Qrule, but have been awkward in the past because the user must be familiar with the data sets generated by Qrule. A series of programmes written in R (Appendix) have been provided here to illustrate and facilitate this process.

An example using data from the piedmont of Maryland has been presented to illustrate the process of hypothesis testing. The question(s) of concern involved the possible effects that increases in urban areas (over an 11-year period) had on the simultaneous loss of forested habitat. A comparison of the two data sets had shown that urban areas had increased and forest areas declined over this time period. Three specific questions were tested: the first question was, "Could the pattern of change be explained by two simple random processes?" Or more precisely, "Were observed changes statistically different from those generated by random processes alone?" The RwC neutral model was developed, random landscapes with identical political and physical boundaries were generated, and a KS test comparing random with actual landscapes showed that the cumulative distributions of random versus actual forest patch sizes were indeed, significantly different. The second question was, "Has there been a significant change in patterns of association between urban and forested areas?" An examination of the association matrix, with a χ^2 of the contingency table, showed that patterns of urban:forest association were less than expected by chance alone, but the loss of forest edge was greater than expected from increases in urban areas and decreases in forest alone.

There are many other hypotheses that could be tested with these data. For instance, we might wish to know "Did urban growth has result in a rate a loss of larger forested areas that was greater than expected by chance alone?" Or perhaps the alternative question would be of interest: "Were smaller patches of forest absorbed by urban growth at a rate greater than expected by chance alone?" An additional set of questions could be developed regarding the effect that changes in agricultural lands might have had on forest loss. The construction of a complex set of tests that compared the relative impact of agricultural versus urban land use change is an intriguing possibility left to the curiosity of the reader. Such inquiries might involve the development of more complex neutral models, or the critical application of the suite of software tools (not just Qrule) that are now available. Whatever the approach, reliable results will only be achieved if the analysis adheres to the statistical principles of hypothesis testing.

15.6 Conclusion

Relatively simple models can often "explain" complicated patterns. Indeed, as Gardner et al. (1987) have demonstrated, real landscapes are sometimes indistinguishable from purely random maps. This result does not argue that real landscapes are produced by simple random processes, but simply defines the conditions in which more detailed explanations of cause-and-effect cannot be demonstrated. When a simple random model fails, then a more complicated explanation (or model) may be tested. For instance, explanations that invoke specific agents of pattern formation due to the physical template (environmental gradients), biotic processes, or disturbance regimes (including human actions). We have illustrated one method for employing a neutral model that considers the effect of landscape constraints on observed patterns. Although the comparisons made here between this model and actual landscapes were dramatically different, variations in the shape, extent, and proportions (p) of land cover types have yet to be fully explored for the RwC model. We hope the combined process of model development and statistical tests will be helpful in the extension of these ideas to other landscapes and landscape questions.

Acknowledgements The assistance of J.B. Churchill in preparing the data set for analysis and the generation of Fig. 15.2 are greatly appreciated.

Appendix

The development of RULE (Gardner 1999) has been an evolutionary process. The current incarnation of this programme, version 4.0, provides extensive changes to the format of the output, data files for statistical summaries, and the inclusion of a new neutral model described in this manuscript (the RwC model selected as option "Y" when executing code). Programme documentation, example input and output files, and the source code for Qrule may be obtained from http://www.al.umces.edu/Qrule. Also available, R4Q, a series of procedures written for the R software (R Development Team 2008). It is hoped that the release of the source code will allow alternative neutral models to be suggested and tested to better understand the relationship between pattern and process at landscape scales. The Open Software License (http://www.opensource.org/licenses/index.php) applies to the distribution, use, and possible alteration of Qrule.

Chapter 16
Stage-Structured Integro-Differential Models: Application to Invasion Ecology

Aurélie Garnier and Jane Lecomte

Abstract Modelling dispersal processes requires a quantitative measure of the amount of individuals dispersed at each distance, which is conveniently summarized in a dispersal kernel. The framework of stage-structured integro-differential models provides spatially explicit population dynamics models that couple a stage-structured life-cycle with a dispersal kernel. This framework is flexible and can notably incorporate density-dependence and stochasticity in demography and dispersal. We first present the general formalism of stage-structured integro-differential models and then provide two examples of application, both from invasion ecology. The first model was developed for the spread of feral oilseed rape along road verges. It is a one-dimensional deterministic invasion model, with a complex dispersal kernel that includes the combined action of various dispersal vectors. The second model was developed for the invasion of grasslands by pines and is a two-dimensional stochastic model with a life-cycle that accounts for the maturity age (10 years) and the dependence of cone production to tree height and tree density.

16.1 Introduction

Dispersal is a major evolutionary force (Clobert et al. 2001) that determines the ability to colonize new habitats, the intensity of competition between kins and the structure of gene flow between populations. Within a population dynamics point of view (i.e. when focusing on individuals rather than genes), measuring and modelling dispersal are crucial aspects in biogeography, metapopulation ecology and invasion ecology (Bullock et al. 2002). Dispersal kernels are known to be a particularly effective tool in modelling dispersal processes because these probability density functions quantify the proportion of individuals dispersing at each distance

J. Lecomte (✉)
Univ. Paris-Sud 11, Unité Ecologie Systématique et Evolution, UMR8079, Orsay F-91405, France; CNRS, Orsay F-91405, France; AgroParisTech, Paris F-75231, France
e-mail: jane.lecomte@u-psud.fr

F. Jopp et al. (eds.), *Modelling Complex Ecological Dynamics*,
DOI 10.1007/978-3-642-05029-9_16,

(contrary to a simple mean dispersal distance). Thus, dispersal kernels can account for the rare but highly important long-distance dispersal events (i.e. the tail of the kernel) which notably determine the ability of a species to colonize unoccupied areas, to find mates located far away or to escape competition (Clobert et al. 2001).

Stage-structured integro-differential models (Neubert and Caswell 2000) are spatially explicit population models that couple a stage-structured dynamics (using a transition matrix, see Chap. 9) and dispersal (using dispersal kernels). A large variety of dispersal kernels can be used, including fat-tailed kernels that model frequent long-distance dispersal events and also mixture kernels that model the action of several different dispersal vectors (Clark 1998). Therefore, the integro-differential population models are an alternative to metapopulation models (which usually consider a simple mean dispersal distance) and to reaction-diffusion models (i.e. partial differential equations, Chap. 7), in which a Gaussian dispersal kernel is implicitly included (Skellam 1951).

In this chapter we first present the formalism of stage-structured integro-differential models and the type of output provided by this type of model. We then provide two examples of application in which an invasion process, i.e. the expansion of a particular species in a new habitat, is modelled. The first example, a simple one-dimensional model, was developed to assess the spread of feral rape along road verges (Garnier and Lecomte 2006; Garnier et al. 2008). The second example, a two-dimensional stochastic model based on a cellular automaton approach, was developed to evaluate the invasion of grasslands by pine species (Boulant et al. 2009).

16.2 Model Structure

Stage-structured integro-differential models are based on an integro-differential equation which computes iteratively the density of individuals within each developmental stage at each time t and on each location x (Neubert and Caswell 2000):

$$\mathbf{n}(x,t+1) = \int_{-\infty}^{+\infty} \left[\mathbf{A}_{\mathbf{n}(y,t)} \circ \mathbf{K}(x,y)\right] \mathbf{n}(y,t)\, dy$$

Bold letters in the equation above correspond to vectors and matrices whereas italic letters are used for scalars (e.g. time, location). The vector $\mathbf{n}(x,t)$ is the density of individuals in each developmental stage at time t and on location x, $\mathbf{A}_{\mathbf{n}(y,t)}$ is the matrix of demographic transitions (which is here dependent on the local density of individuals $\mathbf{n}(y,t)$), $\mathbf{K}$ is the matrix of dispersal kernels and the symbol "o" corresponds to a term-by-term matrix product. The individuals located on y at time t ($\mathbf{n}(y,t)$) contribute to the density on x at time $t+1$ ($\mathbf{n}(x,t+1)$) through recruitment or survival (matrix $\mathbf{A}_{\mathbf{n}(y,t)}$) followed by dispersal from y to x (matrix of dispersal kernels $\mathbf{K}(x,y)$). The integral term represents the sum of all the contributions originating from all locations y.

The Demographic Component: A Transition Matrix

The matrix **A** governs the demographic transitions occurring between each developmental stage at each time step. Each term a_{ji} (row j, column i) of matrix **A** corresponds to the demographic parameter of the transition from stage ***i*** to stage ***j***. For example, let us consider a simple two-stage model for an annual plant species with a soil seed bank where stage **S** corresponds to seed bank seeds and stage **F** to mature plants (Fig. 16.1). The population is censused just before seed release at the end of the reproductive period.

Let us assume that we consider demography on location x. The transition matrix **A** allows computation of the density of individuals in both stages **S** and **F** at time $t + 1$ (population vector $\mathbf{n}(x,t + 1)$) from the density at time t through the relation $\mathbf{n}(x,t + 1) = \mathbf{A}\mathbf{n}(x,t)$, with $\mathbf{A} = \begin{pmatrix} s & fp \\ g & fg \end{pmatrix}$. The matrix **A** can be made space-dependent, for example to model spatial heterogeneity in habitat quality (i.e. direct dependence on location x), or dependent on the density $\mathbf{n}(x,t)$ at location x to include density-dependence, such as in the integro-difference equation above. See Chap. 9 and Caswell (2001) for more details on transition matrices and stage-structured population models.

The Dispersal Component: A Matrix of Dispersal Kernels

A dispersal kernel is a probability density function k that gives the probability $k(x,y)$ that an individual (or a particle, e.g. pollen) moves from location x to location y. This general notation of k does not make any assumption about the spatial homogeneity regarding dispersal. If the space is homogeneous, then dispersal from location x to location y only depends on the distance separating x and y ($d(x,y)$, say $|y–x|$ in a one-dimensional habitat) and thus $k(x,y) = k(d(x,y))$. Most dispersal

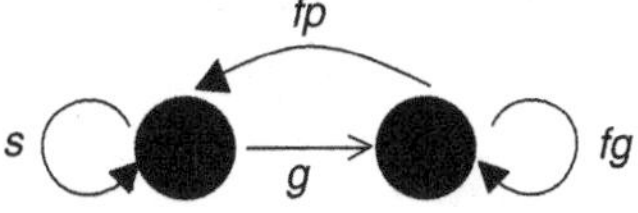

Fig. 16.1 A two-stage life cycle diagram. Stages: seed bank seeds (**S**) and mature plants (**F**). Demographic parameters: annual survival in the seed bank (s), germination (g), rate of seed incorporation in the seed bank (p) and seed production (f)

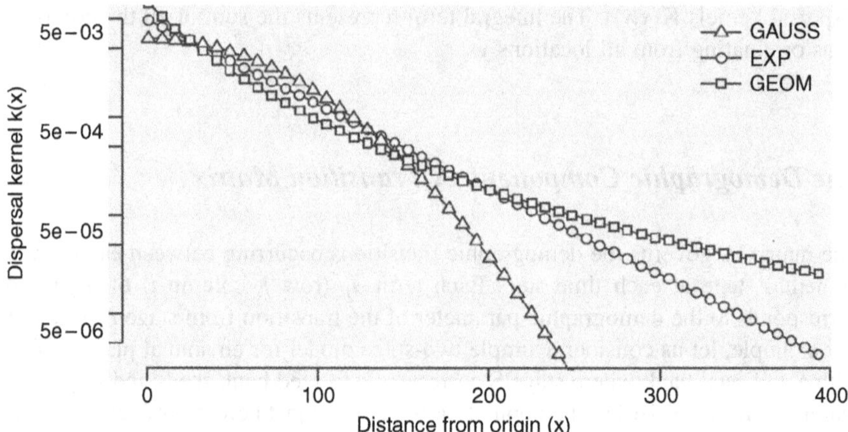

Fig. 16.2 Examples of dispersal kernels: Gaussian function ("GAUSS"; thin-tailed), exponential function ("EXP") and geometric function ("GEOM"; fat-tailed). The *y*-axis is on a logarithmic scale for easier recognition of the shape of kernel tails

kernels are leptokurtic, with dispersal to both short and long distances occurring more frequently than they would under a Gaussian kernel (Fig. 16.2). Moreover, different types of tails exist among leptokurtic dispersal kernels such as thin-tailed kernels (quicker decrease than an exponential), exponential-like kernels (long-range decrease similar to an exponential) and fat-tailed kernels (slower decrease than an exponential). These dispersal kernels generate different colonization patterns (Clark et al. 2001), mixing propagules and gene flow at long distances (Devaux et al. 2007).

The kernel matrix **K** describes dispersal events occurring during each demographic transition in the stage-structured integro-differential model. Each term k_{ji} (row j, column i) of matrix **K** corresponds to the dispersal kernel for dispersal events occurring during the transition from stage i to stage j. The absence of dispersal is modelled by a Dirac delta function $\delta(x)$ which (very roughly speaking) is zero if $x \neq 0$, is infinite when $x = 0$, and integrates to 1. In our two-stage example (Fig. 16.1), dispersal only occurs when seeds are released from mature plants, i.e. during the transitions $\mathbf{F} \rightarrow \mathbf{S}$ (incorporation of dispersed seeds into the seed bank) and $\mathbf{F} \rightarrow \mathbf{F}$ (germination of dispersed seeds). The kernel matrix thus writes: $\mathbf{K} = \begin{pmatrix} \delta(x) & k(x) \\ \delta(x) & k(x) \end{pmatrix}$

Model Output

As in non-spatial stage-structured population models, the asymptotic population growth rate λ in the integro-differential stage-structured model corresponds to the largest of the eigenvalues of the transition matrix **A** at small densities (i.e. in absence of intra-specific competition). Values of λ greater than unity guarantee that population size will grow when small.

In its deterministic form described above, and under some assumptions (no Allee effects, thin-tailed kernel, population growth rate greater than unity) the integro-differential population model generates travelling waves which move at a constant speed c^* (Weinberger 1978; Neubert and Caswell 2000). The invasion speed c^* is determined by the low-density leading edge of the travelling wave and is calculated as a function of the transition matrix **A** for small densities and the kernel matrix **K** (see Neubert and Caswell 2000 for mathematical expressions). Perturbation analyses, i.e. sensitivity (see Chap. 23) and elasticity analyses (see Chap. 9), on λ and c^* can be analytically performed for deterministic integro-differential models [see Neubert and Caswell (2000) for mathematical details]. These analyses generally require simulations when the model is stochastic (however, see Lewis and Pacala 2000 for stochastic unstructured integro-differential models).

16.3 Application Examples

16.3.1 Invasion of Road Verges by Feral Oilseed Rape

Oilseed rape (*Brassica napus* L.) is an annual cultivated species for which feral populations (i.e. populations escaped from crops and established in uncultivated areas) are a common feature along road verges in European and North American farming landscapes (Pivard et al. 2008a, b; Knispel and McLachlan 2009). The persistence and spread of these feral populations are expected to raise both agronomical and ecological issues in the case of genetically modified (GM) cultivars (Hancock et al. 1996; Pessel et al. 2001). Indeed, if the transgene confers a selective advantage to the plant (e.g. herbicide tolerance), and if efficient dispersal vectors move feral seeds across long distances, the GM feral plants could invade uncultivated habitats and thus modify the composition of semi-natural plant communities. Moreover, feral-to-crop gene flow and feral seed surviving within soil seed banks could make the (spatial and temporal) isolation between GM and conventional crops less feasible.

Stage-structured integro-differential models are well adapted to model the spread of GM feral oilseed rape populations along road verges because: (1) the stage-structured dynamics component of the model can account for the different stages of oilseed rape life cycle (seeds in the seed bank and reproductive plants) and for the impact of a selective advantage on specific parts of the life-cycle and (2) the dispersal component can be used to explicitly model the successive action of the different seed dispersal vectors occurring along road verges.

The invasion model we developed for GM feral oilseed rape in a previous study is deterministic and is one-dimensional to mimic the linearity of road verges. The first version of the model (Garnier and Lecomte 2006) was thereafter refined to include experimental data on seed dispersal and to model explicitly the action of various dispersal vectors (Garnier et al. 2008). The demographic component of the final model is reduced to a two-stage life cycle (Fig. 16.1) with seed bank seeds, mature plants and an annual census occurring just before seed release in summer. The dispersal component accounts for the successive action of primary dispersal vectors that disperse seeds from the mother plant to the ground (ballistic dispersal and verge mowers), followed by the action of secondary dispersal vectors that re-entrain fallen seeds (wind and vehicles). Mathematically, the global seed dispersal kernel (k – Fig. 16.3) is thus the convolution of the primary and secondary dispersal kernels (k_1) and (k_2): $k(x) = \int_{-\infty}^{+\infty} k_1(y)\, k_2(x-y)\, dy$. This convolution describes the proportion of seeds dispersed from the location 0 to x via an intermediate location y: seeds are first dispersed from 0 to y by the primary vectors (term $k_1(y)$) and then are subsequently re-entrained from the location y to the location x by the secondary vectors (term $k_2(x-y)$). This process is summed over all possible intermediate locations y (integration symbol). The environment is supposed to be homogeneous.

Because primary seed dispersal along road verges is likely to occur at both short distances (ballistic dispersal, known and measured) and long distances (verge mowers, unknown but assumed to exist at a low frequency), the primary dispersal kernel (k_1) is itself a mixture kernel which sums a small proportion (p_1) of dispersal

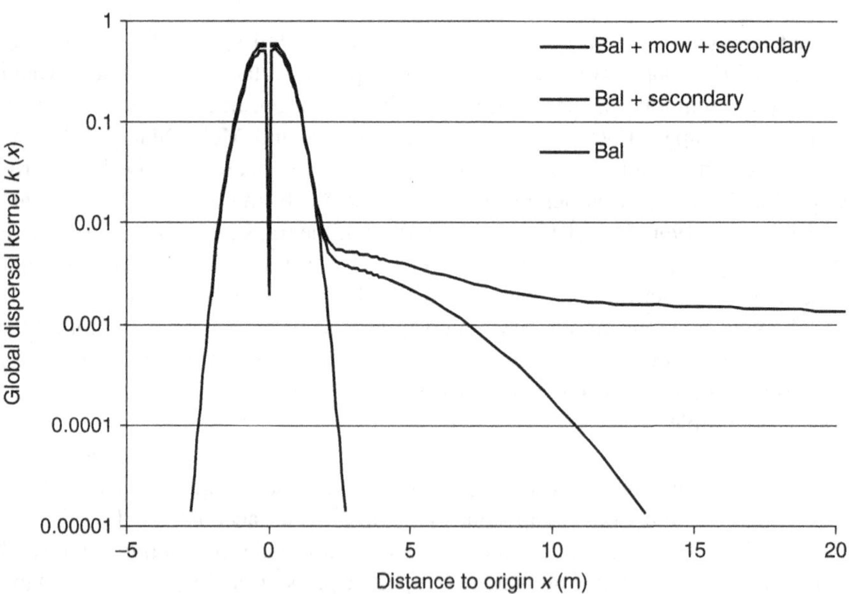

Fig. 16.3 The components of the dispersal kernel for feral oilseed rape seeds along road verges. "Bal": (primary) ballistic dispersal only; "Bal + secondary": primary ballistic dispersal followed by unidirectional secondary dispersal, "Bal + mow + secondary": global dispersal kernel including primary dispersal by ballistic and a small proportion $p_1 = 0.1$ by mowers (mean dispersal distance by mowers $a = 50$ m, unidirectional) followed by unidirectional secondary dispersal

by mowers (k_{mow}) and a large proportion ($1 - p_1$) of ballistic dispersal (k_{bal}): $k_1(x) = p_1 k_{mow}(x) + (1 - p_1)k_{bal}(x)$. Ballistic dispersal is bidirectional in the one-dimensioned road verges (Fig. 16.3) whereas dispersal by mowers is expected to be unidirectional. The corresponding kernel functions are respectively a bidirectional Weibull function for ballistic dispersal [$k_{bal}(x) = m|x|^c exp(-d|x|^{c+2})$ estimated by Colbach et al. (2001)] and a unidirectional exponential function for mowers ($k_{mow}(x) = exp(-x/a)/a$ if $x > 0$ and $k_{mow}(x) = 0$ if $x < 0$ – hypothetical).

The secondary seed dispersal kernel (k_2), that includes the combined (and unidentified) effect of dispersal by wind and by vehicles is a unidirectional mixture of two Gaussian functions that was fitted from a seed dispersal experiment, using a maximum-likelihood method (Garnier et al. 2008).

The invasion model was run with contrasting values of selective advantage (with an effect on both seed germination and plant survival) due to herbicide tolerance. We performed elasticity analyses, which are similar to sensitivity analyses (see Neubert and Caswell 2000 for their application to integro-differential models) to identify the key-parameters of the model. Results showed that the invasion speed was primarily determined by long-distance dispersal (whatever its low occurrence rate). The selective advantage noticeably increased invasion speed, provided some long-distance dispersal events were included.

This study underlined the necessity to obtain relevant estimates of long-distance dispersal of feral oilseed seeds along road verges to make reliable predictions of the spread of (GM) feral populations. However, long-distance dispersal events are known to be very difficult to detect and to quantify because they are rare and highly stochastic (Clobert et al. 2001; Bullock et al. 2002).

16.3.2 Invasion of Grasslands by Pines

The invasion of open habitats by native and introduced tree species is a growing concern in many regions of the world (e.g. Richardson et al. 1994; Dovciak et al. 2005) and is favoured by changes in natural and human environmental factors: fire regime, climate, farming practices (grazing) and forestry practices (large plantations of exotic trees). Seedling recruitment is known to have a key-impact on the population dynamics of several tree species (Harper 1977) and is thus expected to have also a predominant effect on their invasion dynamics. Because seedling recruitment itself depends highly on different environmental factors, simulation models are a useful tool to better understand how these factors interact and influence the spread of trees.

Stage-structured integro-differential models are particularly useful to model the spread of invasive tree species. This is because in slow growing tree species, where maturity occurs after several years, it is crucial to distinguish several age-classes in their life-cycle. Moreover seed dispersal kernels estimated from field data can easily be implemented in this type of models. Stochastic environmental, demographic and dispersal variations are known to have a critical impact on the spread rate of invasive populations and these stochastic factors can be accounted for in integro-differential models (Lewis and Pacala 2000).

Boulant et al. (2009) developed a cellular automaton model (see Chap. 8) for the spread of two pine species (*Pinus sylvestris* L. and *Pinus nigra* Arn. ssp. *nigra*) in Mediterranean grasslands. This stochastic stage-structured integro-differential model allows simulation of pine expansion within a 500m-wide rectangular gridded landscape divided into 25 × 25 m cells. The demographic component of the model accounts for the whole complexity of the pine life cycle: trees begin to produce cones at the age of 10 years and cone production is age-(height-)dependent and density-dependent. Therefore the life-cycle was divided into 15 stages (Fig. 16.4): nine seedlings age-classes (from 1-year old (S_1) to 9-year old (S_9)) and six adult classes. Adults were divided into four 2-m height classes ($A_{2\text{–}4m}$ to $A_{>8m}$) of increasing fecundity. When the total density of adults in a cell reaches the maximal density of isolated trees (D_{is}), adults are considered as dominant in woodland (A_{dom}). Individuals reaching maturity when the total density of adults in a cell is larger than the maximal density of dominant trees (D_{dom}) are considered as suppressed (A_{sup}), i.e. with a reduced fecundity. When the maximum tree density (D_{max}) is reached, no more seedlings can establish. Seedling recruitment was modelled explicitly as a function of tree fecundity and of the interactions between shrub cover, post-dispersal seed predation, grazing pressure, grass competition, and drought (see Boulant et al. 2009 for more details on demographic parameters).

The dispersal kernel of the pine model is a mixture of two exponential functions that combines a small proportion (p) of long-distance dispersal and a larger proportion ($1 - p$) of short-distance dispersal, which was estimated from field data. Since dispersal was supposed to be isotropic, a one-dimensional form of the kernel was used although the landscape was two-dimensional. The probability that a seed produced in a cell i reaches a cell j located at a distance x_{ij} thus writes: $k(x_{ij}) = p\frac{2e^{-2x_{ij}/a_1}}{\pi a_1^2} + (1-p)\frac{2e^{-2x_{ij}/a_2}}{\pi a_2^2}$, where a_1 and a_2 are the mean distances of

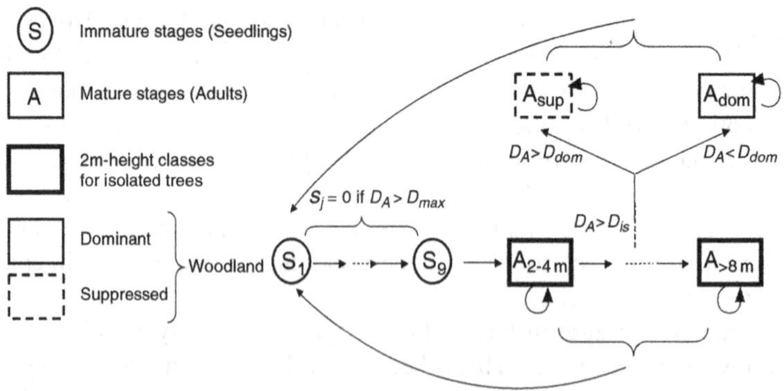

Fig. 16.4 Life cycle graph for the pine model [simplified from Boulant et al. (2009)]. Adult density thresholds (in a 25 × 25 m cell): maximum density of isolated trees (D_{is}), maximum density of dominant trees in woodland (D_{dom}), maximum tree density (D_{max}). D_A: adult density in a cell (including isolated, dominant and suppressed trees), s_j: seedling establishment rate

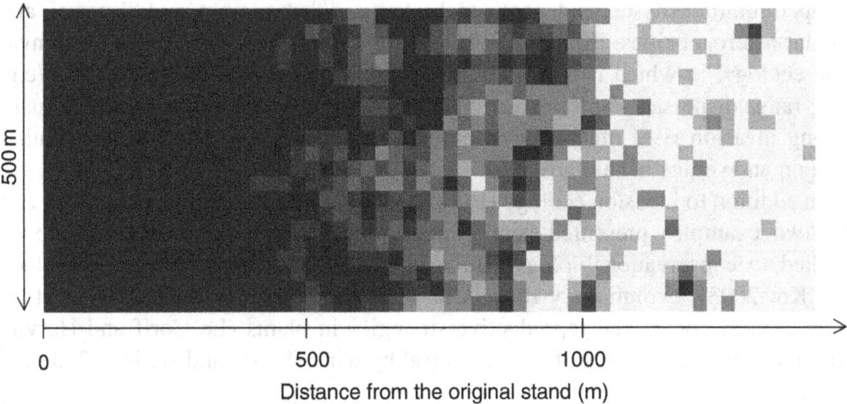

Fig. 16.5 Simulation obtained with the pine invasion model. Each 25 × 25 m cell is coloured in *grey* shades according to tree density (*white* cells = empty cells, *black* cells = maximal tree density). The simulation was run over 100 years

long- and short-distance dispersal, respectively. This function integrates to 1 when summed over all directions and over all distances.

All components of the pine model are stochastic: demographic and environmental stochasticity is included in the demographic component and both the direction and the distance of dispersal are randomly drawn (from a uniform distribution and from the dispersal kernel, respectively). Simulations are necessary to explore the dynamics of such a complex model, which combines a highly structured life-cycle and stochasticity in every component. An example of a trajectory run over 100 years from an initial stand (one column of cells composed of 20% of dominant trees and 80% of suppressed trees) located at the left edge of the rectangular landscape, shows the patchy expansion of the population towards the opposite edge (Fig. 16.5). Invasion speed was computed from the linear regression of wave front location on time (see Kawasaki et al. 2006 for mathematical details).

Elasticity analyses performed on the stochastic growth rate and the invasion speed gave a better understanding of the interactions between the different factors influencing seedling recruitment and notably revealed that grazing intensity was the main factor limiting pine spread. This result has important consequences in terms of management, as it suggests that grazing by domestic livestock could be an effective tool to limit the invasion of open landscapes by pines, which is less expensive than mechanical interventions.

16.4 Conclusions

Integro-differential models are a useful tool to model the spatio-temporal dynamics of a population. They allow sufficient realism in both the life cycle (stage-structure) and dispersal processes. Their flexible formalism permits to include intra-specific competition (density-dependence), stochasticity in demography and dispersal, and habitat heterogeneities. Both biological examples presented here came from invasion ecology, in which it is particularly essential to account for rare events (e.g. long-range dispersal events, catastrophic environmental variations). Indeed, quantifying invasion risks must rely on probabilities associated with scenarios rather than on an average behaviour.

In addition to invasion ecology (Buckley et al. 2005; Jacquemyn et al. 2005 and the two examples presented here), integro-differential models have also been applied to conservation biology (Bullock et al. 2008), epidemiology (Medlock and Kot 2003), evolutionary models in population genetics (Champagnat et al. 2006), studies on mixed reproductive strategies in plants (Le Corff and Horvitz 2005) and on mechanisms of seed dispersal by wind (Soons and Bullock 2008).

Chapter 17
Modelling Resilience and Phase Shifts in Coral Reefs: Application of Different Modelling Approaches

Andreas Kubicek and Esther Borell

Abstract Tropical coral reefs are among the most diverse habitats with great ecological and economic importance. These highly dynamic ecosystems are frequently subject to natural disturbance events (e.g. hurricanes) which can lead to dramatic changes in reef properties if the systems' resilience is already reduced by anthropogenic impacts such as increased levels of nutrients or over-fishing. Due to their high complexity many relations in coral reef dynamics are still poorly understood and ecological modelling becomes increasingly prominent as a tool to close these knowledge gaps. This article gives an overview on different modelling techniques that address the investigation of coral reef dynamics and discusses advantages and disadvantages of respective applications.

17.1 Introduction

Tropical coral reefs are complex ecological habitats, that are the most diverse of all marine ecosystems, with estimates of benthic and pelagic organisms ranging from 600,000 to more than 9 million species worldwide (Reaka-Kudla 1997). Scleractinian corals (stony corals, Cnidaria, Anthozoa) are the main reef builders. They fix calcium carbonate, which produces the majority of the habitat structure for other reef organisms. Coral reefs are dynamic systems within a wider network of closely interlinked habitats such as mangroves and seagrass beds, (Nagelkerken et al. 2002; Mumby and Hastings 2008), which are frequently subjected to natural disturbances (Connell 1997; Buddemeier and Smith 1999). However, the nature and temporal pattern of disturbances have changed severely over the past few decades coinciding with global climate change (Veron et al. 2009) and increasing anthropogenic activities in coastal areas (Mora 2008), often exceeding the regenerative capacity of reef systems (Bellwood et al. 2004).

A. Kubicek (✉)
Department of Ecological Modelling, Leibniz Center for Tropical Marine Ecology GmbH (ZMT), Fahrenheitstrasse 6, 28359 Bremen, Germany
e-mail: andreas.kubicek@zmt-bremen.de

F. Jopp et al. (eds.), *Modelling Complex Ecological Dynamics*,
DOI 10.1007/978-3-642-05029-9_17,

Chronic alterations of a reef environment including increased levels of nutrients, overfishing and the release of toxic compounds can severely undermine reef resilience and thus the ability of reef communities to cope with new disturbances superimposed onto those already existing (Nystroem et al. 2000). Reduced resilience inhibits or delays reef regeneration after a disturbance event, which can lead to long-lasting or even irreversible changes in community structure; so-called phase shifts to alternative stable states (Hughes and Connell 1999; Hughes et al. 2007). The resultant alternative state is manifest in either a new dominant coral species (Aronson et al. 2004) or an alternative life-form, like corallimorpharians (Kuguru et al. 2004), ascidians, soft corals, sponges and urchin 'barrens' (Norstroem et al. 2009) and very often algae (McManus and Polsenberg 2004). Regardless of the nature of these shifts, they generally all culminate in a conspicuous loss of benthic invertebrate and fish diversity as well as a decrease in inorganic carbonate deposition, which in turn reduces reef complexity, overall species richness and increases shoreline erosion.

Important factors that mediate resilience include (1) the degree of diversity within functional groups, functional redundancy and the response diversity within each group, (2) demographic structure of populations, (3) recruitment success, and (4) ecosystem connectivity, i.e. exchange processes among reefs or between reefs and adjacent habitats within a given seascape.

When diversity of coral reef species is high and species interact in a highly structured environment, feedback loops occur over a wide range of scales. Thus, descriptive approaches using mean average measures or starting from reduced statistical assumptions might not be appropriate for analysing the complex structure and underlying processes. Here modelling may help to integrate the multitude of components, relevant variables and parameters to describe and visualize complex ecological processes and the driving forces which shape the resilience of a system. Models may also be used to simulate the behaviour of specific system components in response to a changing environment (Fig. 17.1).

In the following subsections we describe different approaches to modelling reef resilience including examples of a trophic model which is based exclusively on differential equations (Sect. 17.2) and a Cellular Automaton (CA) model which allows spatial explicit analysis (Sect. 17.3). Section 17.4 introduces how Individual Based Modelling (IBM) can facilitate the implementation of direct individual interactions of organisms and Sect. 17.5 gives an example of a grid based community model which combines differential equations and a CA approach. We have chosen these examples to illustrate and discuss the possible advantages and drawbacks of presently applied ecological modelling techniques.

17.2 Equation-Based Modelling of a Coral Reef Food Web

McClanahan (1995) developed a differential equation model to evaluate the impact of fishing on Kenyan coral reefs. The model simulates the food web of a virtual reef ecosystem of undefined spatial extent in which corals and algae comprise the primary

Fig. 17.1 Intact coral reef, Sulawesi/Indonesia (Photo by E. Borell)

producers, herbivores consist of sea urchins and herbivorous fish and predators are composed of invertivorous and certain piscivorous fish as well as humans, i.e. fishermen.

Model relationships and parametrisation were all based on empirical studies and local fisheries data. To keep the model at an operational size, McClanahan considered only a limited number of the system's key components and their interactions and left out other food web pathways such as phytoplankton, detritus or corallivores and top predators, such as sharks. Gross and net reef production is calculated by combining production and respiration for both algae and corals. Although the model considers that both groups fix calcium carbonate from the seawater, corals in the model calcified at rates ten times greater than algae and thus represented the major calcium carbonate depositors. Sea urchins and herbivorous fish competed for algae and in the process of foraging eroded reef structure with the erosion by urchins being tenfold higher than that of fish.

Invertivores controlled the abundance of urchins and in the model switched to an unspecified alternative food source upon depletion of sea urchins. This had the effect that invertivores did not experience bottom-up control, which decreased fluctuations in model dynamics. Herbivorous fish were controlled by piscivores. At the very top end of this web were humans who ultimately constitute a somewhat arbitrary control of all fish present in the system. All processes in this model were described by utilizing a matrix of interlinked differential equations (Chap. 6).

Through variation of the state variables 'fishing experiments' were performed in order to assess the effects of fishing intensity and catch selection on fisheries yield, community structure and ecological processes. McClanahan performed a total of five simulation runs to specifically determine (1) the model's prediction of successional dynamics, (2) the effect of removing all bony fish, (3) the effect of removing only piscivores, (4) the effect of removing all fish except invertivores, and (5) model predictions of a scenario where sea urchins (or fishing) do not have a detrimental effect on live coral. Model simulations of different scenarios revealed that the modification of a single variable in this web of highly interrelated components has important ramifications for reef development. One of the major findings of the model (and later verified by field studies) was that coral reefs are prone to have more than one equilibrium state for realistic parameter ranges (see Chap. 6), influenced by the extent of fishing or the abundance of piscivores (Fig. 17.2). The simulation results showed that if all fish groups were harvested, two equilibrium states could occur, one governed by herbivorous fish and the other by sea urchins. A third ecological state was manifest by high algal and low coral cover associated with the low abundance of either herbivorous fish or sea urchins. The fisheries management strategy that is predicted to produce the highest yields whilst maintaining high primary productivity and calcium carbonate deposition was to harvest piscivores and herbivorous fish and to leave invertivores unharvested. For this case, the model predicted the amount of piscivores to quickly decline, taking predation pressure off the herbivorous fish and channelling the majority of algal production into herbivorous fish, while invertivores kept the sea urchin abundance low. When fishing levels were highest, algal biomass was predicted to increase and

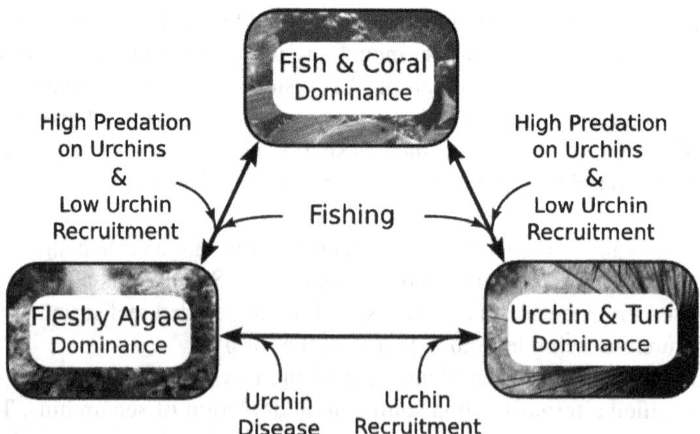

Fig. 17.2 Transition processes between different equilibrium states of the coral reef ecosystem indicated by model simulations and field studies. Adapted from McClanahan (1995)

competitively exclude corals. Calcium carbonate deposition would remain high because of the proportion of calcifying algae and because bioerosion would decrease due to low sea urchin and fish abundance.

McClanahan's model has the general applicability to assess coral reef food web interactions through a somewhat holistic approach rather than focusing on a limited number of organisms and interactions. However, a major shortcoming of this type of modelling is presented by the fact that it is highly aggregated. Treating several components under the umbrella of one variable, i.e. the grouping of various components according to functional types or trophic groups severely reduces natural variability and may therefore easily mask processes which are crucial attributes to the system's dynamics. Considering that different species or functional groups of algae and corals, do exhibit very different types of tolerance to a given environmental condition, the lack of distinction can lead to the oversimplification of a given scenario.

17.3 Spatial Competition Between Coral Species: Application of a Cellular Automaton Modelling Approach

Langmead and Sheppard (2004) designed a classical cellular automaton model (Chap. 8) to assess the effect of natural background disturbance (e.g. sedimentation, predation) on a coral community on a Caribbean fore-reef slope. Each disturbance event can be set to occur over different spatial scales and at varying levels of intensity. The model comprises ten different coral species that compete for space. Each of the species has a specific pattern of recruitment, growth, mortality rate and aggression (i.e. competitive potential).

The simulation area of the model reef comprises a torus with a total size of 9 m^2 that is subdivided into 1 cm^2 cells corresponding to a median sized coral polyp. Each of these cells can contain either bare substratum or one of ten coral species at a time.

The model was exclusively parameterized with data obtained from the literature. Coral growth was based on annual skeletal extension rates of each species and was expressed as radial expansion. Growth was determined by the rules for competitive interactions between corals: Colonies could only grow into adjacent cells if they were either unoccupied (bare substratum) or occupied by a competitively subordinate species (species were ranked according to their aggressiveness). Depending on the differential susceptibilities to disturbance and varying mortality rates, each species was assigned a probability of mortality if impacted by disturbance. Based on data for larval settlement in the study area, each coral species was set to be present at a specific density and the number of potential recruits was then determined annually using a Poisson probability distribution. Recruitment success was determined by larvae abundance and the amount of free space in the plot as recruits are only allowed to settle on bare substratum.

To gain a better understanding of the spatial extent of disturbance (i.e. the fraction of the plot that is disturbed) and the size of disturbed patches, two different scenarios were created: First, the spatial extent of disturbance was varied at five intervals ranging from 0 to 0.6 while the size of disturbed patches was kept constant. In the second scenario, the total disturbed fraction of each plot per year was kept constant and the size of disturbed patches was varied. The sizes of disturbed patches followed a power law model in which frequency of disturbance events is related to their spatial extent; i.e. smaller disturbances were set to occur more often than larger ones.

Recruitment and background disturbances were updated yearly while growth and competitive interactions were iterated once every 3 months.

Simulations were run for 500 years (complete cycle of the model) and percentage cover of each species was taken on an annual basis. The data derived from the model simulations were fed into a Bray–Curtis matrix in order to determine the sensitivity of the variables on species diversity, species composition and mortality in response to each of the model parameters.

The results showed that in the absence of disturbance, the reef was occupied by competitively dominant species and that those species featuring low aggression and low growth rates were lost after short periods of time. Intermediate levels of background disturbance favoured high coral diversity, which supports the classic hypothesis of intermediate disturbance (Grime 1973). Accordingly, the amount of bare substratum increased with higher levels of disturbance and was accompanied by a decrease in biodiversity.

The relative importance of total colony mortality to partial colony mortality changed with colony size class. Total mortality was more important for small colonies while large colonies were most sensitive to partial mortality. For sensitive species, competitiveness, i.e. aggressive potential had the greatest influence on community composition. Growth was also an important factor whereas mortality and recruitment had the least impact on the model.

Model evaluation showed that only five out of the ten simulated species were comparable to actual field observations. Despite this relatively weak congruence (mainly resulting from insufficient data), for coral population size structures, the model was able to accurately represent growth and distribution for seven out of the ten species. The results indicate that size structure of populations is a much more precise indicator for testing the predictive abilities of the model than the simple comparison of coral cover.

Another important aspect of the results was the apparent relationship between the threshold of partial and total colony mortality and modal colony size on a log-scale, which demonstrates that colony size and age may be decoupled earlier than was previously thought (e.g. see also Bak and Meesters 1998). Also, the results indicate that size of disturbed patches was as important in structuring coral communities as the overall amount of disturbance (Fig. 17.3). This poses an important consideration when looking at recovery mechanisms of reefs since the spatial extent of cleared substratum is rarely directly quantified in the field, but is usually determined indirectly through differences in coral cover. Even though

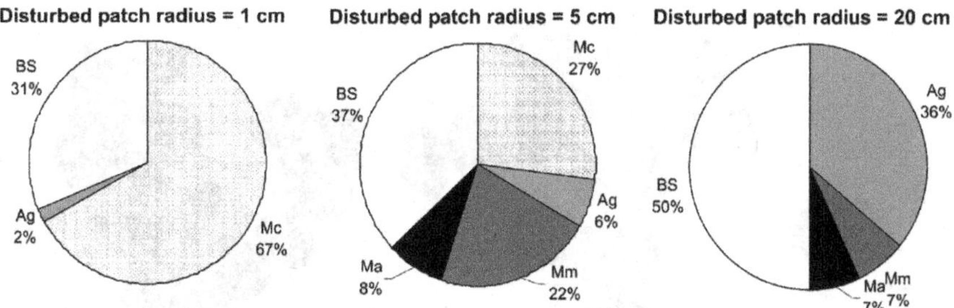

Fig. 17.3 Effect of disturbed patch size on community composition while the total fraction of disturbed reef area was kept constant. Mean percent cover at 500 years for the most abundant species (Mc *Montastraea cavernosa*, Ma *Montastraea annularis*, Mm *Meandrina meandrites*, Ag *Agaricia* spp.) and bare substratum (BS). [From Langmead and Sheppard (2004)]

some interesting information can be deduced from this approach, the overall applicability of the model in a broader context of coral reef dynamics and effects of disturbance events is limited due to its small size and the lack of other key components such as grazers and algae, which can play a crucial role in structuring coral reef communities.

17.4 Macroalgal Growth Patterns Simulated with an Individual-Based Model

Many macroalgae exhibit non-deterministic phenotypic growth, which enables them to thrive under different environmental conditions. Yniguez et al. (2008) designed an individual-based (or agent-based, see Chap. 12) model (SPREAD, Spatially-Explicit Reef Algae Dynamics) to investigate the effects of key growth factors (nutrients, light, temperature) as well as disturbance and mortality on the growth rates and growth morphology of the calcifying algae *Dictyota* and *Halimeda* spp. at four different sites (two inshore and two offshore reefs) within the Florida Keys Reef Tract. Model performance was empirically evaluated with local growth rate and structure for *H. tuna* at these sites.

To determine the effects of different environmental conditions on the growth patterns of the algae, Yniguez et al. (2008) used single modules (see Fig. 17.4) as the interacting components in their model rather than representing whole algal individuals. The emergent properties of superordinate hierarchies such as the whole individual organism, the population or the algal community were thus derived from interactions of single modules. The model environment was composed of a three-dimensional cubical grid (edge length ~30 cm) and subdivided into 1 cm^3 cells. Each cell contained information about light and space availability, nutrients and temperature. Temperature and nutrients were kept uniform for all cells at each time step but changed with season. Light availability was determined for each cell

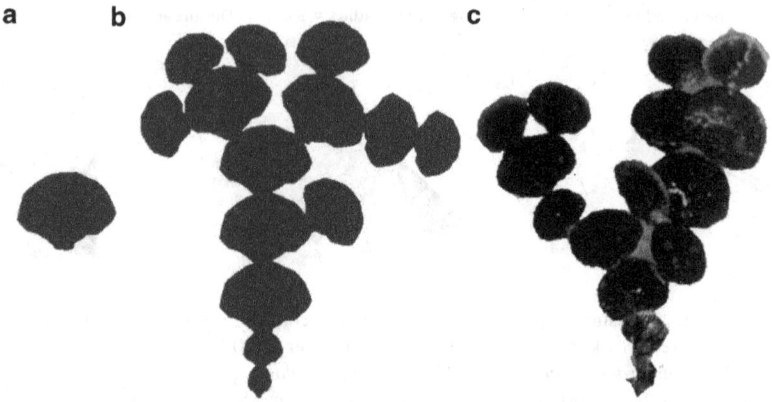

Fig. 17.4 *Halimeda tuna* – a single module as interacting components in the model (**a**) the modelled growth pattern (**b**) and a photograph of the actual alga (**c**). Taken from Yniguez et al. (2008)

individually depending on depth and shading effects of existing modules. At the start of the simulation, the initial values for state variables and the number of model organisms were set corresponding to field observations. The model used discrete daily time steps. Each module had direct information of its position within the grid and was able to retrieve/'sense' environmental parameters for both its own cell and its neighbourhood. Based on this information, probabilities for growth (defined as the production of a new module by an existing one) and growth pattern were calculated and newly produced modules positioned. If existing modules were situated at the edge of an alga they could be randomly selected for fragmentation, which implied an additional mortality probability for the fragment. At the end of an iteration the three-dimensional grid was transformed into a two-dimensional square to calculate the percentage of cover for each species.

A total of 30 replicate simulations (each over a period of 1,000 days) per site revealed striking similarities for morphometric characteristics of model *H. tuna* and field data such as the number of segments per individual alga between different sites and the relationship between growth (segment production) rates and depth. *H. tuna* from the deepest site featured the highest growth rates and the highest number of produced modules per individual, while individuals of the shallowest site had low growth rates and low numbers of segments. Despite small divergences between model prediction and field data, the natural inter-site differences in growth morphometrics were well reflected by the model. Due to its fine-tuned nature this technique is not yet applicable to larger scale simulations mainly due to limitations in computing capacity. Similar methodologies were already applied in the study of plant architecture and morphology by utilizing so called L-Systems (see also Chap. 11). This approach is novel in the marine context and can contribute valuable information regarding small scale processes, which can be fed into larger scale models when depicting natural variability.

17.5 Ecosystem Model for Phase Shifts in Caribbean Coral Reefs

The current collapse of many coral reefs in the Caribbean is thought to be a combined effect (Bellwood et al. 2004) of overfishing of herbivorous fish (Hughes 1994), coral diseases (Bythell and Sheppard 1993), hurricanes (Bythell et al. 1993), coral bleaching (Kramer et al. 2003) and local deterioration of water quality (Littler et al. 1993). Macroalgal blooms on the overfished reefs in the Caribbean were, until the early eighties mainly prevented by a single species of sea urchin *Diadema antillarum*. The mass mortality of *D. antillarum* in 1983 (Lessios 1988), left parrotfish (Scaridae) as the main herbivores on many Caribbean reefs.

To gain a better understanding of the relative importance of fishing of parrotfish and parrotfish grazing for coral-algal dynamics, Mumby (2006) merged different modelling approaches within one application that facilitates the integration of interactions within and between different trophic levels of a typical Caribbean coral reef community. The main model in this study constitutes a grid based spatial explicit simulation model for a hypothetical reef that combines empirical data derived from field studies, experiments and other models. The second model is an equation based approach, which was designed to model the processes of parrotfish grazing in order to parameterize their grazing behaviour in the main simulation model.

The Basic Coral Reef Model

The model addressed the dominant massive reef building coral species (as most branching species were eliminated by the white band disease) on a common Caribbean reef in the *Montastraea annularis* zone at mid-depth (5–15 m) where coral diversity and abundance were found to be highest. The model was parameterized with data from Glovers Reef of Long Cay (LC) in Belize. The virtual reef area comprised a 625 m^2 (25 m $\times$ 25 m) lattice made up of 2,500 rectangular cells (0.5 m $\times$ 0.5 m). The functional organisation inside the cells resembled that of a classical cellular automaton. Unlike a classic cellular automaton however, the cells were able to accommodate more than one distinct entity, i.e. different organisms and/or dead substratum at the same time. The benefit of such an organisation is that the observer is able to split up populations into smaller groups thereby facilitating a semi-individual model behaviour. Contrary to an IBM, organisms in this model did not interact directly with each other but the trajectory of an organism's behaviour and development was determined by the characteristics of its own cell (i.e. composition of components) and those of its neighbourhood.

The model was iterated every 6 months; an interval that is sufficiently long to allow for meaningful assessments of coral growth and coral distribution. Yet, direct algal cover is the outcome of a dynamic balance between algal production (area for colonization, recruitment rate, and growth) and algal removal (grazing), processes that occur over much shorter time spans. For this reason Mumby (2006) developed a second model to parameterize parrotfish grazing intensity, in which he determined the proportion of area grazed within a 6-month period. The calculations incorporate parameters for fish species, abundances, sizes, sex, and feeding activity (bite rates). According to the results, in an unharvested parrotfish community 30% of the total area was grazed, while in heavily depleted populations the grazed proportion comprised only 10% and for areas of intermediate fishing pressure, this fraction amounts to 20%.

The main model included corals, algae, urchins and parrotfish (Fig. 17.5). However, urchins were excluded from most model scenarios since a large scale recovery of *D. antillarum* populations was considered unlikely and outbreaks of diseases could readily reoccur. Corals and algae (see Table 17.1 for summary of attributes for corals and algae) were placed randomly on the grid until distinct

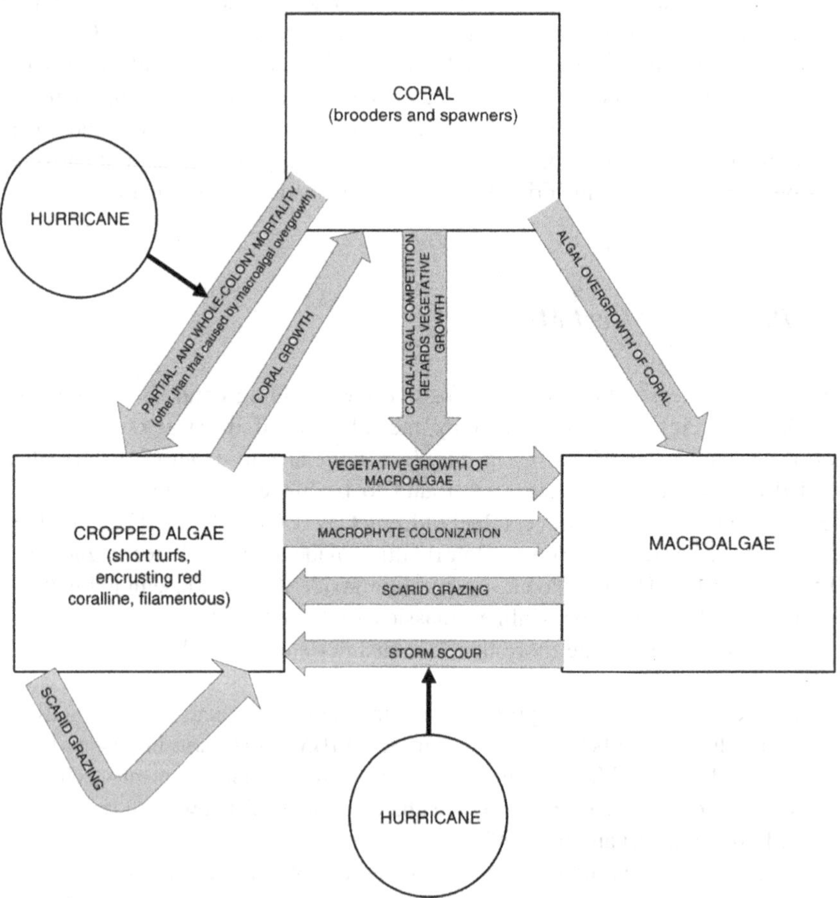

Fig. 17.5 Processes included in the simulation model (*arrows*) that link the major functional categories of reef organisms (*boxes*). Taken from Mumby (2006)

Table 17.1 Major attributes of corals and algae and the main rules utilized in the model

Corals	Algae
General information	
Massive corals Two groups according to reproductive mode: 1) Brooders, e.g. *Porites astreoides* 2) Spawners, e.g. *Siderastrea siderea*	Macroalgae and cropped algae Age classes: Cropped algae 0–6 months Cropped algae 6–12 months Macroalgae 0–6 months Macroalgae 6–12 months
Reproduction/recruitment/dispersal	
Larval production depending on maturity state/size class for high coral cover: juvenile (<60 cm^2) $\rightarrow$ no larvae pubescent (60–250 cm^2) $\rightarrow$ ~50 larvae $\times$ cm^{-2} adult (>250 cm^2) $\rightarrow$ ~210 larvae $\times$ cm^{-2} Number of recruits that can settle per cell depending on settling ground: Bare substrate $\rightarrow$ 4 recruits Cropped algae $\rightarrow$ 2 recruits Macroalgae $\rightarrow$ 0 recruits $\rightarrow$ the density of spawning coral recruits is ten times lower than that of brooding ones	Probabilistic overgrowth of cropped algae (A) by macroalgae (M) within a von-Neumann-Neighbourhood, depending on proportions of macroalgae (M_{4c}) and corals (C) if $C \geq 0.5$ PA $\rightarrow M \rightarrow 0.75 \times M_{4c}$ if $C < 0.5$ PA $\rightarrow M \rightarrow M_{4c}$
Growth	
Constant diametric growth rates: 8 mm $\times$ year^{-1} (brooders) 10 mm $\times$ year^{-1} (spawners)	If cropped algae are not grazed for 1 year they turn into macroalgae
Mortality	
Periodic – annually Smaller coral colonies are more susceptible to partial and whole-colony mortality than larger ones	Periodic – annually Proportion of algae which are eaten by grazers $\rightarrow$ depending on grazer density
Stochastic – hurricanes Proportion of corals is subject to partial or whole-colony mortality; based on mean fraction of reef destruction across the Caribbean	Stochastic – hurricanes Proportion of macroalgae become cropped algae if located on disturbed patch; based on mean fraction of destruction across the Caribbean
Interactions and processes	
If corals reach maximum cell size (2,500 cm^2): $\rightarrow$Larger colonies grow over smaller colonies $\rightarrow$Corals grow over cropped algae $\rightarrow$Corals can displace macroalgae	Macroalgae can grow over cropped algae Macroalgae can grow over corals (depending on coral colony size) and lead to partial or whole-colony mortality

proportions of spatial cover had been reached with the rest being occupied by sand and rubble.

Corals were divided into two functional groups according to their mode of reproduction, comprising either brooders or spawners. Brooding corals produce planula larvae which settle in the vicinity of their parent colony. Spawners (commonly referred to as broadcast spawners) release gametes and the planulae are dispersed in the water column and may migrate many kilometres before settlement.

The model was parameterized with data for *Porites astreoides* and *Siderastrea siderea* for brooders and spawners respectively. Both maturity and reproductive output were expressed as a function of colony-size and the efflux of larvae from reefs was quantified from the size-frequency distribution of coral colonies. Coral recruitment was parametrized with data from an offshore reef in Belize which had high adult coral cover and high biomass of grazing fish. Coral recruitment in the model was set to occur at an initial colony size of 1 cm in diameter with the settlement success being determined by the components of their cell (i.e. rugosity and algal characteristics). A linear stock-recruitment relationship was created based on the assumption of high adult coral cover and optimal larval supply. The massive growth forms of coral colonies were expressed as hemispheres and growth rates were modelled by linear extension rates of the hemispheres (Table 17.1).

Mortality rates of corals were also colony-size dependent where whole colony mortality was generally lower for mature (large) colonies than for smaller ones. Large colonies were able to overgrow smaller colonies in basic interactions once colonies had reached the maximum implied size of a cell. Macroalgae were able to overgrow coral recruits and to cause extensive partial mortality of larger colonies.

Data on hurricane mediated mortality was derived from the impact of Hurricane Mitch on mature colonies of *M. annularis* in Belize where at least 90% of the colonies experienced partial-colony mortality. The frequency of hurricanes could be varied according to geographical area. Since the simulation area of 625 m^2 is relatively small, the chances that a reef would be either completely destroyed or missed entirely by a hurricane were high. For this reason the model used the mean percentage of destroyed reef area for the whole simulation area rather than subdividing it into patches of heavy and light destruction.

Algae were distinguished as either cropped algae (cropped substrata) or macroalgae. Cropped algae included encrusting coralline red algae, fine filamentous algae and algal turfs, which were contained within one category because coral recruitment, i.e. coral settlement and post-settlement mortality is associated with all of these types. If cropped algae were not grazed, spores of macroalgae (here in the model *Dictyota* spp. and *Lobophora variegata*) developed into a fleshy canopy that prevented coral settlement. Macroalgal growth progressed by either of the following pathways: Cropped algae which were not grazed over the period of 1 year turned into macroalgae. Once established, macroalgae were able to overgrow cropped algae in neighbouring cells depending on their relative cover and that of corals within a von-Neumann-Neighbourhood (see also Table 17.1). If coral cover was low, macroalgae could overgrow an area of cropped algae similar in size to the area they occupied, whereby high coral cover reduced this area by 25%.

Simulations and Results

An a priori sensitivity analyses revealed that initial coral cover, grazing and hurricane frequency were all important factors influencing coral cover over a period of 20 years. It is important to bear in mind that the model was only simulated on the reef dynamics of initially 'healthy' reefs. The most important findings of the simulations may be summarized as follows:

First, in the absence of any acute disturbance event and the urchin *D. antillarum*, coral cover always increased when grazing was carried out by an unexploited community of parrotfish. This in turn had a positive influence on recruit survival where highest densities correlated positively with highest coral cover. Second, in the absence of *D. antillarum*, the dynamics of coral cover were highly sensitive to changes in hurricane frequencies. Reefs that experienced hurricanes on a decadal basis showed a net decline in coral cover whereas a hurricane frequency of 20 years allowed for full recovery akin to the initial 30% coral cover. Reefs that were subjected to hurricanes at even lesser frequencies (e.g. 40–60 years) exhibited rapid reef growth. On the other hand, the inclusion of *D. antillarum* enabled reefs to withstand hurricanes on a decadal basis and the results showed that overall diversion in reef recovery (i.e. reef trajectory) between different hurricane frequencies decreased. Third, a reduction in parrotfish biomass lead to substantial changes in reef community. High parrotfish biomass (i.e. high grazing) resulted in a 25% increase in coral cover. Conversely, coral cover decreased from 30 to 7% when parrotfish were heavily depleted. Furthermore, the results demonstrate that grazers (or the depletion of them) is a fundamental and overarching factor in shaping the trajectory of reef development in Caribbean forereefs. All other parameters, such as whole-colony mortality rates, connectivity, larval retention and dispersal had a negligible effect on coral cover. Last, reefs that maintained a healthy parrotfish population showed clear phases of reef growth in between hurricane disturbances which recurred at 40-year intervals. In contrast, reefs with a partially depleted parrotfish population being subjected to the same hurricane frequency exhibited a steady decline in coral cover.

Interestingly, examination of temporal shifts in the relative size-frequency distribution of corals under different disturbance scenarios also indicates that hurricanes and the exploitation of grazers had very contrasting effects on coral populations: Under intense exploitation of grazers, the coral size distribution become bimodal and the population experienced a bottleneck among the juvenile size classes. On reefs with high levels of grazers but frequent hurricane disturbances, populations were characterized by high numbers of juvenile and pubescent colonies while the adult part of the population experienced a bottleneck.

Improvement and Adjustment to Different Questions

This study illustrates that the overall outcome of the model could not have been predicted by simply examining the parametrisation due to the intricate nature of biotic and abiotic interactions across spatial and temporal scales. The results of this model received strong support by proceeding studies using modified versions of the same simulation model. The importance of sea urchins for the ecological balance between corals and macroalgae on Caribbean reefs was confirmed by Mumby et al. (2007) who investigated the susceptibility to and persistence of macroalgal dominance on Caribbean reefs. Mumby and Hastings (2008) extended the model by including vicinity to mangroves as an additional factor and two different depths as additional parameters in order to assess the relative importance of mangroves (they function as nursery grounds for Scarid fish) on the abundance of parrotfish on adjacent reefs. In another recent study Mumby (2009) uses the same model to assess the stability of alternative stable states of Caribbean reefs. Here, the parameters pertaining to external disturbances were omitted and instead the model concentrates on the inherent parameters of coral community dynamics. In contrast to the model discussed here, growth rates, sizes at maturity, overgrowth and mortality were now not set as fixed values but determined probabilistically, which allows for a higher degree of natural behaviour of the model due to a broader range of natural variability. The reef community dynamics were investigated with regard to different levels of grazing intensity ranging from 5 to 40%. Simulations were run in 6 month intervals for a total of 36 single reefs. With this approach, Mumby was able to identify clear threshold levels of grazing intensity (i.e. the level of grazing necessary to prevent the shift to an alternative state) for different sets of initial coral cover and different levels of grazing.

Spatially explicit modelling approaches to understand the impact of grazers on coral reefs have only emerged over the past 5 years. The models by Mumby offer a novel approach to overcome the problem of reef complexity, which to date has complicated experimental studies of the interactions of multiple disturbances. However, a word of warning should be issued regarding the organisation in a lattice. The complexity of the model may lead to complications in the definition of clear rules for interactions between individual organisms which are not located in the same cell. For example, if a coral outgrows its cell, the part that protrudes into the adjacent cell becomes an integral component of that cell and thus 'fragmented' from its original colony. It starts to function as a new and smaller entity with rules and trajectories of a juvenile coral colony, since age is determined via size. The fact that most processes are determined via a von-Neumann-Neighbourhood might compensate for the loss in detail to some extent, however, depending on the questions asked, there may still be a risk to ignore certain important processes.

17.6 Summary and Conclusions

The techniques currently available in ecological modelling all bear certain limitations and the choice of a given approach depends on the question of interest.

The exclusive utilization of differential equations (Sect. 17.2) can provide interesting results for different fishing scenarios in a relatively complex simulation environment, that may help to improve fisheries practices. Often these techniques lack resolution, however, as all components are aggregated into functional (larger) groups, which substantially obscure the inherent natural variability among relevant components. This can be compensated through the employment of a cellular automaton as illustrated by Langmead and Sheppard (2004). It allows for spatially explicit analyses through disaggregating populations into single interacting coral polyps. On the downside, because the rules of this CA model do not address cell aggregations (i.e. the whole coral colony) they cannot change in relation to individual colony attributes.

The grid based approach by Mumby (2006) allows to integrate a suite of important components of a coral reef system, which makes it possible to describe the complex characteristic processes for coral reef communities. This application constitutes a novel approach to the analyses of resilience and phase shifts of coral reefs. It builds on the concept of a cellular automaton by implementing distinct procedures within one cell and allows for dynamic changes of certain rules, e.g. larger colonies are less likely to be overgrown by algae than smaller ones. The combination of different modelling techniques does not only improve model performance but also helps to identify some of the deficits in our current research and may reveal how future experiments could be adjusted in order to fill the gaps. Yet, the structure with distinct spatial entities – the cell – limits this approach in its flexibility. The formulation of rules for several cells or across cells becomes very complicated and could be easier accomplished by utilizing a continuous area.

In contrast, individual based modelling (IBM) is free from such limitations because the model area does not have to comprise spatial aggregation of the acting units. Yniguez et al. (2008) give a good example for an applied IBM. In their model the environment is organized as a grid which holds different states for environmental variables. Interactions either between algal modules and/or algal modules with their environment are possible in all directions with dynamic changes of rules in relation to the component's attributes. The utilisation of IBM offers several useful tools to study resilience as object-oriented programming (Chap. 4) provides the possibility for a detailed description of organisms (as objects) in separate subprograms (see also Chap. 12). This constitutes a very fine-tuned approach to model detailed interactions on small scales. In addition, an IBM allows to integrate all earlier developed modelling techniques, like equation based sub-models or CAs, wherever intended or needed to create an application with highly dynamic performance and realistic behaviour.

Understanding the factors supporting resilience and the characteristics of phase shifts is imperative if we want to understand current and future coral reef dynamics. Both resilience and phase shifts comprise highly complex processes that are not yet fully understood. Over the past few years modelling has become a prominent tool to tackle ecological questions in coral reef science. Modelling has not only contributed a great deal to advance our understanding of potential driving forces pertaining to reef resilience, but also helped to identify the current scientific gaps and research deficits in this discipline. Future modelling approaches that merge past and present information derived from previous models, with data of specific sites will substantially enhance our abilities to identify local driving forces of reef dynamics. This may be employed in management programs that can help to improve the sustainable utilisation of resources.

Chapter 18
Trophic Cascades and Food Web Stability in Fish Communities of the Everglades

Fred Jopp, Donald L. DeAngelis, and Joel C. Trexler

Abstract We introduce the trophic organisation structure of aquatic ecosystems by giving a short overview on some classic landmarks from ecological theory. The concept of trophic cascades describes interactions in food webs that descend the whole structure. They start at the top node of the highest carnivores, the piscivores, by increasing the piscivore's biomass which in turn triggers changes in the successive trophic hierarchical levels. The concept of trophic cascades has long since passed from theoretical into applied ecology. We demonstrate this with an example of a spatially-explicit simulation model that is used to understand the high variability in the aquatic trophic structure of the Everglades marshland. Changes in hydrology of the Everglades over the last several decades have reduced the hydroperiod in some areas and may have diminished foraging fishes and their food base. A key component for restoring fish productivity to historic levels is to understand and to improve the spatio-temporal water patterns in the wetlands. Therefore, by applying the simulation model we investigated the dynamics of an aquatic food web with the following components: primary producers, detritus, invertebrates, fish consumers and nutrients. For this purpose, a hydroscape of 20×20 km was modeled that shows a natural-like elevation gradient. The annual fluctuations in water level were imposed as sinusoidally changing hydrology on the whole system, which resulted in dynamic patterns of flooded and non-flooded areas. We performed long-term simulations over a period of 10 years and examined how the trophic levels reacted to changes in the water level; in particular, how the changing water levels affected trophic cascades. We discuss the consequences of these results for management and restoration of the Everglades aquatic communities.

F. Jopp (✉)
Department of Biology, University of Miami, P.O. Box 249118, Coral Gables, FL 33124, USA
e-mail: fredjopp@bio.miami.edu

F. Jopp et al. (eds.), *Modelling Complex Ecological Dynamics*,
DOI 10.1007/978-3-642-05029-9_18,

18.1 Trophic Organization of Aquatic Ecosystems

From the point of view of functional organization, all ecosystems are characterized by fluxes in matter and energy. They have structural components, trophic levels, which enable and control the transport of these fluxes through the ecosystem. When we speak of a trophic level we mean all biota in a food chain or food web of an ecosystem that are the same number of links away from the ultimate source of energy, which is usually solar radiation. In the lowest trophic level (level 1) are the primary producers, which are normally green autotroph plants, and which support a chain or web of consumers. Depending on their feeding behaviour, the consumers are separated into the group of herbivores (trophic level 2), which feed on the producers, first-order carnivores (trophic level 3), which feed on herbivores, and, often, second-order carnivores (trophic level 4) that feed on first-order carnivores. Whether there are still higher level carnivores is ecosystem-dependent. Decomposers, which enable the recycling of nutrients by feeding on the organic litter, are often classed into the bottom of the food web. In 1941 the ecologist Raymond Lindeman (1941) published a fundamental study on the trophic interrelations of the Cedar Bog Lake, Minnesota, USA, a late stage eutrophic lake. Due to the theoretical implications and the high number of successive studies that were motivated by the work of Lindeman, this study can be regarded as one of the most important studies for modern aquatic community ecology.

Most importantly, Lindeman's work focused on the flow of energy up the food chain from the autotrophs to the top carnivores, and thus on the dependence of each trophic level on the one below it. Biomass usually decreases with each higher trophic level, which is referred to as trophic pyramid. Lindeman explained this phenomenon as a consequence of respiration within each trophic level and of the high losses of energy during transfer from one trophic level to the next, due, in part, to The Second Law of Thermodynamics. Also, he assumed that control of these flows went primarily in one direction, from the autotrophs up to higher trophic levels. In the sense that any given trophic level depends on the conditions of the level below it, Lindeman expressed here the idea of bottom-up control. Figure 18.1 shows such a typical example of pure bottom-up control in a four-level food chain in an ecosystem.

Hairston, Smith, and Slobodkin (1960) reviewed a number of different food chain studies derived from different habitats. They concluded that interspecific competition exists within each of the trophic levels of the producers, the carnivores and the decomposers. They also inferred that the herbivore trophic level is rarely food-limited; instead it appears to be predator-limited. Hence, Hairston et al. introduced the idea that in food chains the controlling effects can also go 'down' the food chain, which is a mechanism that is termed top-down control. These two concepts of bottom-up and top-down control in trophic structures of ecosystems have been and are still widely investigated and seem to be of extraordinary importance for aquatic systems, in particular (DeAngelis et al. 1996). The concepts have been extended in such a way that ecologists encompass not only the impacts of lower on higher

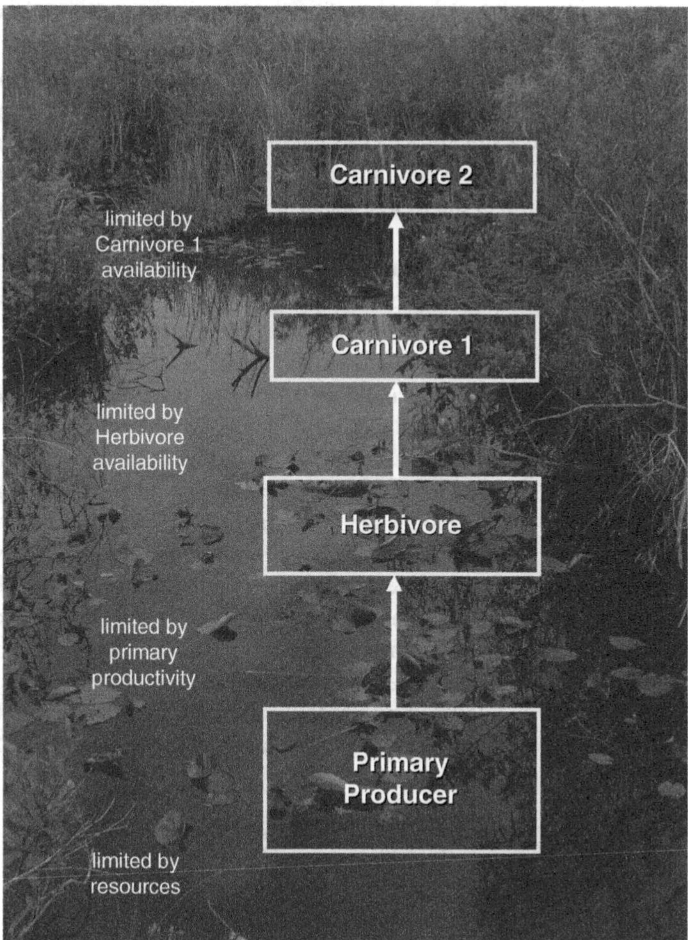

Fig. 18.1 Pure bottom-up control in a four level food chain. Due to the limitations, for every level strong competition effects apply

hierarchical levels with the concept of bottom-up control, but also the impacts of nutrients, physical factors (e.g. temperature) and chemical factors (e.g. water pH, dissolved oxygen). Other major physio-chemical factors that affect lake productivity are light, water-turnover time and vertical mixing. However, it has been found in recent years that even when all of these physical factors are taken into account, they can only explain approximately half of the observed variability in primary productivity and the productivities of higher trophic levels. There is a lot of variation in the production of comparable trophic levels among lakes that have the same phosphorus input or nutrient loading. Carpenter et al. (1986) have offered the hypothesis that much of this variation can be explained by "cascading trophic interactions". These are interactions that cascade down food chains, starting from the highest carnivores, the piscivores. An increase in piscivore biomass causes a decrease in planktivore biomass, which leads to an increase in herbivore biomass (as well as allowing the herbivore community to shift towards larger zooplankton, which are preferential prey of planktivorous fish), and thus to a decrease in phytoplankton biomass. Depending on certain ecological conditions and for certain times of the year the length of the food chains can vary which is paradigmatically depicted in Fig. 18.2.

The ability of predatory fish to control prey populations is well-documented. This can cause suppression of the forage species, which affects species composition and size structure of the zooplankton community, and in turn influences the phytoplankton community. Important earlier work (e.g. Brooks and Dodson 1965) examined the impact of fish predation on zooplankton size structure, showing that the planktivorous fish eliminated the large zooplankton (daphnids), which allowed algae to reach high concentrations. The trophic cascades typically studied by limnologists are similar, but usually extend from piscivorous fish down to phytoplankton. To explain the concept in more detail, consider a simple food web for a typical small lake in northern U.S., as shown in Fig. 18.3. You will note that this web is not quite chainlike, so the trophic levels are not perfectly distinguished. There is much experimental evidence that consumers in this web can control prey. Changes in large piscivorous fish densities (bass, pike, or salmonids) have been shown to cause changes in vertebrate planktivore populations (e.g. bluegills, sunfish, yellow perch). Large numbers of vertebrate planktivores tend to reduce large crustaceous zooplantkon and invertebrate planktivores. In the absence of vertebrate planktivores, invertebrate planktivores deplete small crustaceous zooplankton and rotifers. The effects of zooplankton on phytoplankton are more complex, because zooplankton help recycle nutrients (good for phytoplankton) as well as feed on

Fig. 18.2 Conceptual Model of Trophic Cascades. *P* Producer, *H* Herbivore, *C* Carnivore. The possible impact between the levels is depicted by the signs

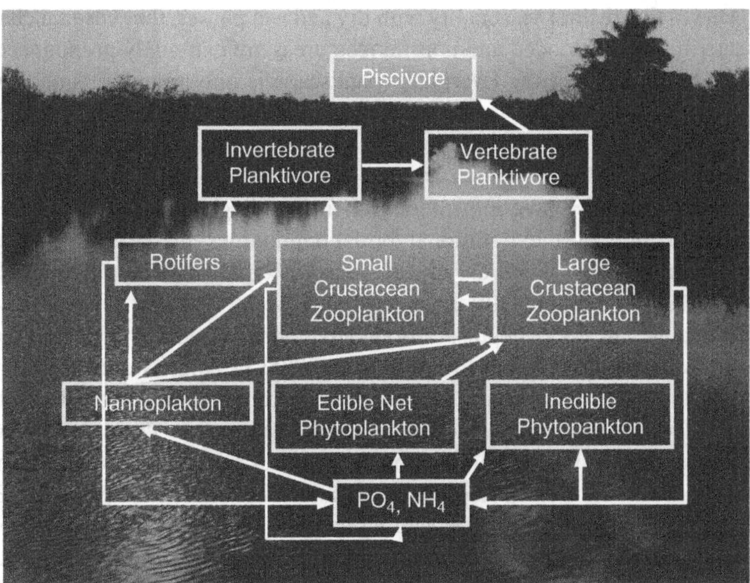

Fig. 18.3 Conceptual model of trophic structure in a typical lake, as described by Carpenter et al. (1986)

phytoplankton. Furthermore, the size structure of the zooplankton makes a difference in what will happen. So we see that the effects of planktivores on zooplankton are complex. A particular planktivore (vertebrate or invertebrate) may not significantly affect the size distribution within the zooplankton (herbivore) trophic level; that is changing the ratio of large to small zooplankton. This can still have a downward effect. Large zooplankton (e.g. *Daphnia*) tend to have a much stronger negative effect on phytoplankton than smaller zooplankton, so cascading effects may still occur. The effect of zooplankton on phytoplankton is also somewhat complex, because (1) some phytoplankton are inedible to zooplankton, and (2) zooplankton may sometimes have a beneficial effect on some phytoplankton by increasing the speed of nutrient recycling into the water column.

In the following, we adapt these theoretical concepts on trophic structuring and cascades to the situation of the Floridean marshlands of the Everglades in the southern U.S and apply them to analyse the implications for the stability of its ecosystems.

18.2 The Aquatic Food Web of the Everglades

The Everglades ecosystem is a subtropical heterogeneous marshland, which offers diverse habitats for a wide variety of species (see Chap. 21), including large rookeries of wading birds. The underlying food web of small fishes and invertebrates are the energy base for much of the biodiversity of the higher trophic levels. Due to the distinct seasonality with dry and wet phases, the seasonal changes in the distribution of flooded and non-flooded areas are extremely pronounced. As the elevation gradient of the Everglades landscape is only minimal, small differences in mean water levels can alter the fraction of flooded habitat drastically. Surviving organisms must be able to cope with these altering hydrological conditions (Trexler et al. 2002), which also determine the available foraging area for wading birds and thus, have influence on their breeding cycles. In this context, a substantial decline of the traditional bird communities in the Everglades has occurred over the past several decades (Ogden 1994). There is empirical evidence that in wetlands large planktivorous and piscivorous fish that are sensitive to seasonal changes in water depths, periodically move in and out of local areas in which the water depth changes (Trexler et al. 2005; DeAngelis et al. 2007). This affects the temporal pattern on which trophic cascades influence the Everglades food web (Dorn et al. 2006; Chick et al. 2008). Within the cycle of annual re-flooding of extensive wetland areas, trophic cascades caused by invading fish can lead to significant changes in the whole aquatic food web structure of the Everglades (DeAngelis et al. 2010). Depending on individual traits of the fish species, they can disperse and exploit different habitat types; e.g. opportunistic fish species can disperse into and exploit re-flooded areas first, while gleaner species, which are good at exploiting resources at low levels, are more successful in dominating permanently flooded wetland areas. The specific combination of heterogeneity in elevation and fluctuations of water level can lead to a community of multiple coexisting species feeding on the same resource (DeAngelis et al. 1998; Jopp et al. 2010).

We now investigate the dynamics of such a food web structure with an annual standard water level fluctuation of 0.6 m amplitude. To do this we introduce a spatially-explicit model framework (for detailed model descriptions, see: DeAngelis et al. 2010; Jopp et al. 2010), that includes the main physical factors, seasonal water level fluctuations and a linearly increasing topographic elevation. The model consists of a simple aquatic food web, as well as rules of movement for certain species populations in the simulated spatial environment. The food web consists of six groups: a functional group of invertebrates, three separate small fish species, which differ in their traits (F1 = an intermediate fish, F2 = a good disperser, F3 = a good exploiter), crayfish, and a piscivorous fish species, which is the top predator. There is also a periphyton functional group, which is a mixture of algae and microbes that serves as a food source for fish and crayfish. Non-living pools exist for detritus and nutrients. For each of these components, there is a set of specific differential equations (see Chap. 6: Ordinary Differential Equations), which describes the interactions between the components. The functional responses of the small fish and crayfish are not Holling type 2, but rather Beddington–DeAngelis (Beddington 1975; DeAngelis et al. 1975). This assumes that the fish are somewhat territorial and thus self-limiting, and helps to stabilize the dynamics of the model. The functional response of the top predator follows a Holling type 2 function. The whole model is spatially explicit, with the food web dynamics occurring on a grid of 100×100 cells. Each spatial cell is assumed to be 200×200 m, which results in an overall simulated area of 20 × 20 km. This two-dimensional topography increases linearly from zero to 2 m along the full distance of 20 km in x-direction, which simulates the slow increasing field elevation gradient. The model also takes into account the seasonal cycle of flooding and drying, which is simulated by a sinusoidal function, producing variability of the water level over the simulation time. For each cell, the potential food web structure is the same, although living biota are assumed present in a cell only when it is flooded, and the fish are present only when water is of sufficient depth. During the execution of the model, the cells show local heterogeneity due to the interaction of different elevation, water stands and biomass levels.

There is seasonal net movement of the fish and crayfish: (a) movement out of cells that are drying, and (b) movement into cells that are becoming flooded. During the period of rising water, a fraction of the population in a given cell is allowed to move up the gradient to an adjacent newly flooded cell, while during falling water some fraction of a population is able to escape being stranded by moving to cells that are still flooded. The fish are also allowed to diffuse among flooded cells and exploit the different habitat resources. To examine how the model components react to annual changes in the water level, long-term simulations over a period of 10 years are performed. In particular, the model predicts biomasses of fish across the heterogeneous landscape over time.

Due to page limitations, full equations and parameters of the model are not listed here, but are documented in Jopp et al. (2010), DeAngelis et al. (2010), and can be found as supplementary material on the MCED webpage (www.MCED-ecology.org).

We use this modelling framework to meet the following aims:

- Describe the resulting temporal pattern of the aquatic food web under the above mentioned conditions
- Investigate how important trophic cascades vary temporally in the model and what this may imply for the southern Florida fish community.

18.3 Model and Simulation Results

We now examine the temporal dynamics of the food web subject to this baseline hydrological regime, with a water level fluctuation of 0.6 m in amplitude. Figure 18.5 shows six snapshots in time through the annual hydrologic cycle along the elevation profile throughout a year for the six higher trophic levels): invertebrates, fish species 1, 2 and 3, piscivorous fish and crayfish. To achieve stable patterns, long-term simulations over 10 years were performed, and the last year, year 10, is displayed and analysed here.

In all sub-figures elevation increases from the left to the right, by about 2 m over a 20-km distance. The first three panels (Fig. 18.5a–c) refer to the dry season, when the water levels are decreasing. The most conspicuous features in these panels are the pulses of fish and crayfish that are retreating towards the flooded lower elevations (towards the left) as water levels decline. Fish 2 (good disperser) moves ahead of Fish 1 (intermediate traits), Fish 3 (good exploiter), and the crayfish. To the right, where the elevation is higher and fish density lower, the invertebrates show a peak, as they are now free from fish predation. In addition, a remnant population of crayfish also survives, as they face less competition from the fish.

As water levels continue to decline, the piscivorous fish will move to lower elevation to the left where the cells are permanently flooded. As the water levels begin to rise again (Fig. 18.5d), some fish begin to follow the water movement, resulting in small population peaks, with the crayfish leading, followed by Fish 3. Subsequently, all species follow the water movement to gradually occupy most of the cells (Fig. 18.5e), except for the last 12 or so cells at the right most (Fig. 18.5f), which are permanently dry. Piscivores occupy some of the cells in the higher elevations when permitted by adequate water depth, but in low density, as they do not have time to build up high biomass. In the simulated food web, there are three distinct food chains observed (see Fig. 18.4): periphyton-fish/crayfish-piscivore, detritus-invertebrate-fish-piscivore, and detritus-crayfish-piscivore. Because of these overlapping food chains, trophic cascades are not clearly discerned. However, some top-down effects are observed in the simulations as seen in Fig. 18.5d, where high density of piscivores is associated with reduced fish and crayfish density, and high density of invertebrates. In the vicinity

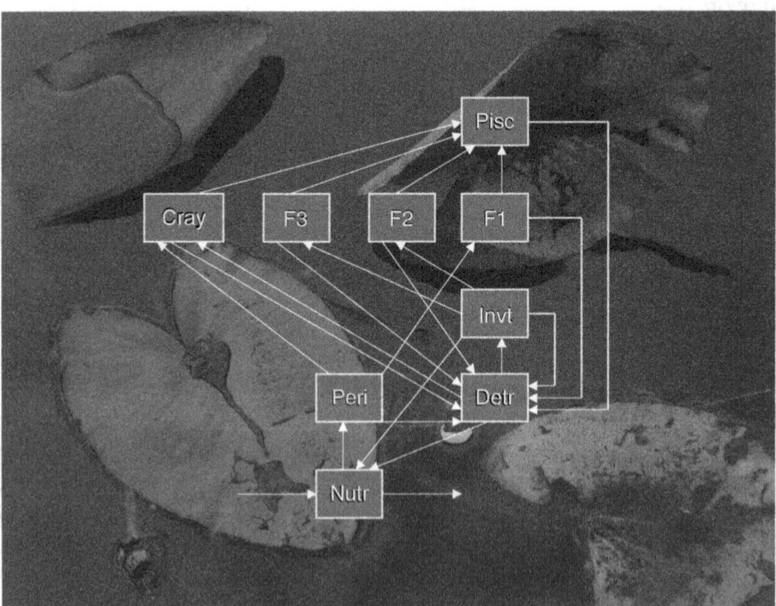

Fig. 18.4 Trophic structure of the basic food web structure of the Southern Florida fish community model. *Arrows* indicate predator/prey relationships. *Pisc* Piscivorous Fish, *Cray* Crayfish, *F1*, *F2*, *F3* Small fish species, *Invt* Invertebrates, *Peri* Periphyton, *Detr* Detritus, *Nutr* Nutrients

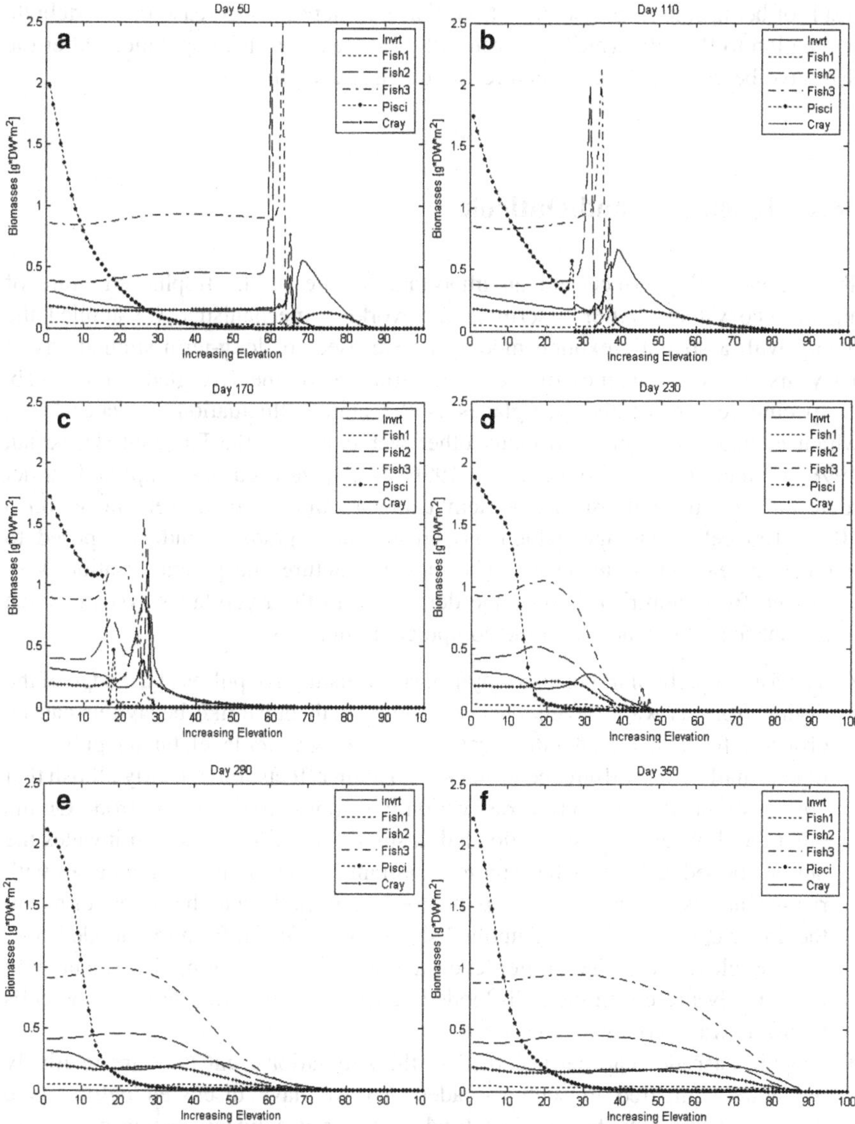

Fig. 18.5 Simulation output of the mean biomass of the food web along an elevation gradient during seasonal variations in water level in the last year (year 10). Starting from day 50 (**a**) each 60 days a 'snapshot' was made until the end of the simulation (**f**). The water level amplitude during this simulation run was 0.6 m

of the wet-dry water margin, these cascading effects become more complex and may be difficult to distinguish from the pulses of invading or retreating populations. In Fig. 18.5a, for example, the invertebrates show a double peak around the wet-dry water margin. The invertebrate's peak observed on the right side is the result of being free from fish predation. Further, there is a decrease in invertebrate population to the left, which is a result of an increase in fish populations after the fish have been freed from piscivore top-down predation.

18.4 Discussion and Outlook

Trophic cascading effects are an important feature of the trophic structure of aquatic ecosystems. After describing the working mechanisms, we applied the theory with a spatially-explicit model that we used for long-term simulations of 10 years for the dynamics of the basic structure of the Everglades food web. A specific feature of the Everglades is the annual fluctuation in water level, which is also fundamental to many other wetlands; e.g. the Pantanal (Heckman 1998), Doñana (Serrano and Serrano 1996). We have used our simplified model representation to study such a system using a minimal food web model on a 100 × 100 cell landscape, which has an elevation gradient and is exposed to fluctuating seasonal water levels. The model structure and parametrisation were suggested from empirical knowledge derived from the Everglades system. Some important features stand out in the computer simulations.

- The fish, in particular, but also crayfish, show distinctive pulses at the edge of the drying front as water levels decline. Some small pulse-like behaviour can be observed following the flooding front during rising water level, but the pulses are much smaller. The difference is due to two main effects. First, nearly all fish that are not stranded move to lower elevation in response to the drying front. On the other hand, when cells are re-flooded, a smaller fraction of the fish invades the newly flooded cells. A perhaps more important reason for the weaker pulses with rising water is that fish moving up the flooding front deplete the supply of fish in the donor cells, which are resupplied slowly by diffusion from the flooded cells at lower elevations. When water levels are falling, the retreating fish pile up with fish already present in the still-flooded cells, which in turn pile up in the cells below as water continues to drop.
- Trophic cascades can be discerned in the simulations, but they are relatively moderate compared to such cascades that we have discussed before. One reason for this may be that our food web is not a linear trophic chain, but contains omnivory as well; that is, piscivorous fish consume all of the fish species, as well as crayfish. Although omnivory is known to work against the strength of trophic cascades (Strong 1999; Nystroem et al. 1996), lake systems contain omnivory as well, so this does not completely explain our results. Another reason, however, is that diffusion is an important mechanism in our model. Smaller fish can move in and out of the zone in which there is a high concentration of piscivores, which may tend to smooth out the top-down effects, which is supported by other studies (Howeth and Leibold 2008; Holyoak et al. 2005; Leibold et al. 2004).

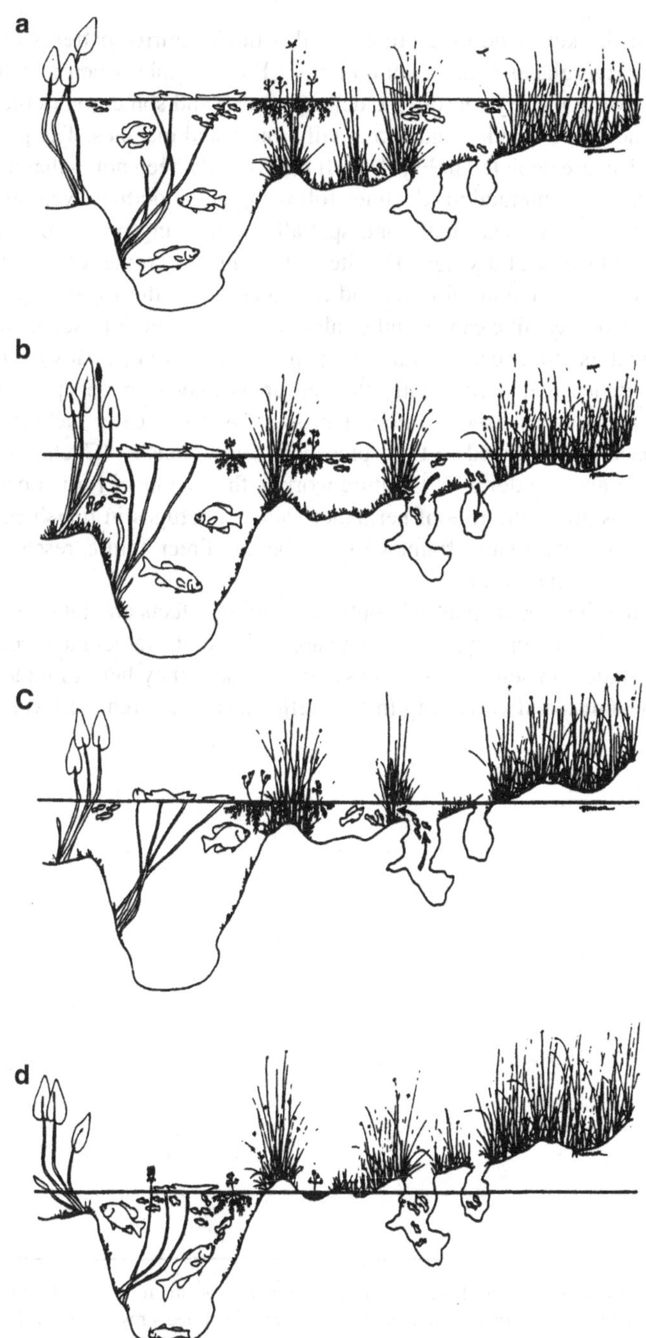

Fig. 18.6 Typical cycle of water levels in an area of Everglades freshwater marsh: (**a**) Water levels are high and both small and large fish are present in the flooded marshes. Trophic interactions

Some of the key patterns captured by this model mirror patterns observed in field data on Everglades aquatic communities. For example, when marshes are re-flooded following the dry season, the density of fish and some invertebrates is low in both the newly re-flooded and continually inundated marshes. The principles of this mechanism are depicted in Fig. 18.6. It has recently been noted that the strength of top-down biotic interactions declines following a re-flooding event and that the interaction web varies seasonally and spatially, depending on the position in the landscape and history of drying at the site and around it (Trexler et al. 2005; Liston 2006). Another major issue for wetland management is the issue of permanently flooded water bodies, like canals and swales. These were built to serve engineering purposes; that is, to expedite water movement and storage. However, there is growing evidence that permanently flooded areas may provide opportunities for introduced piscivorous fishes to persist in systems historically lacking such predators or become stabilized to their presence at low density (Cucherousset et al. 2007; Rehage and Trexler 2006). Future work with this model format can improve our hypotheses about the role of permanent aquatic refuges in marsh ecosystems, particularly in a landscape framework, to better direct future research on this important management topic.

The fascinating mechanisms of trophic cascading effects are not only an enrichment to ecological theory, but can also be applied to management and conservation issues of aquatic and semi-aquatic ecosystems, where they help to understand the complex dynamics and variability in the participating hierarchical levels.

Fig. 18.6 (continued) between these functional groups can occur in the marshes. (**b**) As water levels recede, large fish and then small fish move into refugia of deeper water. (**c–d**) If water levels reach low levels, deep ponds and solution holes may be the only refugia remaining. Intense predation by large fish on small fish may take place in the ponds. When the water levels rise again, the remaining fish can explore the hydroscape and the cycle continues [modified from DeAngelis et al. (1997), illustration by M. Trexler]

Chapter 19
Lake Glumsø: Case Study on Modelling a Small Danish Lake

Søren Nors Nielsen and Sven Erik Jørgensen

Abstract The case study on the small Danish post-glacial Lake Glumsø shows some of the highlights of aquatic ecosystem research over the time span of more than 30 years. The research at Lake Glumsø is closely linked to the development of ecological modelling as a discipline, inspired a wide range of related working approaches as well as advancements in ecological modelling and the development in ecological theory.

In the early working phase, models based on differential equations were used not only to describe the lake processes but also to methodologically identify the optimal formalisation of the described processes. First, modelling applications were continuously improved, thus providing a large body of functional representations for the different compartments. A result of the intense discussions of the early model structures was a new approach called structural dynamic modeling which allowed for varying numbers of variables and a model structure that could change during simulation time. Continuing the search for optimal trade-off between model complexity on one side and efficiency and precision on the other side, machine learning techniques were applied to find out, whether this could help to identify the most reasonable model structures which were tested with Lake Glumsø data and many other lakes world wide. In further contributions to ecological theory, the question was tackled whether overall ecosystem characteristics, like exergy, biomass production or species composition, do converge in a successional process towards optimal values. This led to the development of ecosystem goal functions, which was part of an international ecosystem research debate on the theoretical patterning and the dynamics of ecosystems.

Collaborating research groups on ecosystem research benefited very much from the experiences made and the knowledge gained from the Lake Glumsø studies.

S.N. Nielsen (✉)
ECO-Soft, Kålagervej 16, DK-2300 Copenhagen, Denmark
e-mail: soerennorsnielsen@gmail.com

F. Jopp et al. (eds.), *Modelling Complex Ecological Dynamics*,
DOI 10.1007/978-3-642-05029-9_19,

19.1 Introduction: How a Small Danish Lake Helped to Advance the Development of Ecological Modelling

As new scientific options and approaches emerge, they are frequently tested and applied in a reference case. One such case with a high impact on ecological modelling was the development of the Lake Glumsø model. The lake is a small, shallow, post-glacial lake in Denmark situated on the Danish island of Seeland, 78 km away from the capital, Copenhagen with 266,000 m^2 surface area and 2.4 m maximum depth. This chapter will outline the importance of this lake to the advancement of ecological modelling over a time span of more than 30 years. It also presents an interesting illustration on how the results from a prototypic case may inspire a wide range of other work and developments in application as well as in establishing new techniques.

19.1.1 Why Studying Lake Glumsø?

There is nothing special about this lake. In Denmark, Sweden, Germany, Poland and various other countries throughout the Northern temperate climate zone there are similar post-glacial lakes. Therefore, results obtained for one of them can be helpful to understand the situation of many others: it is a prototypic case. Lake Glumsø, was selected as a case study because: (1) the hydrology was simple, (2) the municipality had planned and decided to invest in waste water treatment to clean up the lake, (3) the lake was highly eutrophic, which made it reasonable to expect a significant improvement following the intended measure, (4) the retention time was only 5 months allowing a fast response of the lake to validate model prognoses. During development of the model for Lake Glumsø it was possible to provide a robust model structure that was widely applicable to simulate dynamic processes requiring minor or moderate adaptation to be used for other lakes, estuaries, reservoirs and near coastal waters. The modifications required the parametrisation of involved processes referring to, e.g., size, depth, and nutrient load. As a prototype, the Glumsø model contained the relevant variables, functions and parameters in a clearly presented and well accessible form.

Starting in the 1970s, lake modelling had become important for several reasons. As a result of an increased use of chemicals in households and diversion of more or less untreated sewage into lakes, as well as an increased use of chemical fertiliser inputs in agriculture, eutrophication of lakes and rivers had become widespread and a major concern in environmental protection. As other works suggest (e.g. Scheffer et al. 1993), specific concepts were required to understand the dynamics of the involved processes, in particular to estimate potential outcomes of management measures to anticipate effort and the achievable results. Important pioneer work in these fields has been done using the Glumsø study site.

19.1.2 The Database

In measurement campaigns from 1972 to 1975 the following data for the state variables was acquired: total and soluble phosphorus, total nitrogen, nitrate, nitrite, ammonium, phytoplankton (biomass and chlorophyll a), zooplankton in the water column, total, interstitial, and exchangeable phosphorus, total interstitial, and exchangeable nitrogen in the sediment. The processes of primary production, and sedimentation were measured in the field. Exchange of nutrients between sediments and water column was determined in laboratory. For the forcing functions inflows of water (tributaries and waste water), outflow rate, as well as total and soluble phosphorus, total and soluble forms of nitrogen, phytoplankton, water temperature, light intensity, and precipitation were determined. Later, the database was expanded with series of frequent measurements, three times a week for the months of April and May. For further details on the available database see (Jørgensen et al. 1973, 1986; Jørgensen 1976)

19.2 Modelling and Simulation: The Basic Approach with Differential Equations

In the early 1970s, differential equation based modelling systems were the predominant approach. From the comparatively few simulation tools that existed and that could cope with complex dynamics, the Continuous Systems Modelling Program (CSMP) developed by IBM was chosen. The program is a pre-compiler and parser that serves as a FORTRAN interface. It was developed for engineers to solve first order time dependent differential equations. Likewise its ability to handle systems with big differences in time constants as well as the possibility of direct integration of FORTRAN programming lines were important in making this choice.

The following are the most important variables and processes that were included in the model:

State variables:

- Minerals: total and soluble phosphorus, total nitrogen, nitrate and nitrite, ammonium
- Minerals in the water column: nitrogen and phosphorus
- Minerals in the sediment: nitrogen and phosphorus total, interstitial, and exchangeable
- Phytoplankton: represented as biomass, carbon, nitrogen and phosphorus
- Zooplankton and fish represented as biomass

Processes are indicated in Fig. 19.1. For further information on model structure, processes and parameters see (Jørgensen 1976; Jørgensen et al. 1978).

In total, the implementation consisted of 19 state variables, and 55 parameters (i.e. constants), which already represented parts of the lake dynamics well (Fig. 19.2).

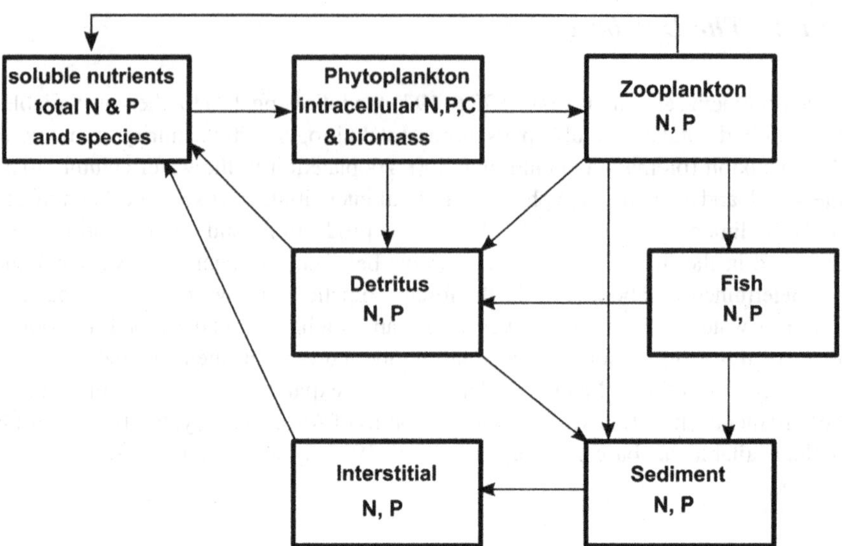

Fig. 19.1 Conceptual diagram of the Glumsø model at the final state of development, where the biogeochemical cycles are aggregated

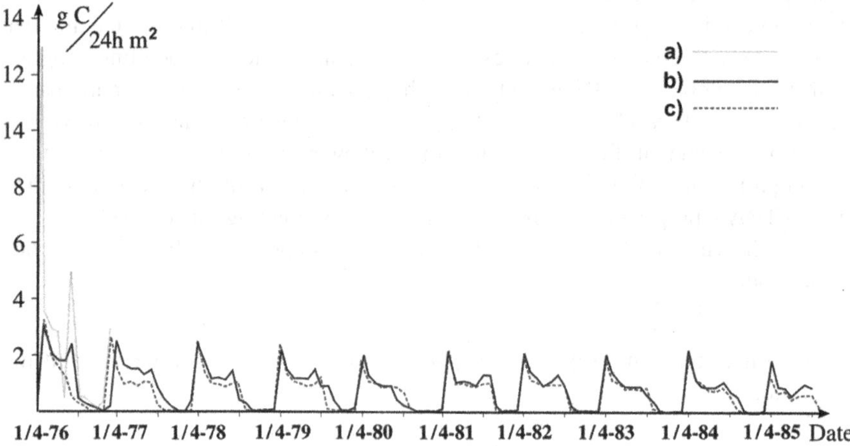

Fig. 19.2 Results from an early version of the model. Productivity in different scenarios [redrawn from Jørgenson (1976)]: (**a**) phosphorus is not removed from the waste water, (**b**) waste water with a discharge of 0.4 mg P/l, (**c**) waste water with a discharge of 0.1 mg P/l

During an early state of development, the model was already applied more widely than previous models as an experimental tool to identify an optimal formulation of processes. Five important observations were made during the development of the model:

- It was necessary to develop a more complex description of the sediment water exchange than applied initially (Jørgensen et al. 1982).
- It was necessary to apply a two-step production sub-model of the phytoplankton involving initial and subsequent uptake and a determination of actual growth by intracellular mineral nutrient concentrations.
- A simple Monod kinetics (in analogy to the form frequently used for enzymatic catalysis) was tested but failed in the verification and calibration phase.
- The presence of various functional types of zooplankton, herbivores, detrivores and omnivores was tested and found not to be important.
- It was necessary to apply a carrying capacity and threshold values for grazing of zooplankton.

While the initial model was able to simulate the correct level in each compartment, the temporal pattern was not sufficiently met. An additional, intensive measuring program was set up for a short period covering the spring bloom in order to calibrate the model involving identification of a proper maximum growth rate and temperature dependency of the phytoplankton. While temporal density of data input can be important, it is not recommended to use a temporal resolution that is too low, as it increases the noise of predictions. In a couple of studies the model was found to perform relatively well even on other lakes, such as Lake Balaton and also to perform well when compared with other models (Costanza and Sklar 1985).

19.3 Further Elaboration of Modelling Techniques Departing from the Lake Glumsø Model

The Lake Glumsø model served as a blueprint for other lake models. Several model descriptions were inspired or advanced by pioneering work done at this lake (see Jørgensen and Bendoricchio 2001). A list of other case studies based on the Lake Glumsø model is shown in Table 19.1.

The many applications that have been possible illustrate that the framework is robust. However, for each case it was necessary to elaborate which additional state variables and processes were of importance and therefore needed to be incorporated in order to simulate the system under consideration.

In addition to its function as a basis for advancement in lake modelling, the case also served as an example to develop new modelling techniques and improve modelling approaches. Some of the developments are briefly described in the following.

19.3.1 Discussing the Model Structure

The observed applicability of the model for a broad number of cases sparked a considerable discussion on what aspects of the lake dynamics were the most

Table 19.1 Case studies using the Glumsø model as platform for development (type of system, system name, level of model development, modifications needed; modified from Jørgensen and Bendoricchio 2001)

Lakes, shallow	System	Modifications made	Level
One layer	Glumsø version A	Basic version	7
	Glumsø, version B	Non-exchangeable nitrogen	7
	Lake Gyrstinge	Level fluctuations, sediments exposed to air	4–5
	Lake Lyngby	Basic version	6
	Lake Bergunda	Nitrogen fixation	2
	Broia Reservoir	Macrophytes, two boxes	2
	Lake Great Kattinge	Re-suspension	5
	Lake Svogerslev	Re-suspension	5
	Lake Bue	Re-suspension	5
	Lake Kornerup	Re-suspension	5
	Lake Søbygård	Structural dynamic model	6
	Lake Balaton	Adsorption to suspended matter	2
	Lake Annone	Structural dynamic model	6
	Lake Balaton	Structural dynamic model	6
	Lake Mogan	Only P-cycle, structural dynamics, competition between submerged vegetation and phytoplankton	6
	Stadsgraven, Copenhagen	4–6 interconnected basins	5/6
	Copenhagen, inner lakes	5–6 basins	5
Lakes, deep more layers/boxes	Lake Victoria	Boxes, thermocline, other food chain	4
	Lake Kyoga	Other food chain	4
	Lake Mobuto Sese Seko	Boxes, thermocline, other food chain	4
	Lake Fure	Boxes, nitrogen fixation thermocline	7
	Lake Esrom	Boxes, Si-cycle, thermocline	4
Estuaries			
	Ringkøbing Firth	Boxes, nitrogen fixation	5
	Roskilde Fjord	Complex hydrodynamics	4
	Lagoon of Venice	Ulva/Zostera competition	6

Level indication: 1 – conceptual diagram selected, 2 – verification has been carried out, 3 – calibration using intensive measurements, 4 – calibration of the entire model, 5 – validation: object function and regression coefficient has been found, 6 – validation of a prognosis for significant changed loading has been carried out **or** the development of structural dynamic model (SDM), 7 – validation of a prognosis **and** development of SDM

important ones and which ones were negligible. Since eutrophication was in the focus, nutrient dynamics and the dominant animal, plant and microbial species involved in the turnover of mineral nutrients and primary production were most relevant. To what extent the inclusion of particular species was necessary or whether an aggregated representation (as a black box) was sufficient, was largely discussed (e.g. Jørgensen et al. 1978). It was already at an early state discussed for how long the "final" structure would be able to simulate the system in a quantitatively reliable manner. Retrospectively, it turned out that with the improvement in the condition of the lake, i.e. the lowering of the nutrient load through the establishment of sewage treatment, the species composition considerably changed. This also influenced the characteristics of the growth dynamics and the nutrient budgets, and thus the overall ecology (Scheffer et al. 1993). The lake responded differently to a comparable impact when the *biocoenosis* consisted of different organisms.

19.3.2 Structural Dynamic Modelling

A further approach where the lake model helped to advance ecological theory was *structural dynamic modelling*. Only a short time after measures had been undertaken, Lake Glumsø showed the first signs of a shift in the species composition of the phytoplankton and zooplankton community. In the first period this appeared not to be important to the quantitative predictions of the model and the validation of the prognosis was performed in an acceptable manner (Jørgensen et al. 1986). This underlines the achievement of the approach as it was the first lake model where it was possible to validate a prognosis. However, the subsequent shift was not captured well by the initial model. This refers to the fact that conventional differential equations have a fixed model structure which describe only the quantitative change of particular variables. What if the number of variables and/or the structure of the equations changes during the simulated time interval? This was investigated in context of structural dynamic modelling. In fact, this approach was useful, since the dominant species had changed during the lake restoration that reduced the nutrient load combating eutrophication. The overall approach of structural dynamic modelling was described in (Jørgensen 1986, 1992). Specific results were also presented in Jørgensen (1995) and Nielsen (1995).

19.3.3 Goal Functions

As an important issue in ecological theory, it was discussed whether certain characteristic features develop towards specific (optimal?) values on the level of the entire ecosystem. Are there coherent trends resulting from the interaction of the components, in form of a specific ecosystem state, that was optimized during succession? At a very early state the exergy function derived from a thermodynamic analysis of the ecosystem (Jørgensen and Mejer 1979, 1981, 1983; Mejer and Jørgensen 1979) was considered as a possible candidate to govern the evolution of ecosystems with time. The function was found to be closely correlated to the buffer capacity of the system (Jørgensen 1982) and it was tested as a goal function for Lake Glumsø by Jørgensen (1986) and later Salomonsen and Jensen (1996). Likewise, its importance was analysed for other similar Danish lakes (Jørgensen 1995; Nielsen 1995, 1997) where the role in selection of species composition of the zooplankton, phytoplankton and macrophyte community, respectively, were analysed. A large variety of approaches were discussed in this context, see Müller and Leupelt (1997) for an overview of goal functions. From different perspectives, several features could be identified, which were possible to be brought together in a theoretic ecosystem pattern (Jørgensen 2002). Furthermore, biomass has been applied in two cases (Straskraba 1979; Radtke and Straskraba 1980), exergy in 21 cases (Zhang et al. 2010). To test and exemplify some of the approaches, the Lake Glumsø model was employed. The overall results supported the hypothesis of the application of maximum exergy as a useful goal function as an emergent property of ecosystem dynamics (Jørgensen 2002).

19.3.4 Equation Discovery

In the early phases of development of the Glumsø model a great deal of work was put into the search for appropriate equations. During this period the search was carried out largely because of the limited powers of computers at that time. In order to save computating time, the most simple equation was looked for that was able to simulate observations within acceptable accuracy. A more complex equation would not only be more costly in computing time but would also likely involve more parameters and eventually introduce a higher uncertainty in the model (cf. Costanza and Sklar 1985). In some of the versions of the Glumsø model up to 7–8 different expressions of various temperature dependencies were tested on process equations for their efficiency, i.e. contribution in improving the precision of model predictions.

But does a model need to be entirely the work of a modeller – or could a computer programme also be used in model development? At least computer algorithms, *machine learning* techniques (see Chap. 14), can help in model construction if it is to identify the most reasonable structure. If a large set of equations exists that lead to nearly comparable results, a computer-based testing of alternatives can be helpful. The approach is to some extent comparable with parameter identification (see Chap. 23), where the value of a parameter is changed as long as it fits an optimization criterion. Here, the number of equations and their algebraic structure is varied and then the simulation results compared with the data. This is what the work of Atanasova et al. (2006, 2008), Todorovski and Dzeroski (2006), Todorovski et al. (1998) and Vladusic et al. (2006) describe using the Lake Glumsø data as an example.

19.4 Overall Contribution

Over the decades, the focus of interest in ecological modelling has successively changed – this went in line with alterations in the environmental situation. In the late 1970s, the environmental impact of excessive nutrient input leading to eutrophication was successively reduced. The installation of sewage treatment units mitigated detrimental effects. Often, as in the case of Lake Glumsø, a simple diversion of pollutants were chosen as it was less expensive and it was still believed that: "dilution was the solution to pollution". With time, lakes are now cleaner than they were 30 years ago. Certainly, modelling has helped to gain an understanding of the nature of the processes involved. These changes are also reflected in the works on Lake Glumsø.

The studies completed on Lake Glumsø influenced and benefited later ecosystem research, such as the German ecosystem research programme from 1988 to 1999. In the Bornhöved Lakes Region with a special focus on Lake Belau (which is largely comparable to Lake Glumsø) the approach was made to investigate aquatic and terrestrial systems in parallel, to see how they influence each other, e.g. through matter and energy flow. Here, models were employed as a primary synthesis tool (Fränzle et al. 2008; Breckling et al. 2005).

Last but not least, the Glumsø studies helped to demonstrate the importance of:

1. Adequate background data, both for verification, calibration and validation of prognosis
2. The use of modelling as an experimental tool
3. The multidisciplinary character of the modelling exercise

Moreover, the case has led to a new type of structural dynamic models that can cope with the shifts in species composition of ecosystems.

Chapter 20
Biophysical Models: An Evolving Tool in Marine Ecological Research

Alejandro Gallego

Abstract Although they have been in use for some time, biophysical models are still a relatively new tool in the study of the ecology of marine zoo- and ichthyoplankton. As the range of specific applications has expanded so has their level of complexity and sophistication. From simple particle-tracking models simulating the transport of zero-drag, neutrally buoyant particles, the field has evolved towards the development of true biophysical models where the "particles" represent biological entities with increasingly sophisticated submodels simulating their development, survival and behaviour. Here I present the results of a modelling experiment to illustrate the effects of increasing model complexity on the trajectory and final distribution of "particles" (e.g. representing early life stages of marine fish). The outcomes are widely applicable and demonstrate the importance of selecting the appropriate level of complexity required for the specific research objectives.

20.1 Background

If we were to pick an environment where physical processes have a major influence on the biological entities that inhabit it, the marine environment would be a prime candidate. Not only are marine organisms affected by the characteristics of their physical surroundings such as temperature, salinity, oxygen and nutrient concentration, to name just a few, but they often live in a highly dynamic three-dimensional world where they can be subject to turbulent motion, wave action, major advective processes resulting from currents generated, e.g. by winds, tides and water density, etc. It should then come as no surprise that biophysical models, i.e. models that represent the interplay between physical and biological processes, are often the

---

A. Gallego
Marine Scotland – Science, Marine Laboratory, 375 Victoria Road, P.O. Box 101, Aberdeen, 11 9BD, UK
e-mail: A.Gallego@marlab.ac.uk

F. Jopp et al. (eds.), *Modelling Complex Ecological Dynamics*,
DOI 10.1007/978-3-642-05029-9_20, © Springer-Verlag Berlin Heidelberg 2011

modelling tool of choice in marine ecology. This is particularly true in studies of zooplankton and the early life stages of fish, which often reside in the water column and have very limited locomotory capabilities. In this context, we generally refer to biophysical models as *Lagrangian Individual-Centred Models*, although *Eulerian* approaches and those that do not focus on the individual are also found in marine ecological literature, as well as a number of cases of hybrid coupled *Eulerian-Lagrangian* models. The definitions of the *Eulerian* and *Lagrangian* approaches used throughout this chapter follow Turchin (1998).

Marine organisms can live within the sediment, on the seabed (the *benthic* layer) and above it, the latter ranging from a close association with the bottom (the *demersal* layer) to living on, or just below the sea surface. A considerable number of organisms, from taxa that span from invertebrates to many species of fish live in the *pelagic* zone, which is not directly associated with the sea bottom. Many of these organisms have relatively limited locomotory ability, particularly in relation to the horizontal flow of ocean currents. However, they are often capable of significant vertical movement and are referred to by the collective term of *plankton*, derived from the Greek *planktos*, which means errant or drifter. This also includes the *ichthyoplankton*, representing the early life stages of many fish. Larval fish have the ability to swim, and directed horizontal swimming has been demonstrated in a number of species, although generally among the more developed and behaviourally capable warm-water, rather than temperate, species (Leis 2007).

The physical environment provides planktonic organisms with their means of transport, their living conditions (e.g. light, temperature, salinity, oxygen) and their food. The direct observation of biophysical processes at scales relevant to planktonic organisms is difficult. Technological advances such as Video Plankton Recorders (e.g. Lough and Broughton 2007), Optical Plankton Counters (e.g. Gallienne et al. 1996), holography (e.g. Sun et al. 2008) and high resolution acoustic methods (Ross et al. 2008) show promising results. In comparison, traditional sampling methods based on, for example, nets or pumps are very limited in their spatial and temporal coverage and resolution, and their sample analysis is time consuming and dependent on increasingly rare taxonomic skills. Biophysical models do not suffer from some of the problems of observational methods and can therefore be used to complement experimental and observational studies in the marine field.

Cod (*Gadus morhua*) populations in the North Atlantic have been subject to a high degree of exploitation from the mid-twentieth century, resulting in some spectacular stock collapses in the western Atlantic (e.g. Myers et al. 1997). Our study area corresponds to International Council for the Exploration of the Sea (ICES) Divisions IV (North Sea), IIIa (Skagerrak) and VIId (eastern English Channel), managed as a single "North Sea stock", as well as ICES Division VIa (the "west of Scotland stock"). Following concerns about the effect of an unsustainable fishing pressure on the cod population (Cook et al.1997), ICES recommended a number of stock recovery measures, which culminated in 2002 in the recommendation of the cessation of fishing in certain areas (ICES 2003). Although the management of the European cod stocks is implemented at coarse spatial scales, evidence from morphometry, tagging, microchemistry and genetic studies (Hutchinson et al. 2001; Wright et al. 2006) points at the existence of a cod meta-population (Hanski and Gilpin 1997), consisting of a number of distinct sub-populations, relatively isolated, but with some degree of exchange between them. Therefore, if the spawning biomass of local sub-stocks reaches a low level, where depensation may threaten recruitment, the extent of immigration from neighbouring sub-stocks may be critical for their recovery. Such exchange may be the result of the advection of offspring or active migration by juveniles or adults but none of these processes have been adequately quantified for the species.

Heath et al. (2008) designed an age-structured population dynamics model to study the consequences of different natal fidelity scenarios on the meta-population dynamics of North Sea and West of Scotland cod. The model followed discrete cohorts (fish born on the same date), originating from ten spatially resolved discrete sub-populations ("natal units"), as their numbers and maturity state (immature or mature) progressed through time (year-classes). Each sub-population had spatially defined spawning and nursery areas (identified from field survey data). The proportion of offspring produced in each spawning area which reached a given nursery area was quantified by Transition Probability Matrices (Paris et al. 2009) calculated from the output of an offline biophysical model embedded within the population dynamics scheme. The biophysical model has been described in detail by Gallego and Heath (2003) and Heath and Gallego (1998, 2000) and it demonstrated that the offspring transport and survival patterns were not temporally or spatially uniform. The correct understanding of those patterns could be critical for the preservation of the cod sub-population structure and potentially for the maintenance of the North Sea and West of Scotland cod stocks (Heath et al. 2008). To achieve the level of complexity in the biophysical model appropriate for the goals of this modelling exercise and the biological patterns that it aimed to represent (Gallego et al. 2007), I carried out a sensitivity analysis whereby I tested the sensitivity of the model results (against a baseline model run) to a number of model components of varying complexity by initially varying one component at time and finally a number of combinations of those.

20.2 Description of the Analysis

The model made use of daily depth-resolved horizontal velocity fields (resolution 0.25° longitude by 0.125° latitude and 11 fixed-depth vertical layers) generated by a statistical model of the north Atlantic circulation (SNAC; Logemann et al. 2004) based on the HAMSOM hydrodynamic model (Backhaus and Hainbucher 1987) and hourly M_2 tidal velocities. Particles were seeded regularly (at the same horizontal resolution as the HDM) into the model domain within the areas identified as spawning locations (see above), which were resolved at 1.0° longitude by 0.5° latitude rectangles. For these sensitivity analysis runs, year 2002 flow-fields were used and the simulations were run from the start of egg production (see below) until the end of August. By this time, all surviving juveniles were expected to have become demersal, following settlement to the seabed. The model time-step was 1 h.

Of all the possible components of the biophysical model outlined in Fig. 20.1, a summary of the different model setup combinations used in the sensitivity analysis runs is presented in Table 20.1.

The baseline simulation had a single particle release date within the spawning areas (defined for each of the ten sub-populations), corresponding to their specific date of peak egg production (ranging from day-of-year 62–69). The simulation was purely deterministic in relation to horizontal transport (horizontal diffusion

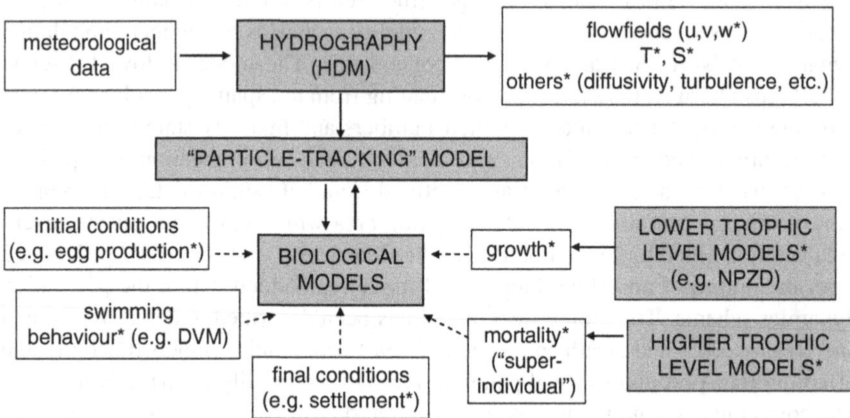

Fig. 20.1 Schematic diagram of the components of a biophysical model of fish early life stages. *Asterisks* indicate non-critical components; *solid arrows* are inputs/outputs and *dashed arrows* indicate constituent components. Abbreviations: HDM: hydrodynamic model; u, v and w are the three-dimensional velocity components, where w is the vertical component; T: temperature; S: salinity; NPZD: nitrate, phytoplankton, zooplankton, detritus model; DVM: diel vertical migration

Table 20.1 Summary of the model setup combinations used in the different runs (see text for an explanation)

EP	H.diffusion	Behaviour	Mortality	Settlement
Single date	No	Fixed	No	Time
Single date	No	Fixed	Yes	Time
Single date	10 ppsp	Fixed	No	Time
Single date	50 ppsp	Fixed	No	Time
Single date	100 ppsp	Fixed	No	Time
Gaussian	No	Fixed	No	Time
Single date	No	DVM	No	Time
Single date	No	Ontogenetic	No	Time
Single date	No	Fixed	No	Complex
Single date	50 ppsp	Ontogenetic	yes	Complex
Gaussian	No	Ontogenetic	yes	Complex

The top row describes the characteristics of the baseline run

EP egg production, *H.diffusion* horizontal diffusion, *ppsp* particles per start position, *DVM* diel vertical migration switched off; see below) as well as vertical movements (fixed depth of 25 m, which is approximately half the typical mixed layer depth in the area at the time), and all cod larvae settled to the seabed when they reached a given age (90 days). The particles initially represented cod eggs, and their hatching time and subsequent (post-hatch) larval growth were modelled as functions of their individual temperature history. Egg development is largely dependent on temperature (e.g. Fox et al. 2003) and temperature, in addition to its direct effect, can be used as a proxy for other co-varying factors influencing larval fish growth in the field, such as food availability, day-length, water column stability, etc. (Gallego et al. 1999). Daily temperature fields were linearly interpolated from monthly mean three-dimensional temperature fields derived from a statistical model of observational data (ICES statistical rectangle horizontal resolution: 0.5° latitude, 1.0° longitude and five layers in the upper 150 m; Heath et al. 2003). In the baseline run, the particles were not subject to mortality. Each particle was considered a "superindividual", i.e. an assemblage of identical individuals. The weighting of each particle was given as a fraction of the sub-population's egg production (total egg production by the sub-population divided by the number of particles released within its spawning area). When applied, mortality was implemented by reducing the weighting of each individual particle (see below).

The effect of mortality was investigated by applying a mortality rate parameterized as a function of length whereby faster growing superindividuals experienced lower cumulative mortality over the simulation period.

The effect of horizontal diffusion was tested by subjecting the particles to an additional stochastic horizontal velocity component. This was computed at particle positions as a function of horizontal velocity shear from the hourly residual plus tidal current fields, following a method derived from Oey and Chen (1992). To achieve statistical stability (see Brickman et al. 2009), multiple particles must be released from each start position so I carried out runs with 10, 50 and 100 particles per start position.

The effect of particle release pattern was tested by comparing the baseline, where all annual egg production was released on the date of peak spawning for each subpopulation, with a run where the start and end dates of the spawning season were fixed for each spawning area, based on field observations. The proportion of the annual egg production shed each day was represented by a Gaussian curve with mean at the date of peak spawning and standard deviation equal to 1/6 of the spawning season duration.

It is generally accepted that planktonic organisms are passively advected by the horizontal flow but are capable of vertical migration. To test its effect, the baseline simulation (vertical depth fixed at 25 m; see above) was compared with a simulation where particles migrated with a sinusoidal diel vertical migration (DVM) pattern between 5 and 45 m depth. It was also compared with a more complex pattern where the vertical position of cod eggs and larvae varied ontogenetically, using a model (Heath et al. 2003) derived from observational data reported in the literature.

At the end of their pelagic phase, cod juveniles become demersal. This transition period is likely to be one when density-dependence can influence year-class strength

(Gallego and Heath 2003). To test the effect of a density- and habitat-dependent settlement scenario, a simulation was carried out where juvenile cod could settle within an age window (90–100 days) but only if there was sufficient free settlement "carrying capacity" ($\leq$0.5 individuals m^{-2}) within a suitable depth range (20–325 m). Pelagic mortality (where implemented) increased exponentially within that window.

Finally, computing hardware constraints prevented me from carrying out a combined simulation with Gaussian egg production, horizontal diffusion with an adequate number of particles per start location ($\geq$50), ontogenetic vertical migration behaviour, mortality and the more complex settlement scenario. Rather, two simulations with a partial combination of those characteristics were carried out (see Table 20.1), although their results will not be presented here in detail.

20.3 Model Results

The sensitivity analysis presented here will concentrate on the effect of varying model complexity on the *patterns* of transport and settled juvenile distribution, rather than the absolute values. The comparisons between the baseline and more complex runs will be largely qualitative. Although I will present a metric of average model performance, the mean absolute error (MAE; Willmott and Matsuura 2005), I will illustrate differences in transport patterns with particle trajectory plots (Fig. 20.2). I will also illustrate differences in juvenile distribution patterns with gridded (1.0° longitude by 0.5° latitude) maps of the percentage of the total (summed over the model domain) settled juvenile population present in each grid cell (Fig. 20.3).

A more detailed quantitative analysis has been carried out. However, it is less relevant in the context of the application of the biophysical model in the overall modelling system of Heath et al. (2008) and it cannot be adequately summarized in the space available to this Case Study. However, in general, although a number of "biological" techniques can be employed to validate biophysical model results (see Paris et al. 2009), the appropriate ones for this type of sensitivity analysis are best taken from the physical fields such as oceanographic modelling literature (see "model error quantification techniques" referred to by Lacroix et al. 2009) and atmospheric/climate research (e.g. Taylor 2001).

20.3.1 Mortality

There should not be noticeable differences in the particle trajectories between the baseline run (Fig. 20.2a) and its equivalent with mortality (not shown), as its only effect would be to "cut short" certain tracks where the weighting of superindividuals became $\leq$0. Overall, the effect of mortality on juvenile distribution patterns, as implemented here and at the spatial resolution of this analysis was quite subtle

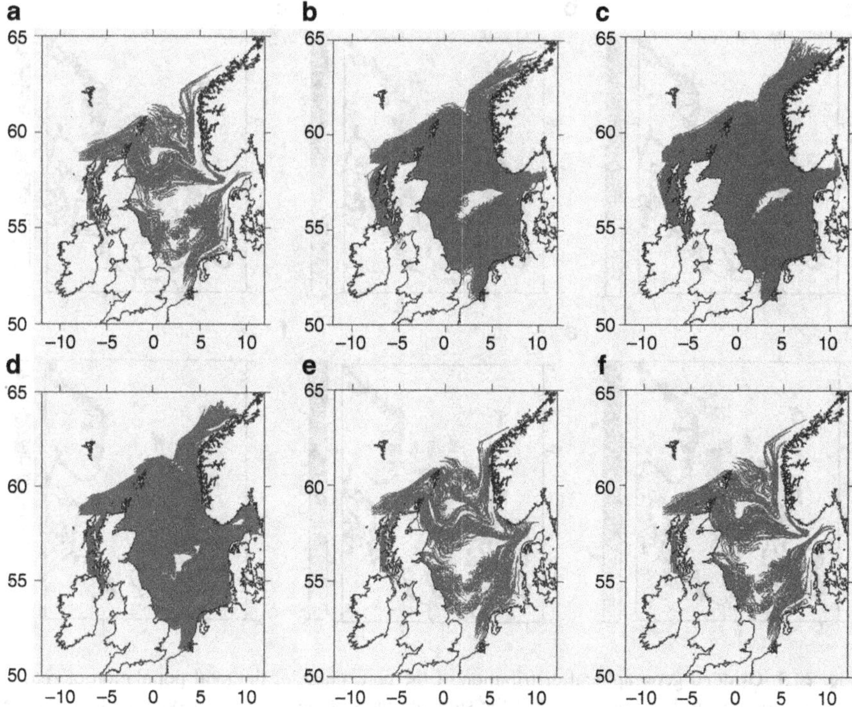

Fig. 20.2 Particle tracks over the duration of the simulation runs. (**a**) baseline run; (**b–c**) runs with horizontal diffusion with 10 (**b**) and 100 (**c**) particles per start location; (**d**) run with a Gaussian particle release (spawning) pattern; (**e**) run with an "ontogenetic" (see text) vertical migration behaviour pattern; and (**f**) run with a complex (see text) settlement rule

(Fig. 20.3b). Mortality slightly smoothed out the juvenile distribution (max. % within any given grid cell was 7.8 and 3.8 in runs without and with mortality, respectively), without a dramatic effect in geographical patterns. The largest differences were along the areas where juveniles accumulate at the end of the simulations (some of which correspond with actual cod nursery areas), as it could be expected. The MAE between these 2 runs was 0.0480.

20.3.2 Horizontal Diffusion

Horizontal diffusion increases the horizontal spread of the particle tracks (e.g. Fig. 20.2b–c) compared with the purely deterministic tracks of the baseline run (Fig. 20.2a). As the number of particles per start location increases, so does the relative horizontal dispersion of the settled juveniles up to a point when increasing numbers of particles do not result in significantly different patterns, i.e. the simulation results become independent of stochastic effects. The biggest differences occur

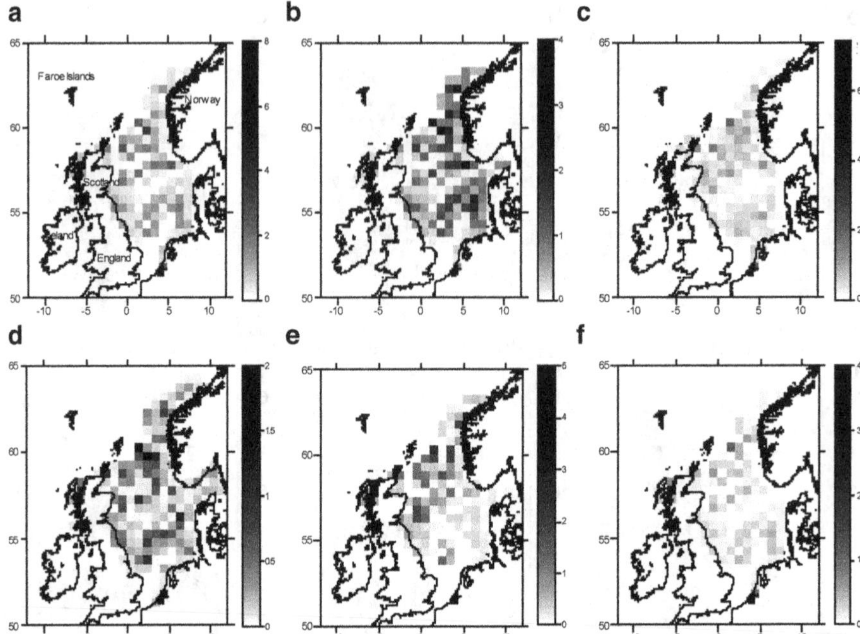

Fig. 20.3 Gridded geographical distribution of the percentage of the total population of settled juvenile cod per grid cell at the end of the simulations, summed over the model domain, for the baseline run (**a**) and absolute differences between alternative runs and the baseline (**b**–**f**); (**b**) run with mortality; (**c**) run with horizontal diffusion with 100 particles per start location; (**d**) run with a Gaussian particle release (spawning) pattern; (**e**) run with an "ontogenetic" (see text) vertical migration behaviour pattern; and (**f**) run with a complex (see text) settlement rule

between runs without and with horizontal diffusion while, at the relatively coarse grid scale used to illustrate these results, the effect of number of particles is hardly noticeable and only the case of 100 particles per start position is presented (Fig. 20.3c). At finer gridding resolution (results not shown), the differences between 10 and 50 particles are more obvious, suggesting that the appropriate number of particles per start position should be somewhere between 50 and 100. This is supported by the pattern of change in MAE values (0.0787, 0.0755 and 0.0759 for differences between runs with 10, 50 and 100 particles and the baseline, respectively).

20.3.3 Particle Release Patterns

The particle release pattern corresponded with the daily proportion of the annual egg production (represented by a Gaussian curve) at each spawning location, compared to the baseline where all particles were released on the sub-population-specific peak spawning dates. The runs were purely deterministic but the effect of day-to-day flowfield variability results in differences in the trajectories of particles released from the same location on different dates (Fig. 20.2d, compared to 20.2a). Indeed, the effect of particle release pattern on particle trajectories is similar to that of horizontal diffusion (see Fig. 20.2b–c), although the effect on the distribution of juveniles is more spatially variable than that of horizontal diffusion (compare Fig. 20.3d with 20.3c but note the different colour scales). The MAE of this simulation compared with the baseline is 0.0539.

20.3.4 Vertical Behaviour

The particle tracks of simulations with particle vertical migration behaviour (Fig. 20.2e; run with DVM pattern not shown) were quite similar among themselves and to those of the baseline run, where all particles were kept at a constant depth (Fig. 20.2a). However, there are some absolute differences between the end distribution of juveniles at the end of these runs (Fig. 20.3e; run with DVM pattern not shown) compared to the baseline (Fig. 20.3a). The DVM pattern has larger absolute differences than the "ontogenetic" one but the latter is more geographically variable. Their MAE values were 0.0763 and 0.0823, respectively.

20.3.5 Settlement Rules

The effect of additional settlement rules on the trajectory of the particles was very small (compare Fig. 20.2f with 20.2a). This was expected since the consequence of the additional constraints is just to prolong the pelagic stage of any juveniles ready to settle (where there is either no spare demersal carrying capacity or suitable depth) until suitable conditions are found. The only appreciable differences between the baseline (Fig. 20.3a) and the present runs are in some limited areas in the west coast of Scotland and central northern North Sea (Fig. 20.3f). The MAE was 0.0344.

20.3.6 Combined Factors

Since both horizontal diffusion and extended particle release patterns had the effect of "spreading out" the particle trajectories, it is not surprising that those were quite similar among the two runs with multiple changes to the characteristics of the baseline run (the two bottom lines on Table 20.1). However, horizontal diffusion had a greater effect on the spread of the final distribution of settlers than a Gaussian particle release pattern, as demonstrated by a greater MAE (0.1039 and 0.0908, respectively).

20.4 Conclusions

The sensitivity analysis presented in this Case Study serves to illustrate a number of issues regarding how to decide on the appropriate level of complexity for biophysical models in marine ecological applications, although the specific lessons derived from this exercise are only applicable to the model that was being examined, i.e. one designed to investigate the *relative* distribution and survival of North Sea/west of Scotland cod early life stages at reasonably coarse spatial and temporal resolution, to provide input to a coupled model studying the metapopulation dynamics of cod in the area. Also, as I discuss at the end of this section, increasing the model complexity introduces an added element of uncertainty, associated with the chosen functional form and parameterization of the additional processes explicitly modelled. An in-depth quantitative analysis of such uncertainty is beyond the scope of the current Case Study. Instead, the functional relationships and parameter values used in the more complex simulations represent the best available knowledge of the processes modelled, based on the relevant ecological literature and our own research. Nonetheless, the next step should be to evaluate quantitatively the more complex models against the uncertainties introduced by their extra parameters.

In the present analysis, settlement scenario was the component the model was least sensitive to. However, this analysis was carried out at a coarse spatial resolution and the more complex settlement scenario was still considerably simplistic so we should not interpret these results to lessen the potential relevance of settlement. On the contrary, as our knowledge of the cod settlement ecology increases, we should test the model sensitivity to more realistic scenarios.

Applying size-dependent mortality to the model did not have a large effect on settlement distribution patterns. Mortality is one of the main biological processes where fundamental information is still lacking in the field (Gallego et al. 2007). Characterizing it as a simple size-dependent function is unlikely to capture the appropriate degree of ecological realism. As above, we should test model sensitivity as new information becomes available.

The particle release pattern had a similar effect to that of horizontal diffusion, although of a lesser magnitude. Realistic egg production models (Heath and Gallego 1998; Scott et al. 2006) are relatively rarely used in biophysical models but over-simplistic particle release patterns can artificially reduce variability if flowfields are sufficiently variable through time.

Biophysical models focus on advective processes and diffusion is often overlooked, even though it can have a major effect on particle distribution. This is of crucial importance in studies of connectivity between different populations. Sensitivity analyses like those carried out here will not only illustrate this effect but are also necessary to determine the appropriate number of particles to be tracked, thus avoiding stochastic bias.

Finally, this sensitivity analysis endorses the importance of modelling realistic vertical behaviour patterns compared to a fixed depth position. Of course, their effect is dependent on the vertical flowfield variability. Even the more complex ontogenetically-varying behaviour was relatively simplistic. With the development of our ability to model the physical and biological environment (e.g. light, temperature, salinity, predator and prey fields, etc.) at increasingly finer resolution, we must continue to test the sensitivity of our results to increasingly complex models to decide on the right level of complexity.

Biophysical models for the study of marine ecological issues are becoming more common and more sophisticated (Miller 2007). This, in part, has paralleled the developments in computing power and sometimes it is unclear whether increased model complexity is necessary, rather than giving in to the temptation to virtually reproduce nature. Higher complexity does not always result in a better model because of parameter uncertainty and variability – often functions and parameters used in a model are taken from different species, developmental stages, environments, etc. The output of more complex models is more difficult to interpret and "external" issues such as higher computational costs complicate necessary steps like a comprehensive sensitivity analysis. Biophysical modelling research would benefit from the type of analysis to investigate the effect of varying degrees of complexity that has been carried out for trophic ecosystem models (e.g. Anderson 2005). Methods such as using the output of a more complex model as a baseline to evaluate the performance of simpler alternatives (e.g. Fulton et al. 2003; Raick et al. 2006) have provided a very useful insight and some general guidelines (Fulton et al. 2003). In the absence of a more comprehensive exercise, the obvious conclusion, as we wrote (Gallego et al. 2007) is that biophysical "models should be as simple as possible but as complex as necessary" and "the level of complexity should be adjusted to the model objectives and the observed biological patterns that it aims to reproduce".

Chapter 21
Modelling the Everglades Ecosystem

Fred Jopp and Donald L. DeAngelis

Abstract The Everglades represent a delicate ecosystem in Southern Florida and is the largest sub-tropical wetland system in the USA. It provides home for a wide variety of unique biodiversity and wildlife. Due to its vicinity to human settlements, the Everglades have been under threat since the beginning of the twentieth century. With the dangers of Global Change on the horizon, this pressure will increase in the near future. The Comprehensive Everglades Restoration Plan (CERP) is a major environmental restoration effort, which is under way and will profoundly affect the Everglades and its neighbouring ecosystems in southern Florida. In the CERP, ecological modelling plays a central role for science-based decision making. In this paper, we introduce the general strategy of modelling for the purpose of ecosystem restoration. We also present two special modelling frameworks and show how they are being used for ecosystem and population-level modelling to help in the planning and evaluation of Everglades restoration.

21.1 The Threatened Ecosystem of the Everglades

The subtropical wetlands of the Everglades are located in southern Florida in the United States (see Fig. 21.1), and extend over an area of ca. 6,100 km^2. It is a delicate ecosystem that comprises the largest subtropical wilderness in the U.S. and provides habitat to a unique compilation of biodiversity. Many threatened species, like the Florida Panther, the American Crocodile, and the greatest diversity of wading birds in Northern America, contribute to this local assemblage of species. Because it lies in the vicinity of large human settlements in southern Florida, the integrity of the Everglades faced increasing threats during the twentieth century. After catastrophic floods in the 1940s a regional management plan, the Central and Southern Florida Project for Flood Control and Other Purposes, was set up with the following objectives: to control the hydrology and possible flood events, to provide

F. Jopp (✉)
Department of Biology, University of Miami, P.O. Box 249118, Coral Gables, FL 33124, USA
e-mail: fredjopp@bio.miami.edu

F. Jopp et al. (eds.), *Modelling Complex Ecological Dynamics*,
DOI 10.1007/978-3-642-05029-9_21,

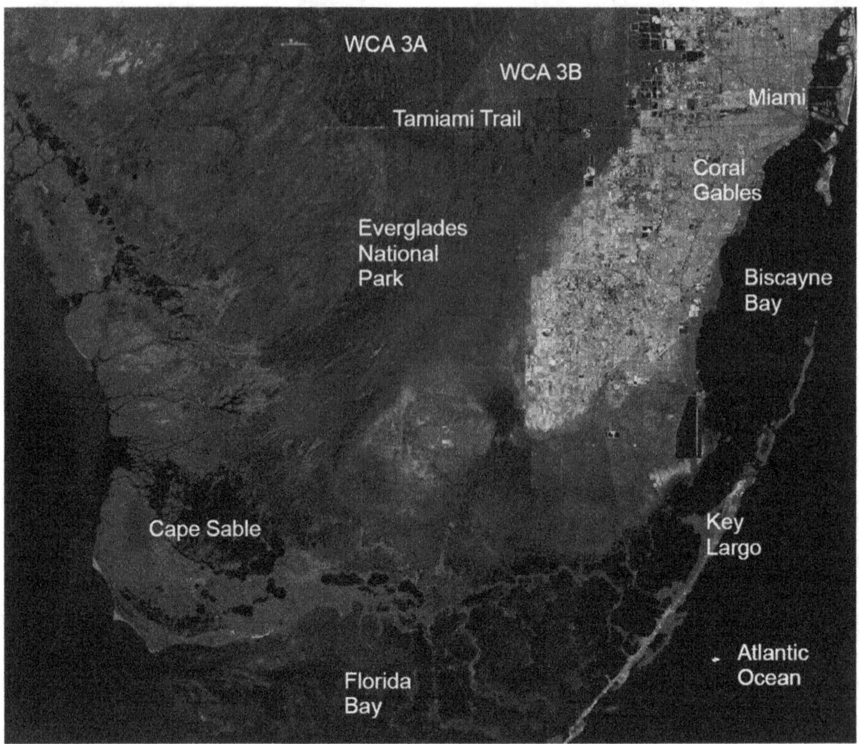

Fig. 21.1 Map of the Southern Everglades, MAP I-2742 by Jones, J., Thomas, JC and Desmond, G.; courtesy of U.S. Geological Survey (WCA: Water Conservation Area)

water supply for the big cities, including Miami and Ft. Lauderdale, and to develop the area for residential purposes and for agricultural use. Consequently, a giant canal system was established to control and carry off water during periods of heavy rainfall, which altered the water flow from its natural condition.

This system of integrated water regulation and flood control makes use of Lake Okeechobee as the central water reservoir, which allows large-scale agriculture but also has drawbacks. Not only do the consequent human-induced major fluctuations in Lake Okeechobee endanger the aquatic and semi-aquatic communities directly in the lake, but due to its position upstream of the Everglades, those habitats are also heavily affected. As side-effects of agriculture, the Everglades ecosystem receives heavy loading of nutrients, which have had impacts on the vegetation and wildlife in parts of this originally oligotrophic system. Today, because of the loss of land and changes in the flow of water, the unique biodiversity of the Everglades is under strong pressure: the wading bird rookeries and colonies have drastically declined within the last decades, the Florida panther is endangered, and the effects are felt by alligators and the fish communities, as well. With the dangers of Global Change on the horizon, which will bring with it sea level rise, more extreme weather events, and invasive species, the future of the Everglades is precarious and ecological restoration is becoming more and more relevant. After concerns were first expressed by the early environmentalist Marjorie Stoneman Douglas, the Everglades were given the status as an International Biosphere Reserve, World Heritage (1979) and protected wetland area, thus, Everglades restoration has had a long history. The focus of the restoration efforts in the Everglades has been an ecosystem approach rather than protecting just one designated geographical area. Such an ecosystem approach of Everglades restoration must not only respect the particular features of the diverse landscapes, like hammocks, pinelands, mangroves, freshwater sloughs, marshes and estuarine habitats, but must also understand how these interactions form patterns and processes on the landscape level (Fig. 21.2).

After much effort, in the year 2000 the Comprehensive Everglades Restoration Plan (CERP) was approved. The focus of this plan is to care for "restoration, preservation and protection of the South Florida ecosystem, while providing for other water-related needs of the region". The plan aims to spend $10.5 billion over 30 years and combines more than 50 different projects; in this sense it can be surely called the biggest restoration project, worldwide. The central idea of CERP is "getting the water right", which means restoring, as much as possible, the quantity, timing, and spatial distribution of the historical water flows. Hence, many projects of CERP are dealing with re-directing billions of gallons of water daily to enable enhanced water deliveries. Another aim is the purchase of land at the borders of the current protected natural areas for ecological buffer zones. Within the whole of CERP, modelling has a central role for science-based decision making.

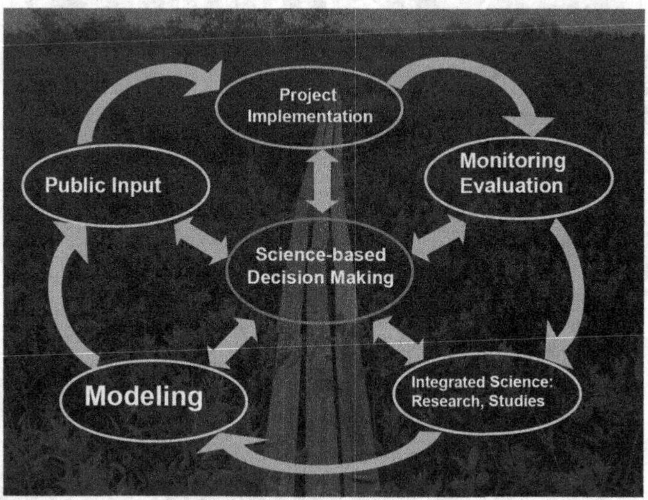

Fig. 21.2 Adaptive Management Restoration Approach, where modelling plays an important part. Background picture was taken on a small vegetation island close to Cape Sable. Board walks like this are maintained by the Everglades Park Service

21.2 The Everglades Integrated Modelling Approach: General Strategy

Due to its unique ecosystem and to the creation of Everglades National Park, the interest in the ecological integrity of the Everglades has a long history. The amount and variety of the collected data is in keeping with the importance of this ecosystem, and ranges from climatological and meteorological data, to biotic data, sociological and economical data. The last two deriving from opinion polls or related economical census, as far as they are relevant for the ecosystem. The data bases in these areas are very good in many respects and the modelling aspect of the Everglades research makes intensive use of all of those objects and qualities, concentrating on sources that are important for Everglades restoration on the landscape level. The goals of this restoration process are twofold: (1) maintaining the ecosystem-specific traits and processes in situ, and (2) maintaining viable populations of all native species in situ (Science Sub-group 1994).

In trying to achieve these goals, models and modelling tools are intensively used to increase understanding and are applied on nearly every hierarchical level of the research process to integrate the available knowledge (see Fig. 21.3).

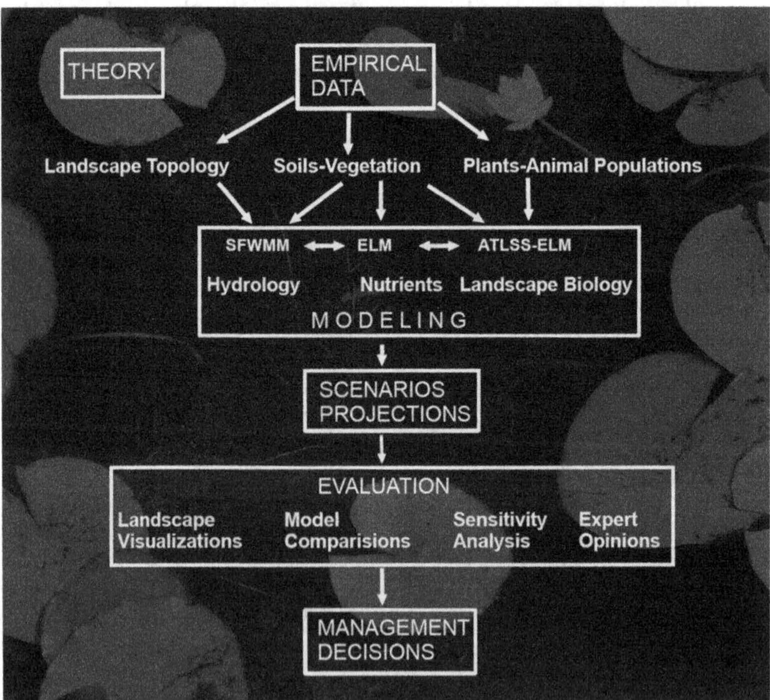

Fig. 21.3 Interplay of data acquisition, modelling, evaluation and final management decision within Everglades' research

In the following section we explain the general strategy of the integrated modelling approach for Everglades restoration. As a first step, the empirical data are analysed and checked for consistency, which helps to fill-up knowledge gaps. The empirical sectors involved are "classical" ecologically: landscape data (e.g. topology, hydrology, soil, vegetation), and population data (e.g. plant and animal populations and their interactions). Then, in a second integration process, sector-specific model-frameworks are applied, which are used for forecasting purposes. Examples of such model-frameworks are the South Florida Water Management Model (SFWMD 1992) and Natural Systems Model (Fennema et al. 1994), the Everglades Landscape Model (ELM; Fitz and Sklar 1999) and the Across Trophic Level System Simulation Complex (ATLSS; Gross and DeAngelis 2001). The last two model frameworks and their working principles will be explained in detail in the next paragraph.

After the developmental process of representing the data and describing the system dynamics is accomplished, the model frameworks can be used to generate computer scenarios and make projections. In computer scenarios, a characteristic set of conditions and traits is grouped within a special case. The procedure referred to as computer scenario is when different sets of conditions are sensibly grouped into alternative cases and the reaction of the model components to the variation of these characteristics is assessed through computer simulation. The scenario technique is used to evaluate the outcomes of possible future situations on the target variables. A possible goal of such an approach could be to generate scenarios on breeding and foraging behaviour of animals under different conditions of hydrology and vegetation types (DeAngelis et al. 1998).

In providing input for ecosystem restoration, special emphasis is put on model evaluation (see also Chap. 23), which normally begins with a sensitivity analysis and visualization processes, to display model outcomes on the landscape level (see Fig. 21.3). If more than one model or model type is available for the same or close related issues, these model attempts are thoroughly compared and their relative advantages and disadvantages are evaluated. This procedure guarantees that the different "powers" of the model systems are systematically evaluated. For all steps of the evaluation processes described here, expert advice and opinions are obtained and integrated before the outcome is finally communicated to stakeholders on management decisions on Everglades restoration.

Figure 21.4 focuses on the integrating power of modelling within an ecological project. Here, even at the starting point of data and information collecting, (see Fig. 21.4 left, upper corner) modelling techniques can help to overcome possible difficulties. Basically, these difficulties have to do with aggravating circumstances during data collection or with intrinsic characteristics of the natural phenomena in focus, such as heterogeneities and variabilities that can lead to data gaps and, later, to inconsistencies. Appropriate modelling tools can help to evaluate these first data structures, understand the heterogeneities, and fill-up the data gaps. When dealing with complex ecological phenomena, the following aspects of knowledge visualization are important: (1) Providing for spatially explicit model representations is more and more requested, which also helps to encourage the communication

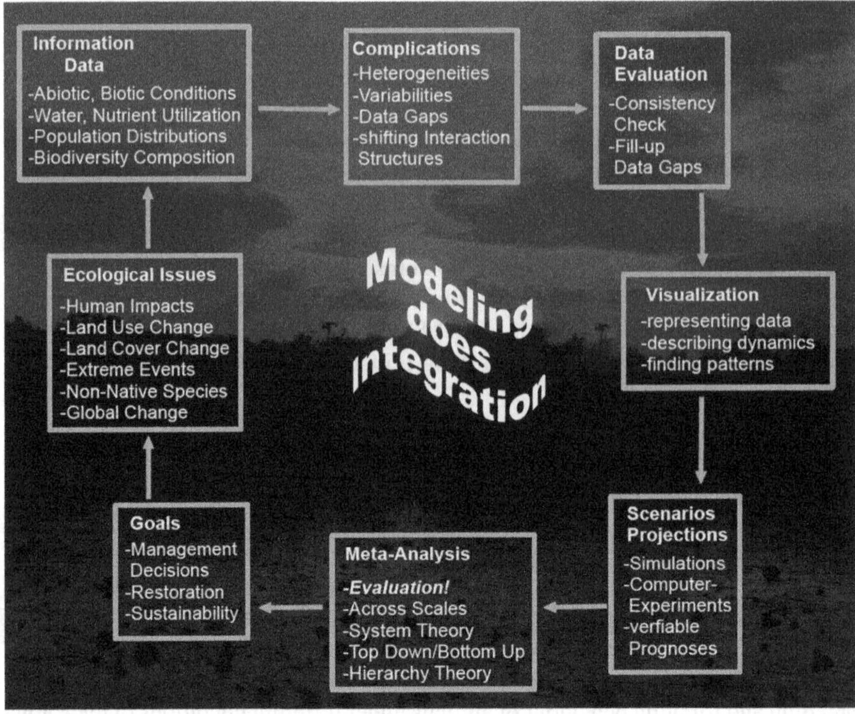

Fig 21.4 Integrative capacity of modelling

process within the project; (2) it is useful as a consistency check and can help to find relevant patterns within the data collections. In particular, models play a most important role when it comes to analysing across-scales phenomena (Reuter et al. 2008), which is one important part of a meta-analysis approach. Finally, stakeholders have to make management decisions on restoration and conservation for the Everglades ecosystem. Because these decisions have to be science-based, also in this step models help to integrate and communicate understanding and viewpoints about the system.

21.3 Examples of Everglades Dynamic Landscape Model Frameworks

Within the Everglades Restoration project, the spatially-explicit modelling frameworks of the Everglades Landscape Model (ELM; Fitz and Sklar 1999) and the Across Trophic Level System Simulation (ATLSS; Gross and DeAngelis 2001) are closely interlinked. In the following, we will explain their working principles.

Providing the Landscape Basis: ELM

The Everglades Landscape Model ELM (Fitz and Sklar 1999) is a regional-scaled, grid-based model framework, that brings together different landscape layers, like vegetation, hydrology and topology, and combines these layers to a GIS-like output representation. The vegetation types used here derive originally from the Florida Gap Analysis project (Pearlstine et al. 2002), providing a spatial resolution of 28 × 28 m, which are aggregated in ELM into cells of the size 1,000 × 1,000 m. The ELM model uses its own hydrologic modelling, but uses boundary conditions from the larger scale South Florida Water Management Model (SFWMD). ELM also contains simple modelling of the nutrient (e.g. phosphorus) fluxes of suspended material, and growth and competition of major vegetation types. A key objective of ELM is to project the movement of phosphorus from the agricultural areas into the Everglades.

Modelling Animal and Population Distributions: ATLSS

The Across Trophic Level System Simulation program, or ATLSS, is an integrated set of computer simulation models representing the biotic community of the Everglades/Big Cypress region and the abiotic factors that affect it (DeAngelis and Gross 2001). The major objective of the ATLSS models is to estimate the effects of hydrologic scenarios on key biota of the Everglades. The spatial extent of the models is the entire Everglades/Big Cypress region and some surrounding areas, and the spatial resolution is generally 500 × 500 m cells, though sometimes finer. Relevant abiotic quantities, such as hydrology, fire, and major storms are modeled. Hydrological scenarios from the SFWMD are used for this purpose.

The output of this model at the 2 × 2 mile grid scale, is converted to a 500 × 500 m grid scale. The ATLSS models are spatially-explicit, using GIS map layers of topography, soil, vegetation type, etc. The biotic community is represented by a hierarchy of models, beginning with the process models of the biota constituting the energy base, including vegetative biomass, lower trophic level invertebrates, and decomposers (Fig. 21.5). Models that contain some relevant details on size and age structure simulate several important functional groups, such as fishes, macro-invertebrates, and small reptiles and amphibians, which utilize the production of the energy base and provide food for some of the top consumers (for size/age structure models see Chap. 9, for an example on local models and process models see Chap. 18). Several individual species that are highly valued because they are unique or threatened, or are regarded as indicators of the overall conditions of the ecosystem, are modeled in much greater detail, using individual-based models (see Chap. 12). These species include the American alligator, the American crocodile, several species of wading birds, white-tailed deer, the Florida panther, the Cape Sable seaside sparrow, and the snail kite. The objectives of the ATLSS program over the long term are to aid in understanding how the biotic communities of South Florida are

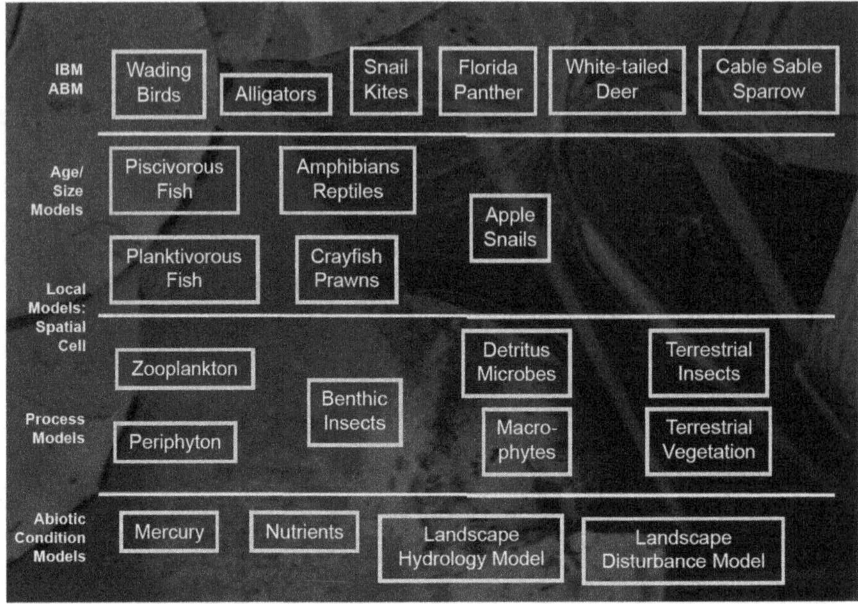

Fig. 21.5 Applied model types within the ATLSS project, following DeAngelis et al. (1998)

linked to the hydrologic regime and to other abiotic factors, and to provide a predictive tool for both scientific research and ecosystem management. The distribution, volume, and timing of water flow influences the energy and material transfers among ecological components within and across the trophic levels of these systems. The ATLSS integrated models mechanistically simulate the causal relationships between hydrology and the biotic components of the Everglades/Big Cypress region.

The ATLSS Program has produced model output used in evaluations conducted by DOI agencies (USFWS, NPS) of Everglades restoration scenarios proposed under the Central and South Florida Project Comprehensive Review Study (Restudy). Starting in the autumn of 1997, a set of simulations was produced each month or less and disseminated both in hard copies and on the web to agencies engaged in the evaluations of the Restudy plans. This output was used in making the final decision on a restoration plan, and at the same time was used in a scientific publication (DeAngelis et al. 1998).

21.4 Future Modelling Contributions for Everglades Restoration

In the subtropical wetlands of the Everglades, ecological modelling plays a central role in science-based decision making for restoration purposes, a function which is becoming more and more important as the delicate ecosystem is drastically confronted with the side effects of non-sustainable practises. Like any other relevant scientific tool, modelling has to be evaluated and to be adapted to new conditions.

Regarding the development of the ATLSS program, we can state, that since the Restudy plan, the whole framework has continued to develop in several ways. First, the models have been applied to other scenario evaluations in South Florida, such as the Modified Water Delivery Program. Second, the number of models within ATLSS is expanding to cover crayfish and apple snails, with other models in the planning phase. Of particular importance will be a model of vegetative succession that will forecast changes in numerous vegetation types under changing hydrology and nutrient inputs. Third, the geographic range of ATLSS is expanding to cover the mangrove estuary areas of South Florida. Mangrove estuary vegetation, fish and other fauna are being modelled. Fourth, a GIS-based ATLSS Visualizer has been developed to allow agencies to view and analyse all ATLSS output data. Fifth, sufficient data now exist for improved calibration and validation techniques to be applied to the ATLSS models, which is in progress now, along with uncertainty analysis.

What are the future demands that the discipline of ecological modelling has to face when dealing with Everglades issues and what, from the methodological point of view, is needed to overcome these? For the Everglades, in the times of the "Grand Challenges" (NRC 2001), the need for positive actions is becoming more urgent. Without doubt, a landscape that is totally surrounded by coastal margins is most at risk from "Global Change" in the form of sea level rise. Here, a projected rise of the sea level of 2–8 ft by 2100 (Overpeck and Cole 2006; Overpeck and Weiss 2009) would in the least, strongly affect, and at worst, devastate most of the positive efforts that have been accomplished for restoration in the Everglades to date. Consequently, more positive actions must be taken. Following Gardner et al. (2008) these actions should include:

- Identifying endangered landscapes/habitats and broadly communicating the essential reasons for preserving these critical resources
- Linking numerous and specific case studies across countries, biomes and landscapes to develop robust criteria for sustainability and adaptive management
- Increasing the awareness within financial sectors of the fundamental relationships between economics and landscape management
- Working out new formal techniques to visualize the rates and the consequences of landscape change

Realizing these actions will depend strongly on the development of adequate theoretical tools and modelling tools to tackle the issues that stand before us. Along these lines, we expect developments in which originally separated areas of modelling grow together. An example for such a development can already be seen with the integration of GIS-like systems into broader modelling frameworks (see Chap. 22). Another important keyword for future ecological modelling is model "coupling", which can be performed in two ways: (1) the output of one model is used as input of a hierarchically higher located model. So far, there have been excellent outcomes, in terms of data evaluation with this technique in the ATLSS project; (2) model frameworks consist of different sub-model types, in which, for instance, one ODE-based part determines the resource for a population, while the individuals of this population are represented with an IBM-like model type (more on model coupling see Chap. 20).

With the approaching of the Grand Challenges to the Everglades ecosystem, the terms of reference for ecological modelling also expand. In a kind of modelling disclosure procedure, we, as modelers, have to make clear, exactly what models can accomplish, what are the conditions under which models can work, and particularly, what they still cannot do. By informing stakeholders of what really can be achieved with theoretical and modelling tools, confidence in our scientific methods will be strengthened, and there will be an increased understanding of the usefulness of modelling in determining how the Everglades ecosystem may change in the future.

Chapter 22
Model Integration: Application in Ecology and for Management

Dietmar Kraft

Abstract Integrated Environmental Modelling (IEM) is a prospering and multifaceted field of science. Many research projects employ several individual models in combination in order to describe entire ecosystems. *Integration* is the skill to join single models, data, and knowledge on a technical as well as on a conceptual level. IEM has become a prerequisite for modern environmental management and has been promoted by fundamental changes in environmental politics in the last decade. The goal and strength of IEM is the interpolation of information on local ecosystem behaviour and the extrapolation and transfer of results to other locations, to different scenarios and into the future.

This chapter deals with the integration of models, data and information to represent and analyse dynamics of complex ecosystems by using Geographic Information Systems (GIS). GIS can either act as a framework to combine data and execute separate (and incorporated) models, or can simply be applied to manipulate data used by autonomous models.

Three specific application examples illustrate possible structures and functionalities of IEM and highlight the benefit of using GIS to join data, models and information. Along with many challenges concerning the conceptual details of model joining and integration of information and knowledge, the use of GIS to support IEM is extremely beneficial, and constantly drives advancements in ecosystem research.

22.1 Introduction: Integrated Modelling

The complexity of a model tends to increase with the extent of the spatial scales and the number of organizational levels examined. Not only does the number of elements that need to be considered increase, but also there is an expansion of

---

D. Kraft
Institute for Chemistry and Biology of the Marine Environment (ICBM), 26111 Oldenburg, Germany
e-mail: dkraft@icbm.de

F. Jopp et al. (eds.), *Modelling Complex Ecological Dynamics*,
DOI 10.1007/978-3-642-05029-9_22,

internal coupling of elements and of the degree of aggregation of the components (WGBU 1997). To combine different scales and levels of an ecosystem, models have to include horizontal as well as vertical interactions, interdependencies and reciprocities of the examined entities as well as human activities (Schellnhuber et al. 1999). These relations exceed the boundaries of the subunits of ecosystems and geographic features. They transcend socio-economic sectors and scientific disciplines. Consequently, "holistic" models require the expertise (and the models) of different scientific disciplines. In this context, integration has two main objectives: (1) developing interdisciplinary, networked approaches of *joining* disciplinary data, models and methods – joining understood in the sense of conceptually and technically combining them, and (2) finding methods to handle the resulting complexity. These objectives include both technical challenges as well as contextual questions, the latter resulting from new aspects following upon data, models and methods integration. Integrated Environmental Modelling (IEM) addresses both objectives: IEM couples models from different disciplines to find answers for holistic questions and helps to aggregate complex results to make the models operable and communicable.

Integration is a driver for the development of different tools, concepts and approaches of modelling. As a result, environmental research, which has concentrated on specific topics and a centred set of methods for quite some time, is now broadening its methodological approach. Knowledge that was quite centred at first, may now be more broadly linked with different models to deal with a particular subject and thus widen the scope of investigations and enable results that were not achievable before. Consequently, in the context of this broader consideration IEM is a sophisticated object under investigation. Based on a combination of models, IEM have the potential to answer newly arising questions by interlinking particular disciplinary knowledge. The analysis of complex and abstract phenomena like dynamic interactions of inhomogeneous socio-ecological systems rely especially on sophisticated methods that might be offered by IEM. The management of catchments, coastal zones, and marine landscapes where elements with differing dynamics converge, present many examples for trans-disciplinary approaches dealing with this complexity.

Geographical Information Systems (GIS) offer two general techniques to support this kind of integration and therefore are an important addition to IEM. GIS are able to link data based on their location and to aggregate them. Data linkage can be performed by intersection of the respective geometries or just by overlaying different thematic layers in a map. Based on their spatial reference, GIS may serve as an interface between data, models and paradigms. In this way, GIS has catalysed integrated modelling and environmental research over the last decades.

The needs of management, policy and planners increase the demand for easily accessible, manipulable and presentable spatial information. GIS have strongly supported a spatially-explicit way of looking at natural systems: Location, understood as the habitat of organisms, is increasingly used as a level of integration in ecosystem research, providing the environment for organisms and defining territories and expansion of biocoenoses. Any information on a habitat having a spatial reference can be processed and linked with other spatial data in a GIS.

In addition, requirements of resource management and environmental policy have strongly favoured the development of IEM (Dolk et al. 1993). Considering global change, politicians, spatial planers and environmental managers require predictions describing possible future developments in scenarios, forecasts, and projections. Information should be regionally differentiated, comprehensible and immediately available. Environmental management assesses the consequences of changes, developments or shifts of the systems described. Weighting the relevance of impacts, assessing states, and evaluating ecological functions of entities, animals, plants, as well as communities, are central objectives (Costanza et al. 1998).

Most dynamic simulation models have limited ability to represent spatial processes, as they focus on specific aspects of the processes themselves (see the chapters on PDE, CA, IBM, and SDM – Chaps. 7, 8, 12, 13 respectively). GIS offer methods to handle temporal dynamics, analysing the changes of the relevant parameters across the area. In contrast to many other approaches that represent space in terms of square boxes or simplified grids, GIS operates with spatial references in terms of geodetic coordinate systems, e.g. by using geographic or metric based coordinates. Together, GIS and IEM have the ability to address both challenges: joining complex data and aggregating them to manageable information.

IEM follows a holistic, area-wide and trans-disciplinary approach. Data or models describing the status of different subsystems are interacting in a joint software-framework describing the status of the whole (considered) system, using adequate indicators and appropriate geo-referenced datasets. IEM simulate the observed behaviour of an entire system and based on scenarios, intend to provide projections of future states.

This chapter has the following outline: General technical and conceptual aspects will be described, which have to be considered when developing and discussing the results of an IEM. The third section focuses methodically on the stepwise process of model integration via GIS. The main exemplary GIS-techniques for joining data layers (and hereby models respectively) and aggregating complex interactions are described. Section 22.4 describes three examples of IEM coupled by GIS. Finally, I conclude with general strategies of model coupling with GIS and discuss future perspectives.

22.2 General Aspects of IEM

Interoperability, dealing with the problems of assembling and interrelating data and/or models, is one of the central objectives of IEM (Argent 2004; Goodchild et al. 1997). The technical ability to exchange data and the conceptual capacity to make use of information intrinsically ties models and GIS to each other (Abel et al. 1994). There are two general options to follow when developing a (integrated) model:

1. Monolithic Approach: Writing an over-all model including more or less detailed representations of subsystems. This straightforward type of integrated model is appropriate when focusing on a clearly distinct aspect, describing an ecosystem based on limited factors and indicators. Monolithic all-in-one-models require a deep knowledge of the system – profound software skills. Monolithic IEM have many advantages: they tend to be high-performance and profit from enhanced internal data exchange. They follow a coherent and regularly well documented concept with every component directly adjusted to each other and the whole system. Typical disadvantages of monolithic IEM are elaborated manageability and a high risk of conceptual errors hidden deep in their code.
2. Modular Approach: Using existing models and combining them into an integrated model. Figure 22.1 (left) shows an example for this using GIS as an integration tool. The GIS manages spatial data in a database, supports model(s) with input, retrieves results from it, and passes the information to tools, which apply specific methods of aggregation and evaluation. Finally, the GIS executes the visualization of results.

Although a number of established monolithic IEM exist (Argent et al. 2006), this chapter focuses on modular approaches (Hinkel 2009; Voinov et al. 2004) as they highlight the process of integration much better. The technical side of an integration process necessary to actually integrate different models, teaches a lot about the conceptual side of how to integrate the contents. Modular approaches can easily demonstrate both aspects of integration (see Section on Conceptual and Technical Aspects).

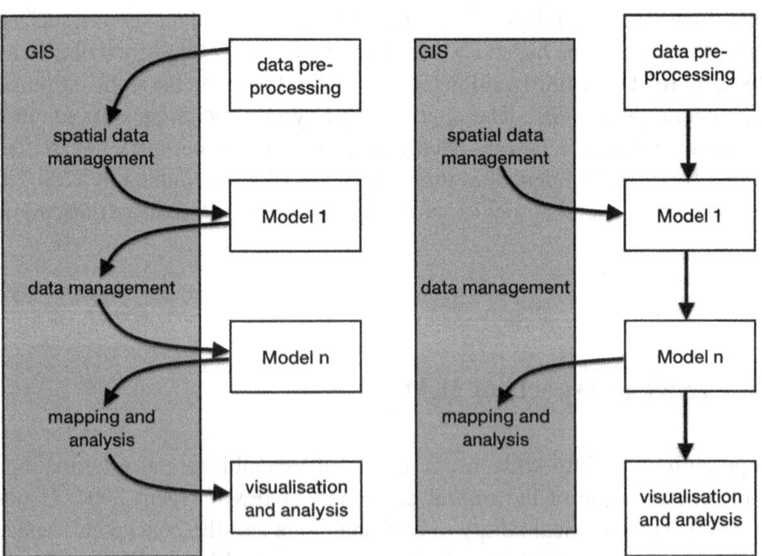

Fig. 22.1 Using GIS as a framework to couple data and models (*left*) or as a tool to manage, map and analyse data for a cascade of models (*right*) (modified from Argent 2004)

Basic Approaches

Figure 22.1 illustrates the two basic, structural approaches of using GIS for model integration (Brandmeyer et al. 2000; Argent 2004): (1) The integration proceeds with the GIS being used as a control instance that manages, analyses and maps the spatial data. Retrieved by the GIS, the modelling components create this data. (2) Alternatively, a cascade of models and tools run independently from the GIS, using the GIS-services to get, manipulate and push data. While the first approach is using GIS to execute models with all the limits of a GIS, the second uses the abilities of the models running a GIS as spatial database and service-provider, with all the limits of access and performance.

On a low level of integration (Fig. 22.1 right), the GIS is limited to managing the data generated by one or more independent models. In this context, loose coupling would be done by plain interfaces allowing data exchange. Tight coupling via programming might allow a dynamic use of the components as a "running" model, while loose coupling is limited to static "put and get". On a high level of integration, GIS-components build a functional unit within a model framework (or vice versa). Having flexible programming environments or macro-languages available is a precondition for tight coupling of models and GIS, while effective software-components are the basis of an embedded modelling approach (Sui et al. 1999).

Conceptual and Technical Aspects

Expanding on the basic approach above, there are two contextual approaches for integrating data, models and methods from different disciplines:

The vertical path, overarching the different sources of data, focuses on the more technical aspects of integration. Especially in a GIS framework, vertical integration interconnects information from different disciplines on one common level. It couples data, models and tools with the same content on a spatial level. Vertical integration follows the idea of trans-disciplinarity: using information from a tributary submodel on specific aspects to support a comprehensive model addressing an overarching question (Oxley et al. 2007). Vertical integration of data and models to a coupled framework strongly supports fundamental investigations on the characteristics of the system described. It enables the analysis of interactions of entities on the same organizational level based on information from different levels.

The horizontal path generalizes complex information gained from the coupling of models and connects it to the knowledge gathered from the analysis of the coupled data (Yokozawa et al. 1999). It aggregates data, combines different information and uses them for conceptual generalizations. Horizontal integration builds on the results of the vertical approach, which is the first step of the integration process within the model framework. The horizontal, conceptual orientation is necessary to connect information from different levels of complexity to address comprehensive questions on the behaviour of an ecological system.

Combining models from different disciplines into integrated applications poses technical as well as conceptual challenges:

On the technical side, legacy models (i.e. software that has been outdated but is difficult to replace because of its wide use) are not often developed to interact with other models, neither within nor between disciplines. These models are well known techniques, especially in disciplines with many popular models like hydrology, forestry and ecology. However, limits are noted and problems of integration with other models are obvious (Loucks et al. 1984; Mackay 2000; Zalewski 2002). Consequently, the topic of interoperability, the task of joining data, models or methods receives a lot of attention (as a start see Goodchild 1999).

On the conceptual side, integration has to deal with a range of problems. In addition to ontological discrepancies between scientific disciplines, as well as between science, management and politicians, the concepts of the different sciences are often "hidden" in ancient code. This is true at least for the use of typical units, for the classification of systems and for definitions and specific disciplinary terms that might have a different meaning in each discipline.

Obviously, both aspects are relevant; they are both sides of the same coin, which are necessary to integrate data and models. During the development of an IEM, technical as well as conceptual aspects are crucial for the success of the approach.

22.3 Model Integration by GIS

GIS are powerful and established tools in sciences dealing with spatially referenced data. Common GIS (like ESRI ArcGIS®, MapInfo®, GRASS and QGIS) offer a variety of tools to manage, visualize, analyse, and interpolate data. Handling different coordinate systems and projections and visualizing information in detailed maps are commonly used features. Maps are useful to present and discuss results, especially with decision makers dealing with environmental issues (Aspinall et al. 2000).

GIS: A Brief Definition

A GIS is a software that manages, analyses, and presents data that are spatially referenced by their location.

GIS depict any information system that enables the user to capture, store, integrate (join, merge and unite), edit, analyse, share, and display geographic information.

In a more common logic, GIS-applications allow users to analyse spatial information, edit data (aggregation, generalization, and projection are most typical editing actions), create maps, perform logical or spatial queries, and present the results of all these operations in maps or tables.

From the application-oriented view, GIS is an information platform used as a interface to obtain thematic maps, which have to be distinguished from the underlying software providing tools for the management of spatial data. In this sense, modellers are developers, picking methods and resources from GIS-applications and using them as a toolbox.

By providing data management, (statistical) analysis and visualization tools, GIS were among the first frameworks that made wide-ranging methods of integration available.

Capturing spatial data in GIS follows the general idea shown in Fig. 22.2. Graphical layers or *features*, which could be entities with a geographic location, consisting of vector structures like points, lines, polygons, any kind of raster-files or objects from a database model, are tied to a geodetic coordinate system, defining their position (see Derby et al. 2005). The relevant descriptive records are linked to the features by unique identifiers.

Using geographical information for integration enables modelling with precise spatial data and condensing respective results in maps (Livingstobe et al. 1994). This is advantageous since information in maps is spatially referenced. GIS can

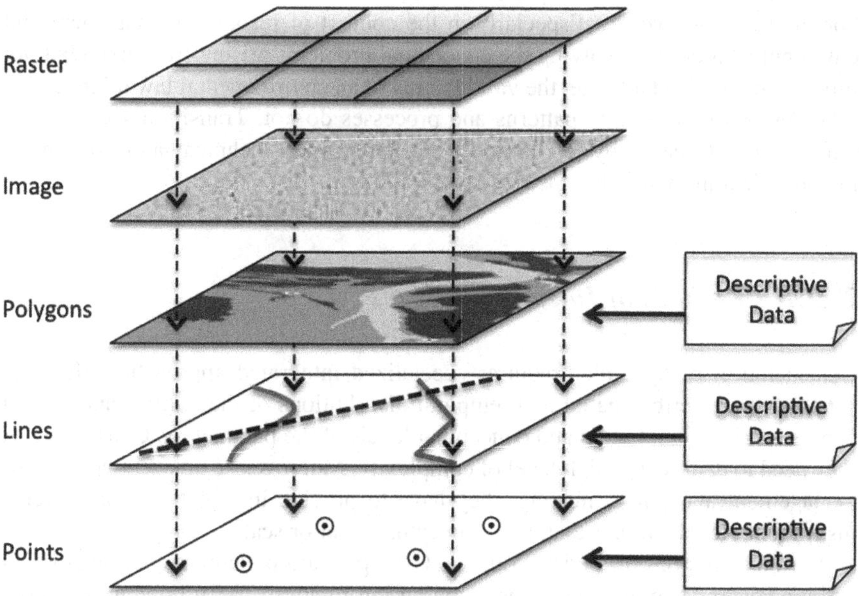

Fig. 22.2 General structure of GIS: geo-position as connection between different layers of data. Vector data (*polygons, lines, points*) are linked to attribute tables with descriptive data (modified from Derby et al. 2005)

support embedded models with more realistic spatial characteristics of the entities, than simple box models. The capabilities of a GIS are eminent: it can calculate dimensions of objects (even if they are dynamic) represented in different coordinate systems and projections. Together with information on the velocity of processes, the according rates can be calculated. Being spatially explicit, it is possible to connect data and models based on their determined location.

Nevertheless, attention has to be focused on the spatial fitting of the data, models and methods. Distribution of abiotic as well as biotic phenomena tend to be heterogeneous and inconsistent. The representativeness of the measurements has to be secured. Especially in complex systems, driving forces and resulting impacts might act on different spatial levels (Levin 1992) underpinning its importance (Schreier et al. 2002). Naiman (1992) argues a number of issues related with the possible mismatch of scales. Important issues concerning management aspects are:

1. The mismatch between local controls and system-wide needs and priorities.
2. The discrepancy between the scale or level of available factual information and the dimension of a phenomenon, e.g. between the catchment level and the level at which important decisions are made (Aspinall et al. 2000).
3. An absence of systematic investigation of processes on a suitable scale.

Furthermore, on a more institutional level, misfits between natural system boundaries and legal units can exist (Young 2003). Institutional arrangements like administrational units or protected areas are not necessarily based on natural borders (and vice versa). Especially in the context of resource management and environmental politics, solving the conceptual problems arising from fixed but not fitting borders is a challenge: the validity area of an environmental law might end at a border, but ecosystems' patterns and processes do not. Transition sections and buffer zones (created by a GIS tool) are appropriate technical and conceptual methods dealing with this aspect of spatial fitting.

22.3.1 Methods of Integration

Consideration of dynamics requires specialized integrated approaches. IEM are supposed to describe spatial and temporal distributions of the entities and should consider the different units and conceptual levels of the parameters. Consequently, they need to deal with a high level of complexity. Moreover, to answer questions of management and policy making, they have to provide information from diverse institutional levels, rather than any conceptual level or scale.

Various functions offered by common GIS-applications support coupling of data and models and aggregating of the resulting information. The following workflow focuses on the main steps in a common software framework.

As an example, Fig. 22.3 shows the stepwise integration of a simple model on land use. Integration aims to identify areas that are based on the intensity of land

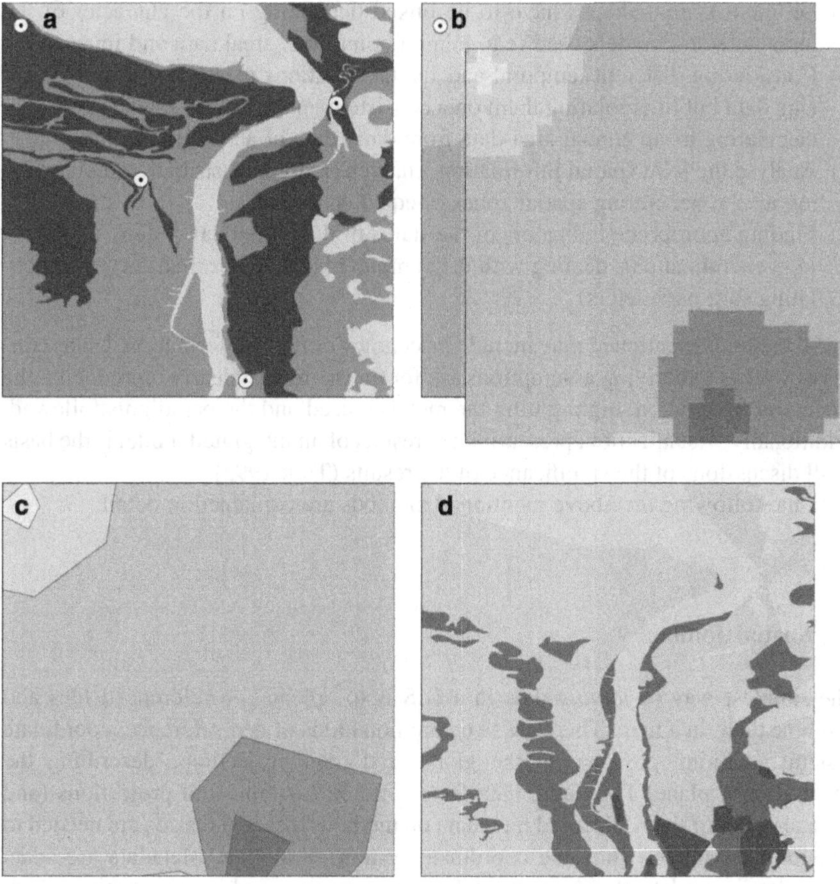

Fig. 22.3 Procedure of stepwise integration of a simple model on land use to identify areas sensitive to high rainfall: (**a**) joining land use and rain gauge, (**b**) interpolation of precipitation, (**c**) classification of high rainfall, (**d**) generalization of sensitive areas (see text for more details)

use, and are sensitive to high rainfall. Using different socio-economic scenarios, the land-use model thus calculates the suitability for a specific agricultural land use. Data from a projection of precipitation are aggregated to zones of high rainfall. All information is combined to identify sensitive areas.

The first step of any model integration is to make all data and model-interfaces available within the technical framework and to the software (by file managing systems etc.). Fortunately, common operation systems do not bother the user with data connectivity problems anymore. Standard file formats should be accessible to the GIS directly, as long as they are geo-referenced.

Basic methodical requirements of model integration are multifaceted (see Fig. 22.3):

(a) Different spatial shapes have to be linked depending on the character of the input data and models used (e.g. joining point data, areal data and images)
(b) Considering different temporal and spatial situations means not simply assessing data but interpolating them on a consistent temporal and spatial level (e.g. calculating mean annual area-data from daily data of a station)
(c) Analyse the joint spatial information, characterizing its distribution and typifying it (e.g. generating spatial zones of equal character)
(d) Finding appropriate indicators of the status of the individual system, undertaking generalizations, dealing with the complexity of the information (e.g. calculating sum parameters)

All methods mentioned may include inaccuracy or fuzzy information. Uncertainties as well as underlying assumptions ask for transparency, clearly reproducing the process of integration, highlighting the methods used and the paradigms followed. Additionally, visualization, presenting the results of an integrated model is the basis of all discussions of the significance of the results (Turk 1992).

In the following the above mentioned methods are explained in detail:

a) Spatial Joining

The simplest way of *joining* data in a GIS is to upload geo-referenced files and combine them in a map. There are two common kinds of geo-reference: coordinate systems, defining position on the globe, and map projections, describing the position on a plane. To couple data from different systems and projections (and there are a lot of them) detailed metadata on methods, dates and sizes are needed to transform a data set from one coordinate system to another. Metadata are "data about data", used to describe the definition, structure and administration of all contents of a data file in its context. Finally, combining geo-files that establish and maintain metadata is a basic requirement.

For *joining* non-georeferenced data, GIS offer two kinds of functions: Either, by connecting data by known identifiers like attributes of places (or names of locations as towns or water bodies), or by an ID or name of e.g. a sample station. Rules are offered to either join objects that intersect each other, or where one object completely falls inside another one. Both are feasible techniques for coupling data and models. The result of joining attributes or locations adds to the input layer.

Joining data into a GIS by their coordinates is a common procedure. Assuming that the position of a station is given as geographic longitude and latitude in degrees, minutes and seconds, this information can be used to add an x-y-feature in a GIS. As a rule, GIS uses decimal degrees, so minutes and seconds have to be converted into decimal degrees. To fit the points from a coordinate system (e.g. survey data from the investigation area) to a given layer in the GIS that is available as a map projection, the positions have to be transformed. The coordinates of the points are projected to the map, considering details like date, zero meridian, and excursion. Transforming data from one grid to another is a critical job of GIS and a challenge for the administration of data and their meta-information. A fitting coordinate system and a fitting projection of the data respectively, is a core issue of spatial data especially at the boundaries of spatial entities.

b) Spatial Interpolation

Transformation, conversion or the translation of data from one unit or level to another is the next step of coupling: e.g. transform frequencies to densities, summarize number and abundance of species to the occurrence of communities in relation to land cover patterns. Most important for the process of integration is the interpolation of single data-points to area-wide information. Interpolation creates a surface, a regionalization, from point-data by geostatistical operations. There are several types of interpolation methods, each with distinct features that treat the data differently and depend on the special characteristics of the data. When applying interpolation methods, one must check whether the source data will change, or the method is subjective (hence, a human interpretation) or objective, and whether the changes between points are abrupt or gradual. Besides triangulated irregular networks used to construct digital elevation models, there are other interpolation techniques such as IDW (Inverse Distance Weighted), kriging (interpolate a random field), and spline (approximate complex shapes) which are widely used (Fortin et al. 2009). In a more ecological context, nearest-neighbour interpolation is a simple method of multivariate interpolation in one or more dimensions. The final result of an interpolation process is a new layer.

It is common to have interpolated point measurements of annual rainfall (see Fig. 22.3b) with IDW. IDW assigns values to unknown points (of an area) by using values from a widespread set of known points. The value at the unknown point is estimated by weighting the sum of N known values. Most GIS-tools tend to dump interpolations as a grid-file, bound by given lines like edges of the investigation area.

c) Spatial Classification

A spatial analysis characterizing distinct areas by information from other layers is one of the core features of a GIS (see Fig. 22.4). The values of one layer are separated by patterns from another layer and then statistically analysed. Thus, the average character of a larger discrete object can be calculated based on regionalized data. This function results in new attributes added to the analysis layer.

GIS offer a variety of geo-processing features, all dealing with the management of the data:

- *Dissolve* aggregates features with the same value of an attribute
- *Clip* cuts features out of a input layer without joining the attributes

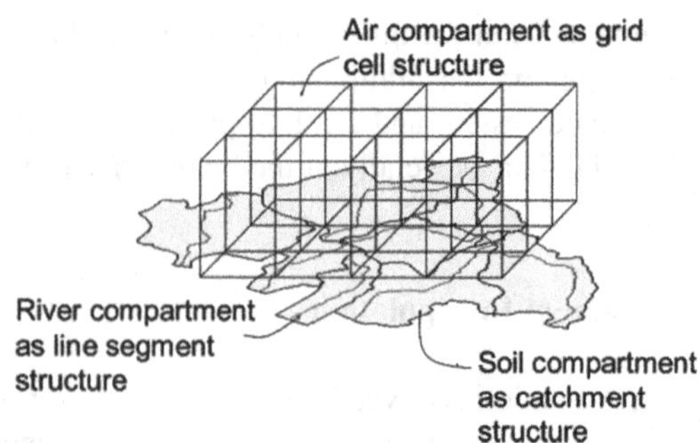

Fig. 22.4 Basic design of the media integration of *G-CIEMS*. Data from different media and different geographical shapes joined by projecting areas or length between the grid cells, catchments and river segments (Suzuki et al. 2004)

- *Union* combines features from different layers
- *Merge* appends features of different files by attributes
- *Intersect* cuts features out of an input layer and joins the attributes

The latter is commonly used to aggregate information from one layer by objects to another. Having interpolated precipitation as a grid and the geometries of the catchments in an area, the information in a GIS can easily be intersected. The GIS either calculates the mean value within (underneath) a geometry, adding this result to the attributes of the file, assuming that the separate catchments will be modelled as a whole; or *cuts* the intersection of the catchments and iso-surfaces of values from the grid, assuming different allocations. The latter aspect requires that the raster (or grid) is being converted into a feature file, following adjustable preferences on classification of the values.

d) Spatial Aggregation

Aggregation aims at the simplification of data. It joins existing data based on their characteristics by calculating parameters, for e.g. mean direction of currents, maximum wave height.

Generalization aggregates objects by *combining* or *composing* them, based on specific (spatial) attributes. Data from different sources or dissimilar measurements can be calculated to a characteristic factor or typified in classes or ranges. Results of these functions are new layers, larger than the input layers.

22.3.2 Handling Dynamics

Maps, the primary output of a GIS, are static representations of ecological systems and landscapes, using the topography of their components. In contrast, dynamic data contain changing numbers, positions and states of entities, altering their quantities and qualities in space and time. Typically, series of point measurements are stored in a database, while two-dimensional quantifications are saved in one (or several) image file(s). In both cases, dynamics (the changing character of the objects) are not originally included in the files. The description of the movement of a water body is as well the result of joining single measurements similar as the changes of a habitat or air circulation. Calculations of changes of quantities over time result in speed, rates and directions of changes.

If and how a model is dynamic (just using a pre-calculated file describing the dynamics of an entity, or continuously executing calculations directly with the model) depends on the architecture of the GIS model framework of the IEM: the tighter the coupling between model, GIS and data is (see Sect. 22.2) the easier GIS and model(s) can exchange data. A loose coupling via data files might be limited to a stepwise calculation of a dynamic process, especially considering the lack of (standardized) appropriate file structure. A tight coupling might evoke the tight embedding of the model within a GIS or vice versa.

Actually, GIS need some technical and conceptual extensions to deal with dynamics (Derby et al. 2005). Again, GIS uses methods of integration to access *continuous* data. The location of the underlying data is the key to (spatially) join (see Sect. 22.3) time series as well as to merge periodic data sets. By spatially matching multiple layers or periodic data GIS are able to calculate, to analyse, and to model spatial relationships over time, identifying trends, patterns and changes within a defined location. The GIS matches *layers* of data from the same area by their geo-position; one data set laying on top of the other. The difference between the superimposed data layers yields the *delta* needed to calculate parameters of dynamics. Besides the simple calculation of the difference between the quality of two (or more) layers, GIS offers the calculations of distances, length and areas of the objects that changed.

Temporal/Spatial Aggregation

In addition to one-dimensional point data files from measurements and two-dimensional data from mappings and remote sensing stored in raster or vector files, model data most commonly are stored in n-dimensional data arrays. Platform-independent data interfaces like netCDF (*Network Common Data Format*) offer a sophisticated method to store data as well as metadata. Time series in multi-dimensional files accessed via interfaces, enable the GIS to *cut* single temporal layers from the file. Processing complex dynamic data requires the aggregation of information from multipart files (Shao et al. 2000). Integration necessitates methods to quantify and qualify the changes and to adjust the parameters of models capturing field data.

Data analysis generalizes changes within the arrays, analysing speed, location and extension of the changes. Beyond the single change from one time-step to the next, general trends and directions are relevant to understand the system observed. Boundary layers, diffusion of pollutants and movement of organisms are resulting topics.

Identifying the characteristic aspect that most likely describes the tendency of a data series is one of the core challenges of data analysis. In contrast to time, space enables for-and-back dynamics: an entity apparently might move aimlessly but is heading in an overall direction. Plotting single observations (or a calculation from a model) of the movement of an entity in a GIS-map (by joining them), makes directions of spatial trends obvious and calculable.

Scenarios

Modelling dynamics, especially projecting known behaviour of an ecosystem to a probable future, must deal with inaccuracies and uncertainties (see also Chap. 2 on Model Development). Methodically immanent inaccuracies are in the end the result of the complexity of the systems we observe: the more parameters a monitoring involves, the more variables a model calculates, the bigger the methodical error is likely to be. Calculating an area-wide parameter based on a limited number of measurement points increases the propagated error, although each measurement might be accurate. That especially applies for projections of the development of highly dynamic components of ecosystems like meteorological events. Describing scenarios is an appropriate method in dealing with these uncertainties and increases the relevance of a projection. Descriptions can be used to estimate the spectrum of possible future scenarios, and determine boundary conditions of the simulations scenarios, which fix the edges of complex IEM. Comparing the results of calculations based on different scenarios gives a hint about the possible dynamics of an ecological system.

22.4 Example Applications

A wide range of IEM exists worldwide. The spectrum of IEM, in a broad sense, spans from global world models via regional landscape models up to sophisticated decision support systems. Classic examples for world models are the computer simulations world3 – more system oriented than spatially explicit – used by the Club of Rome (The Limits of Growth, Meadows et al. 1972) and the spectrum of models used by the IPCC to describe and project the ongoing climate change (IPCC 2007). With the increasing power available with GIS and the desire for integrated approaches for management, many example applications have been developed that address problems in atmosphere–surface systems, hydrology, forestry and biology (see e.g. Basnyat et al. 2000; DeVantier et al. 1993; Goodchild 1993, 1996; Haan 1996; Su 1997). The development of IEM, now using the GIS capabilities to be more spatially explicit, does not end with an increasing number of decision support systems dealing with questions of resource management and environmental policy support (Denzer 2005; Freda 1995; Keen 1978; Lam et al. 2004; Struss 2009).

As integrated approaches are an integral component of e.g. European politics and legal frameworks, the demand for IEM increased significantly in the last years. This is highlighted by diverse DSS for river basin management plans resulting from the legal requirements of the Water Framework Directive (WFD). Due to historic development as well as good funding and early need of management, hydrodynamic models acted as a cutting edge of model integration. As many ecological models take hydrology into account, hydrodynamic models are an integral part of environmental modelling and many approaches in IEM are derived from hydrological applications. GIS-based models dealing with emissions in catchments are good and common examples for IEMs.

MONERIS

MONERIS (Modelling Nutrient Emissions in River Systems) is an example of a model calculating the water quality in catchment areas. The model addresses three goals:

- Identifying the sources and pathways of nutrient emissions at the smallest calculation units level.
- Analysing the transport and the retention of nutrients in river systems.
- Providing support for examining management scenarios of different adaptation measures.

MONERIS evaluates emissions of nutrients from point sources as well as from diffuse sources into surface waters. As Fig. 22.5 illustrates, it integrates many sub-systems: Beginning with the atmospheric deposition, paths of nutrients via urban areas, overland run-off as well as effects of erosion, drainage and groundwater are represented.

As point data (e.g. waste water treatment plants), areal information (e.g. soil data), and administrative information (like statistical data for districts), are integrated, the application of geographic information systems (GIS) is crucial. The GIS integrates the results, although the *MONERIS* system uses only classical database management to join the data and simple spreadsheet processing for calculations.

For scenarios the model evaluates the efficiency of management measures, assigning the measures applied on analytical units to catchment level. The multiple measures implemented by the user can be based on analytical units or cover larger areas, predicting the effect of measures on loads in the whole catchment. *MONERIS* has been applied to numerous river systems: the Axios, Danube, Daugava, Elbe, Odra, Po, Rhine, Vistula, all of Germany and river catchments in Canada, Brazil and China (Behrend et al. 2005; Behrendt et al. 2000; Behrendt et al. 1999; Schreiber et al. 2005; von Sperling et al. 2007: Xu 2004). *MONERIS* is an example for a loose coupling approach, using GIS preferentially to aggregate both, input data from measurements and information calculated by the (sub)models.

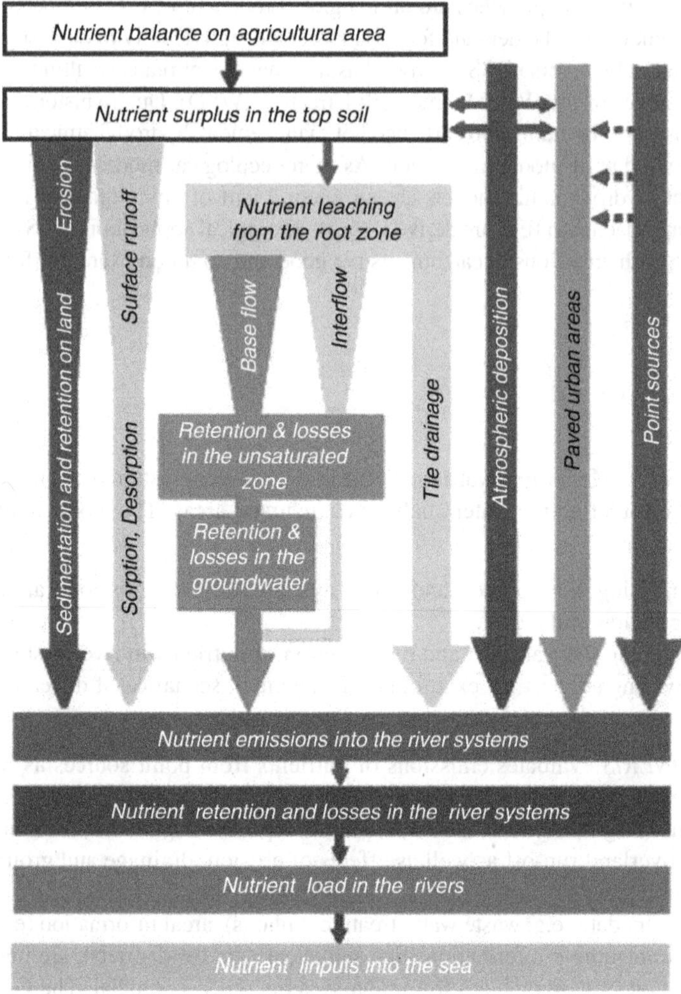

Fig. 22.5 Paths of nutrients within MONERIS (Behrendt et al. 1999)

G-CIEMS

Geo-Referenced Multimedia Environmental Fate Model (G-CIEMS) is a dynamic forecasting model. Suzuki et al. (2004) applied a GIS to make the model spatially resolved, georeferenced and multi-medial. The fate model projects explicit information of the distributions of chemicals in the media air, waters, and soils. Transports between different media with topographically different shapes are calculated based on the projected area or distance. The case study for Japan was based on air grid cells, catchments and river segments performing the projection for air pollutants. *G-CIEMS* calculates exposure-weighted averaged concentrations in air to approximate the

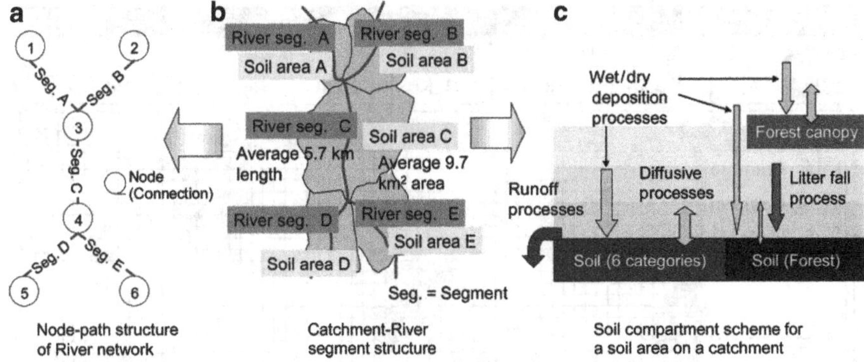

Fig. 22.6 Transformation of the geographical structures of the sub-catchments and river segments (*centre*) to the river network (*left*) and the soil compartments (*right*) within G-CIEMS (Suzuki et al. 2004)

exposure of the population. In this case, the implication for the use of an IEM approach was to provide more accurate exposure estimation with distribution information, using generally available data sources. Suzuki et al. (2004) highlight with the *G-CIEMS* approach that a geo-referenced and spatially highly resolved all-encompassing approach can result in higher (within the factor of 2–3) accuracy of exposure estimation as compared to the results from a monitoring approach. Figure 22.6 illustrates the general information flow within the approach. *G-CIEMS* is an example for a tighter coupling of models and databases, using GIS to join data from different media (see Fig. 22.4).

WadBOS

The information system *WadBOS* developed by RIKS (Research Institute for Knowledge Systems, Buuren et al. 2002), offers support to decision makers in the Dutch Waddenzee area, a coastal zone of tidal flats located in the southern North Sea. This DSS features an integrated model representing the ecological and economic functions of the Wadden Sea system. The submodels represent processes operating at different time scales, varying from daily to annually, and they characterize processes operating at three different spatial scales: the entire sea, 12 homogeneous compartments within the sea, and small cellular units. The *WadBOS* system relies on GIS information for its inputs, but its models use economic, demographic and ecological data from other sources. Its different temporal and spatial scales correspond to the different organizational levels: e.g. the short-term, local impact of the disturbance of breeding birds by boats on the lower, more detailed scale of habitats projected on the higher temporal and spatial scale of management and policy, to generalize effects per breeding season and different protection zones.

As RIKS developed *WadBOS* for laypersons, it is notably user-friendly. It has remarkable interactive capabilities like many editable parameters. GIS capabilities

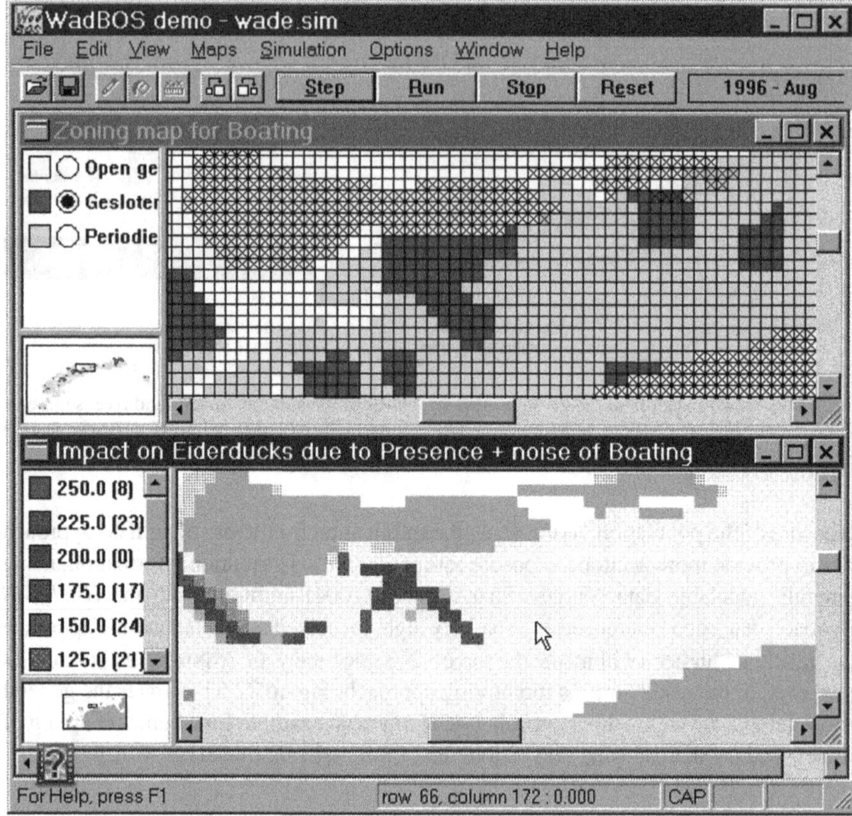

Fig. 22.7 Screenshot of *WadBOS*: Zoning maps for boating in relation to the impact of boating on Eiderducks
Source: http://www.riks.nl

integrated in the *WadBOS* framework allow the (geo)graphical representation of its dynamic output (Fig. 22.7). In addition, the system is highly transparent, as a documentation is available for any module of the system. *WadBOS* can perform different spatial analyses, however, its communication and learning capabilities are at least as important. *WadBOS* is an example for a tight coupled IEM, were GIS and submodels are embedded in an enclosed software environment.

22.5 Conclusions and Future Prospects

GIS enable the assessment of data and models thus facilitating many technical requirements of model integration. Today, different software and modelling frameworks offer remarkable capacity for the technical (and conceptual) development of the Integrated Environmental Modelling (IEM) systems (Filippi et al. 2004).

Thereby, they improve the recognition, description, and evaluation of the characteristics of the ecosystems in focus. The progress made in the technical development of GIS, model environments and frameworks, and the technical and conceptual problems that are being overcome in IEM imply a positive future development for these methods (Pullar 2004; Rebolj et al. 1999). GIS may be applied as a form of decision support systems capable to manage diverse scenarios based on their location, showing options for actions and the potential consequences, and highlighting developments.

Using typical GIS-Software as a main part of integrated models has its pros and cons: On the one hand, GIS provide flexible and approachable techniques to join scientific (simulation) models with management applications, which allows to easily add new data, models or knowledge. Expertise gained in a certain model is portable to another site, time, or political and legal framework. Common GIS are not yet primarily dealing with dynamics like changing size of an entity or changing entities over time. Nevertheless, analysing dynamics by describing allocation, dispersal and distribution as well as patterns, boundaries and shifts is a main aim of the application of GIS. On the other hand, the intense use of GIS for model integration leads to a fixed technical (and conceptual) framework. To some extent GIS-applications define and delimit integrative approaches, giving quasi-standards of data, their format and the means of manipulation and exchange. Rather than contributing services and tools for analysis, management and visualization of spatial data to model integration, GIS are often understood as *the* technical and conceptual frameworks for environmental modelling. However, dealing with 3D and 4D data is still a bottleneck of actual GIS.

Still, developing widely accepted technical structures and conceptual issues across disciplines is a challenge (Hoch et al. 1998). Once the development and application of IEM has become a routine tasks, we can have a closer look at our modelling essentials itself. Finally, we can dare to think about approaching some of the technical and conceptual conventions in different ways.

The issues arising from new environmental directives (e.g. Kay et al. 2006; Borja et al. 2006) show the relevance of integrating GIS and modelling approaches and point to what is expected from models if they are to offer adequate information to support decisions. Many of the issues refer particularly to spatial scales and organizational levels. They serve as important conditions of monitoring programmes and the construction of integrated models and may specify the relevance of information, knowledge and decisions.

Additionally, integration also means joining information and knowledge, and – with the above-mentioned background of new legal frames – sharing it. Transparency, awareness raising, and participation are crucial objectives of sustainability and integrative science. In this context, transparency is understood as clearly communicating the background and methodology of any model to the public and to make accessible the underlying data and evaluation approaches. Awareness raising could mean to enable the stakeholders to use a model to learn about the system (Wang et al. 2005). Participation in IEM allows stakeholders to join the decision-making process by e.g. defining their own model parameters, scenarios and options for calculation.

Part IV
Integrative Approaches in Ecological Modeling

Chapter 23
How Valid Are Model Results? Assumptions, Validity Range and Documentation

Hauke Reuter, Fred Jopp, Broder Breckling, Christoph Lange, and Gerd Weigmann

Abstract In the previous chapters we have described a spectrum of different modelling techniques and application examples. Now we return to overarching aspects which are relevant for model developments of different types. Model validity and considerations which conclusions can be derived from model results are presented jointly for different modelling techniques and application fields. Furthermore, we discuss adequate means of communicating the models to others. The overall views are largely theory-based and show that knowledge on the theoretical background can present an important guidance to making the most appropriate use of ecological modelling results.

23.1 The Last Stage of Development: Limits of Reliability

In the first chapter of this book we started the introduction of ecological modelling by pinpointing the possible discordance in the relationship of formalized models and real-world ecological objects. When assessing the quality of models and model results, we have to keep this in mind. The structure of an ecological model does not represent the structure of the real ecological objects one-to-one. It is a formalized extract to represent ecological interactions. When discussing the limits of reliability for the developed construct, it is important to be aware of this difference. A computer model is not a neutral, objective compound. It is based on the underlying assumptions of model development (Chaps. 1 and 2). A relevant but frequently forgotten part of model development derives from the abstraction concept which underlies the modelling approach. The first step to evaluate the

H. Reuter (✉)
Leibniz Center for Tropical Marine Ecology GmbH (ZMT), Fahrenheitstraße 6, 28359 Bremen, Germany
e-mail: hauke.reuter@zmt-bremen.de

F. Jopp et al. (eds.), *Modelling Complex Ecological Dynamics*,
DOI 10.1007/978-3-642-05029-9_23, © Springer-Verlag Berlin Heidelberg 2011

validity of the model therefore is to estimate how well the given ecological context conforms to what the chosen modelling approach actually can capture. Hence, three questions need to be considered when starting to assess the quality of a computer model (1) what are the underlying assumptions of the model and (2) how do these assumptions relate to the model results of the specific case, and (3) do the boundary conditions hold?

The importance can be well illustrated with ordinary differential equation (ODE) models (Chap. 6). Characteristic for ODEs are functional relations of a limited number of variables. When using these relations within an ecological context, in most implementations it is assumed that heterogeneous structures and activities can be functionally represented in form of homogeneous variables. The degree to which this assumption holds is very important for the validity of the results. ODEs are frequently used to describe population growth or density. Starting from an *Eulerian* perspective (see Turchin 1998), for a population study a point in space can be focussed and the fluxes or the numbers of dispersing population members are counted for this point. The counts (f.i., the bypassing individuals of the population) are summarized and averaged over the inspected time interval. In physics the density of particles in a liquid may be measured this way. A question that needs to be considered is how reasonable it is in a particular application to assume that population density can be measured with such an approach? This crucially depends on species characteristics and behaviour. What might be operational for certain unicellular organisms in an homogeneous medium may be inappropriate for colonial birds – with many situations in between where the degree of adequacy is an issue to be carefully examined.

When we have assessed these questions we can proceed to step 2: in what way do the specific assumptions interfere with the model outcomes? What we have at hand is a formal abstraction, not the biological objects themselves. That is, the model does not give us information about the area in which the population was recorded, specific properties of the individuals, their possible interactions, spatial heterogeneities in the environment or temporal/seasonal differences – unless specifically implemented. Model representation focuses, e.g. only on the temporal changes of the sheer numbers as they were recorded in the point of interest.

The third aspect refers to the problem delimitation. In constructing the model, decisions have been made as to which aspects to include, and which ones to leave out of consideration. In a specific application or repetition of a situation it may well be the case that unexpected external influences play a role which were not considered during development. In most of the practical cases where model predictions do not hold, this implicit *ceteris paribus* condition was not valid (see Sect. 23.4.1).

In all further steps of the assessment of specific model qualities, e.g. how it fits to empirical data and situations, the answers of the questions about assumptions and interactions will be of great value to judge the reliability limits of a developed model. Biological relevance and plausibility are crucial aspects in model evaluation.

23.2 Working with Model Results

Within the process of constructing a model, ecological processes usually have to be quantified in extensive measurement activities, handbooks have to be consulted, code has to be written, debugged and sometimes, the reasons for unexpected model behaviour have to be analyzed, understood and eventually corrective action has to be taken (Chap. 2). Once the primary part of the developmental work is done and after achieving first results, the work enters a further stage of development: securing the specific correctness of the model results and then, continuing to work with the assessed model and its results. Now, we need to understand what the model results mean, how robust and reliable they are. This process is frequently referred to as *validation*. Depending on the model complexity this part can be as demanding and relevant as the primary model development itself.

Validation of differential equation-based models has a long history (see e.g. Power 1993). For other model types specific evaluation approaches exist, which are accompanied by a large body of literature on specific aspects of model analysis (e.g. Rykiel 1996; Klepper 1997; Sargent 1998; Jager and King 2004). On this basis, some general steps of quality assurance for models can be deduced for most approaches (Jakeman et al. 2006). In this context we discuss the principles of parameter identification, sensitivity analysis and the process of model validation. Finally, we will give some suggestions on how to scientifically communicate model structures and model results. We explicitly do not emphasize the usage of the term verification which is occasionally applied in the context of model evaluation (Oreskes et al. 1994; Mitro 2001). The term verification derives from the Latin *verificare* which stands for *making true*. Philosophers are very careful in using the term *truth*. Can a model represent the truth? In itself it can be formally and mathematically correct which is the reason that the term verification is sometimes used for ensuring a mathematical correct formulation of the model (e.g. Oreskes et al. 1994). There is always a necessary deviation between model representation and represented reality, between *explanans* (the statement that explains) and *explanandum* (the context that is explained). Modellers as well as those who apply models must be aware that a formal construct developed for a system representation and real things *as they are*, do not necessarily coincide. They are not identical. A model can capture eventually some of nature's interesting properties, but not on the basis of a true identity. The important point is to find out to what degree a similarity can be expected. In our opinion, this is indicated better in the term validation than in the term verification, though the latter sometimes is favoured in the literature (e.g. Sargent 1998; Manson 2003).

23.3 Assuring Correctness of Model Results

The protocol for model analysis follows certain steps, in particular, the specification for model parameters, their possible ranges, and the analysis of the model behaviour under specified assumptions (Sect. 2.4 and Fig. 23.1). These steps are based on

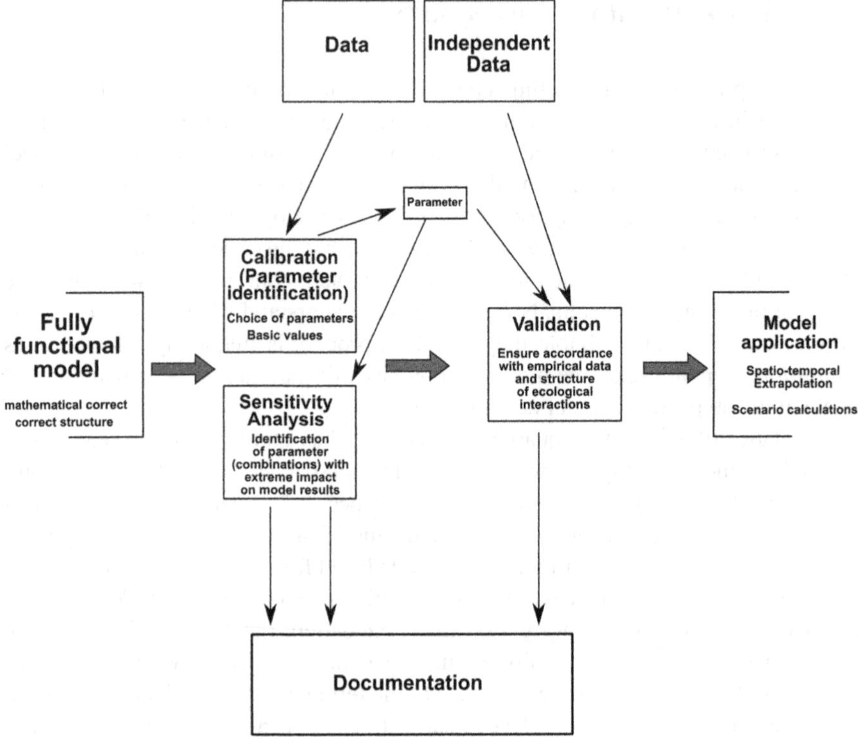

Fig. 23.1 Basic steps to ensure model accuracy. The sequence of steps may be changed depending on model purpose and assessment routines

a preceding assessment of the specific range of contexts for which the model can provide reliable results. Discussing the numerical correctness of the result is an important point of model evaluation. In most (practically all) of the ecologically relevant cases, the mathematical correctness of the executed simulation cannot be mathematically proven (Hawking 1988; Oreskes et al. 1994). Therefore, it is a matter of plausibility analysis to estimate whether the results appear sufficiently reliable. This is of specific importance for differential equation models, but also plays a role for other modelling techniques. The following approaches are frequently used to investigate the numerical correctness – they give hints to increase our trust in the model results but without representing a proof of correctness:

- A relatively laborious way would be to implement the same model on a different hardware- and software platform (on a different computer and/or with a different simulation software, compiler).
- It is less laborious to change step width of model calculation and/or the integration method. This will usually lead to slightly different results. If the differences are not meaningful, the numerical approximation can be considered as appropriate for the aim of the investigation.

- What should be done in addition is a consideration of the expected dynamics: Does the model contain conditions under which critical phases occur – are there extreme changes in very short time intervals? Are there higher order nonlinearities which are important during particular phases of the simulated time? If this is the case or it could occur under particular boundary conditions this represents a challenge for numerical correctness, and all indicated phases should be investigated in detail.
- For critical phases the model behaviour in extreme (but still plausible) situations should be looked at, e.g. by running the model with a higher resolution for the critical conditions. The time span when a population outbreak or collapse occurs can be an example for a critical phase of model development.
- Are there hints for the occurrence of chaotic dynamics (e.g. the occurrence of continuing endogenous changes of amplitude and frequency without an implemented stochasticity)? For chaotic dynamics infinitely small changes are successively amplified. These models usually are not suitable for long-term projections other than statistical interpretations.

In ecology, most of the dynamic processes we deal with do not involve numeric extremes and run relatively smoothly, however, relevant differences of a correct and a numerically approximated result can even be observed in quite simple cases. The Lotka–Volterra equations for a predator prey interaction (see 6.11) can be used for a demonstration. The "true" solutions of the equations are accessible through mathematical integration – unlike most of the other more complex cases. Therefore, we know that the model result is a closed trajectory with constant amplitude. Using the Euler integration routine for simulating the equations, errors accumulate and yield an oscillatory pattern with successively increasing amplitude. Reducing the integration step width improves the situation only gradually. Even for extremely small intervals the effect is still observable. Changing to the Runge Kutta 4th order integration, the results are considerably better. For beginners it can be an interesting and instructive exercise to start such a simulation experiment and to recognize that numeric artefacts are not only a myth and model results must be analyzed carefully before drawing conclusions.

23.3.1 Model Structure and Parametrization: The Issue of Aggregated Parameters

Model parameters do not necessarily derive from direct measurements. In particular, in equations which describe population dynamics (e.g. Chaps. 6, 7, 9), parameters usually represent aggregated averages of a larger set of phenomena, e.g. the overall lifespan, mean population increase or specific tolerance ranges. Sometimes, they can be determined empirically for a specific set of individuals but rarely for the whole population. Regardless of their origin, either from a field campaign or even from further assumptions ("educated guess"), such parameters for aggregated phenomena, can play a crucial role for the overall system dynamics. When these parameters aggregate many external and internal influences, the resulting high context specificity will limit the global applicability of the model and its outcome. If an underlying mechanism can be assumed and measurement data are available, it is frequently possible to determine which parameter value would be best to minimize the differences between modelled and observed data. In the sensitivity analysis it is possible to resolve which parameters are most critical for further adjustments.

23.3.2 Calibration

Calibration (also referred to as "parameter identification") is a procedure in which model parameters are changed to minimize the difference of the model output and a given set of measurement values. For this purpose, a fully developed, executable ecological model must be available. In addition, a target dataset is required that demarcates the output which the model should generate in case of an ideal fit. It is important to ensure that this dataset is independent of the measurements which were used for model specification, because otherwise it would strongly limit any conclusion on the reliability of the identification process.

Systematically, parameter values are varied, applied to the model and the outcome is compared, to see whether the fit was improved. The direction and extent of the change is then used to calculate a new set of parameter values which is again tested in an iterative procedure in which the quality of the fit can be expressed quantitatively (Janssen and Heuberger 1995).

As with other optimization processes, it is not guaranteed that the iterative procedure always finds the best overall fit when stopping at an extreme value. Therefore, it is usually advisable to start parameter identification as close as possible to the assumed values, and repeat the procedure with a number of slightly different starting points to assess whether the results remain comparable. Often we can find a strong dependency between model complexity and data requirements. Lack of data and the use of over-parameterized models may also limit the success of model calibration (see, e.g. Marsili-Libelli and Checchi 2005).

Although most of the established optimization techniques have been originally developed and applied for differential equation-based models as these regularly operate with aggregated parameters, there are also adapted calibration processes being developed for individual-based models IBM (e.g. Pereira et al. 2008), which operate in analogy, i.e. change quantities in model specification and compare the results with regard to an optimality criterion.

The approach of machine learning (see Chap. 19) conceptually expands the approach beyond a variation and adaptation of parameter values. In addition also functional expressions (e.g. different forms of nonlinearity) are changed and then tested, as to whether this would lead to an improved approximation of the target values.

23.3.3 Sensitivity Analysis

After identifying the most adequate model parameters, the procedure of sensitivity analysis provides information about how the model results depend on specific parameter values. In further model development procedures these parameters should receive the highest attention and precision in acquisition effort.

The sensitivity to a given parameter change constitutes an inherent property of the developed model. Its analysis implies a systematic variation and combination of parameters. For a given standard model setting the simulations are repeated with one of the parameters marginally increased or decreased by some defined amount, e.g. ±10% of the starting value. The larger the deviation caused by the minimally changed value, the more sensitive the model is with regard to this parameter (see Fig. 23.2). Such a sensitivity test can be done successively for any chosen set of parameters.

Usually it turns out that many parameters have only relatively small influence, while only a few drastically change the model outcome. Relatively inert parameters can then be considered for elimination in a following model simplification process. For differential equation models it is not uncommon for the fourth or fifth decimal of a sensitive parameter to change the results by fifty percent or more. Many statistical approaches have been applied for sensitivity analysis or have explicitly been developed for specific fields of model testing (e.g. different multivariate approaches, Klepper 1997; spatial aspects of sensitivity analysis, Jager and King 2004).

It has to be noted that the results of a sensitivity analysis are not globally valid: they can be applied only to the given set of parameter values for which the procedures were performed. If more than one parameter is changed simultaneously, model sensitivity may significantly vary between the different settings. Frequently, the result

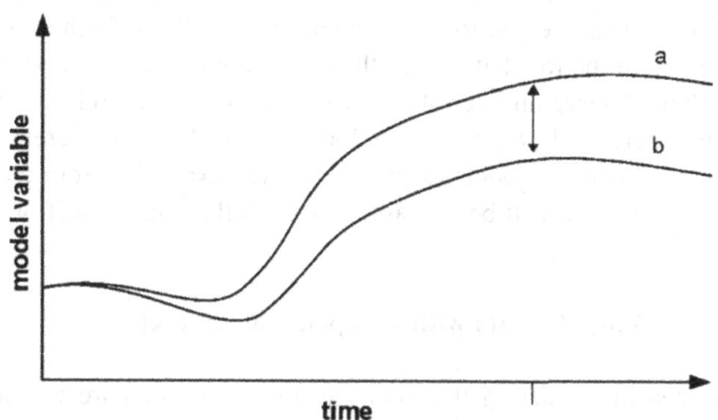

Fig. 23.2 Sensitivity analysis: If a standard model run yields the *upper curve* and a slight deviation of one parameter yields the *curve below*, then the difference of both at a selected point in time (drawn along the x-axis) represents the sensitivity of the model with regard to the changes of the according parameter. Sensitivities of the parameter used in a model can be compared for the sensitivity of the model to parameter changes. Usually, the sensitivity is calculated for a specific point in time. For another example see e.g. Jepsen et al. (2005, Fig. 4)

for one set of parameters provides some orientation also for the surrounding, even if, in a strict mathematical sense, sensitivity analysis has a zero-dimensional validity with respect to the parameter space. Thus, it should be re-applied for any introduction of new parameters or other changes to the original model system.

To cope with the context specificity of sensitivity analysis, multi-parameter approaches for analysing the mutual influence of parameter combinations on model results are applied (e.g. Van Griensven et al. 2006; Makler-Pick et al. 2010). For models with a low number of parameters it may be feasible to analyse most of the possible combinations. For more complex models this could by far exceed the number of possible setups. Thus, it is very important to explicitly control the necessary number of selected parameters for sensitivity analysis and employ techniques to reduce the number of required model runs and replicates.

The concept of sensitivity analysis can also be extended to models where dynamics are substantially determined by the inherent rules and less by external parameters: when models depict individual interactions this could lead to the change of a behavioural rule or it could even involve the structure of the model itself (Jakeman et al. 2006).

23.3.4 Model Validation

Model validation tries to identify how reasonable, reliable and precise model results are with respect to the scientific focus and the intended use and purpose. Not for all desirable situations such a concept can be realized in the strict sense (Konikow and Bredehoeft 1992). Besides assuring numerical correctness, specific investigations of model properties and of their relation with the ecological context (i.e. the appropriate degree of ecological realism) can help to decide to reject or accept a model. If a model was created to depict a specific situation, further assessment is mandatory to determine to what degree the model application is generalizable (see e.g. Rykiel 1996). A large number of approaches exist to test model predictions and validity ranges (Sargent 1998; Troitzsch 2004; Martis 2006). Different approaches are applied for validating models (Table 23.1 gives examples from recent modelling literature) which we will be explained in the following subsections.

Comparison of Model Results with Independent Datasets

An essential step in validating the model results is to compare the outputs with independent datasets which previously have not been used in the model development process. Such a comparison is considered a standard procedure (see e.g. Fig. 23.3). This comparison can be based on a split dataset using the parts not previously used in parametrization and calibration, or may apply cross validation techniques, in which iteratively different subsets of the data are used for analysis (training set) and validation (validation set, see also Chap. 14). Aside from cross

Table 23.1 Recent examples of model analysis approaches. The table illustrates the large range of techniques and different approaches used in model validation

Topic and modelling approach	Model evaluation approach	Reference
Movement and distribution of elks; Individual-based Model (IBM)	Multi-step approach which comprised testing of the developed model and the external ecosystem model SAVANNA Visual accordance of spatio-temporal patterns of habitat use between model and observation on different organization levels Comparison of density maps and accordance of movement corridors Correlation of space use with environmental factors (snow cover) for simulated animals and independent samples with simple statistical tests	Rupp and Rupp (2010)
Suitability of habitats for forest passerines; Spatial Distribution Models (SDM)	Analysis of Habitat suitability in relation to demographic data Habitat suitability index was integrated into a general linear model framework. Testing and validation was done between the different models	Rittenhouse et al. (2010)
Prediction of copepod community dynamics, biological–physical coupled population model	Genetic algorithms for parametrization of copepod properties Statistical comparison (root mean square difference) of model predictions with independent datasets from different years Despite varying results from genetic algorithms, data fit may be equally well	Record et al. (2010)
Investigation of causal factors which facilitate the spread of plague; IBM	Structural validation which focuses on the correctness of process representation and not exclusively on the accuracy of dynamics as data to compare results are scarce Qualitative assessment of key model properties (e.g. secondary infections, contact frequency) to variation in input parameters	Laperriére et al. (2009)
Testing models for prediction of the distribution of invasive plants; Different modelling approaches	Comparison of model results with data previously not used in model parameterization. Statistical evaluation e.g. using presence and absence data, maximized Kappa to measure the proportion of correctly classified points and the area under Receiver-operating characteristic (ROC) curves (AUC) for threshold-independent values	Evangelista et al. (2008)

(*continued*)

Table 23.1 (continued)

Topic and modelling approach	Model evaluation approach	Reference
Sustainable production of mussel aquaculture; Box ecosystem model coupled with a hydrodynamic model	Set-up of 2 year experiments to compare key variables (mussel growth, nitrate, phytoplankton concentrations) from model results with experimental data Visualization of plotted data from experiments and models	Grant et al. (2007)
Hatching of eggs and survival time of herbivore soil-dwelling insects; Partial differential equations	Application of the model to other insect taxa Visualization of plotted data from experiments and models	Johnson et al. (2007)
Dynamics of a woodpecker population; IBM	Qualitative assessment of model performance as location used by the birds are not completely independent (e.g. part of time series) Comparison of secondary model predictions (e.g. natal dispersal distance, population structure) with descriptive statistics	Schiegg et al. (2005)
Prediction of distribution and density of badger sets; (SDM)	Statistical validation with ROC curve using density data from two independent sites Model accuracy as a proportion of correctly classified cells (locations)	Jepsen et al. (2005)

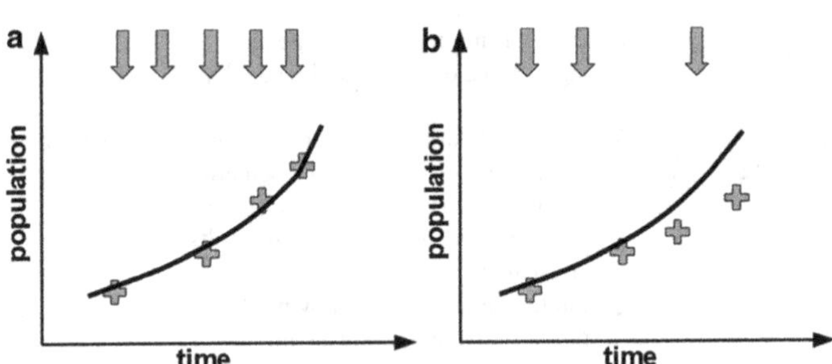

Fig. 23.3 Illustration of the development of a fictional plant population. *Crosses* represent measurement data, the *solid line* model output and the *arrows* precipitation events: (**a**) (*left*) The measurement points were used for model development, the fit it relatively good. (**b**) (*right*) shows an independent dataset with a different precipitation regime (*arrows*) which was not used during model development. The fit is less good and indicates, that for model validation the precipitation response might be reconsidered to improve the application range of the model. For instance, Rupp and Rupp (2010) (Fig. 7) illustrate the application of the approach with empirical data

validation also other approaches which make privileged use of independent datasets for validation can strongly reduce the risk of asking the wrong questions or following hypotheses which seem to be reasonably suggested by the original data (i.e. getting the right answer to the wrong question, type III errors, see Mosteller 1948).

The task of comparing the model output with independent datasets can be focused into the substantial question of statistics, namely whether two samples (the result of the model and the independent data) are derived from the same population. Evidently, this is a further source of uncertainties and even errors in the qualitative assessment of the model. Applying statistical tests to data can be viewed as modelling of the underlying properties of their distributions. Therefore such testing implies an additional step of abstraction and reduction of information of data. In fact, what one does is to model the independent data and extracting certain test statistics which are then compared with the corresponding model of the data as generated by a model using ecological data. The more links such a chain of data manipulation has, the more caution is necessarily concerning the interpretation of their final results.

In applying statistical tests for model evaluation the user is prone to the usual errors and pitfalls of statistics. Among those might be the risk to underrepresent the actual shape of the data distribution with the test statistics and, in consequence, to assume a common population of the samples. This phenomenon has been elaborately demonstrated by Anscombe (1973), who has shown that critical qualities of distributions, like mean and variance of a dataset's x and y, the correlation between x and y, as well as the linear regression line can be identical in four vastly different datasets.

Another usual problem arises from datasets that violate fundamental assumptions of the applied statistical tests. For parametric tests this is mainly the assumption of normality of data. Normality might be achieved by certain transformation techniques, but sometimes one has to refer back to non-parametric tests, e.g. when testing categorial data. Though there is no normality assumption for non-parametric tests, there still are assumptions about data distribution that can be violated (see e.g. Jopp and Lange 2007). Therefore, a generalized "assumption of no assumption" for non-parametric statistics is not appropriate.

Among users of statistics there seems to be a firm belief in a $p < 5\%$ for significant test results. While this is far from being irrational, there are more important lessons to be learned for the proper application of statistics, especially regarding natural science data: Nothing in statistics is unquestionable, not even the desire of a $p < 5\%$ (Stoehr 1999). In contrast, it is far more important to always refer back to the ecological sources of the data and to interpret them according to the available biological knowledge. There are examples where using any statistics at all will lead you to the wrong conclusions, because statistics is inherently ignorant of the involved scientific disciplines. This is the reason why, in some cases, alternative approaches, like structural model validation, might be more appropriate than a straight statistical validation.

Structural Model Validation

Structural validation investigates in how far the model mechanisms reproduce the proposed characteristics of the studied ecological context as described by the conceptual model. Thus the model should not only reproduce the observed system behaviour but also reflect the causal mechanisms and processes in which the real system operates to generate this behaviour (Troitzsch 2004). This approach is mostly applicable for model systems which allow to represent self-organization processes and where higher-level properties emerge from the interaction of lower-level entities (IBM, Chap. 12). When it can be ensured that no type III error is made (see above), such a consistency check for key processes and dynamics on different hierarchical levels increases the probability that the system behaviour has been represented correctly and results are reliable within the applied conceptual system.

An example for a structural validation was presented by Laperriére et al. (2009) who modelled the spread of plague with an IBM. Although the processes leading to disease outbreaks and the role of the different components of the infection system are well known, the data for the investigated case on the island of Madagascar are scarce with low temporal resolution, which is not uncommon for ecological data (Jopp and Lange 2007). Therefore, the model validation process was performed more in a conceptional way rather than testing a one-to-one accuracy: various input scenarios (e.g. prevalence levels in recipients) were linked to model output and a comparison of sufficiently general characteristics of the target systems (e.g. threshold levels) was performed.

Structural validation allows to test theory-driven models. It is often performed with expert knowledge or when the direct participation of stakeholders is desirable. In the latter case, model results are evaluated by experts for their consistency. Another way of doing this is to perform a Turing test, where the experts have to decide whether the results derive from simulations or from real-world field investigations.

23.3.5 Limits of Validation and Validity Range

The described approaches help to learn more about the conditions under which a model can be used and how eventually critical points can be avoided. Usually there are no sharp boundaries of a model's validity range. It is more reasonable to consider a gradual range of certainty depending on the given conditions. Therefore, frequently models inform more about potential outcomes rather than providing a strict prognosis. Experience shows that ecological prognoses are difficult. Models usually are stronger in providing `if-then` information: `if` the model assumptions hold, `then` a specific result can be expected. However, whether the given structure and parameter values are actually the optimal description for a specific case remains difficult to foresee.

23.4 Assessment of Higher-Level Model Implications

Validation attempts to directly relate model output with ecological observations. It is also possible, and for some model approaches useful, to include more indirect and conceptual analyses which investigate model implications on higher organizational levels, and thereby, extend the possible range of validation applications.

Many ecological situations depend on very specific local conditions and are hence difficult to reproduce: an example for this would be a pronounced spatial heterogeneous distribution which is partially amplified through biotic activities. Or as in standard models of plankton algae population dynamics: it is difficult to meet all responses to the influencing conditions and thus to simulate specific site conditions. However, the quality of a model could be assessed by comparing the frequency of certain situations (e.g. population outbreaks or occurrences of the maxima of a particular species) in a larger time span and analyse whether repeated model applications with randomly varying input would lead to comparable distributions. Thus, for many complex situations it is possible to develop reasonable models which might not be able to directly describe a particular development of a process but at least capture and formalize certain general characteristics. In this case, validation procedures which operate on higher-level characteristics can be of great help (see e.g. Sect 12.4).

23.4.1 Scenario Calculations and the Validity Window

In situations where it is not possible to predict precise developments it can be useful to investigate scenarios (see Chap. 22). In the context of ecological modelling this means to consider a set of assumptions and then use a model to calculate the implications of the considered assumptions. In other chapters you find interesting examples for scenario calculations, e.g. in Chap. 2 a model shows the implications of carabid beetle dispersal. This model analyses the colonization success of beetles for several different scenario settings with different stepping stones in an adverse landscape surrounding (Fig. 2.6). Another example relates to the nutrient retention capacity simulated under the assumption of different agricultural processing intensities (Fig. 2.7, nitrogen leaching in industrial agriculture and in green agriculture simulated with the WASMOD modelling system).

A seemingly negative result of a scenario calculation would be to find out that certain expected model results are not compatible or consistent with particular input conditions. This frequently limits the window of expectation in an interesting way, as it helps us to understand more details of the investigated ecological processes and relates them to a larger context.

In other cases of model validation we can find out how far the conditions can be varied before the output becomes ecologically unreasonable. For temporal and spatial extrapolations it is important to note that, again, a *ceteris paribus* condition is used. This means that all basic relations and processes which are not explicitly changed in the scenario setup are kept unchanged, and that all external influences which are considered only implicitly, remain the same, hence ignoring potential changes in processes, reactions of organisms or structural relationships. It is in the responsibility of the modeller to consider and document this thoroughly when extrapolating the results.

For complex ecological situations it is interesting to see that to a surprisingly large extent the degree of freedom is relatively limited. Selecting biologically plausible parameters within a certain range often leads to reasonable results without elaborated fine-tuning. A reason for this is that, on the level of the single organisms, processes operate with a certain robustness, as depending on constricting extreme precisions would frequently conflict with the sustainability of the general conditions for existence. On the other hand, the occurrence of extreme sensitivities frequently indicates that highly aggregated process descriptions were used. Ratz (1995) illustrates in a forest fire model that model assumptions are likely to be unrealistic if the result depends on extremely fine-tuned parameter (in this case within a linear model), whereas other assumptions (here, the introduction of a nonlinearity) makes the model considerably less sensitive to changes in parameterization. Hence, we can state that beyond limits of plausibility for biological parameters the results frequently turn unrealistic.

23.4.2 Result-Probability Distributions, Parameter Ranges and Phase Transitions

Often, not just one specific simulation result is of interest, but instead we have to deal with a whole class of simulation output resulting from systematic tests of varied initial conditions, parameter ranges or external influences. Such an approach is frequently used for dynamic systems in all fields of application. When for a given system the outcome cannot be predicted with a high precision, frequently it is possible to specify a probability range which can be derived through a large number of model re-runs with varied parameters or initial conditions.

Frequently, an unevenly distributed probability density structure can be found and characteristic spatio-temporal probability structures emerge. Structures can comprise heterogeneous output densities, phase transitions between different dynamic or structural regimes and gaps in which it is improbable to find the model's state. Without specific reference to this concept, it has been used in a previous chapter (see Fig. 1.4). It shows the result of re-iterations of a small equation-based system with a successively changed input parameter and shows the response space which exhibits a complex fractal structure.

23.5 Model Documentation and Communication

Modelling involves complex interaction networks (Reuter et al. 2008). While a short formula can be relatively easily surveyed, it is a challenge of its own for most of the practically relevant ecological applications to bring them in a form that conforms to scientific standards, including an independent confirmation of the results. This marks the difference between art and science: while an artist is happy with a unique product, a scientific result is not valid if it is not independently reproducible by other researchers.

In ecological modelling we are faced with the problem that often the model code is too long to fit into the usual format of a scientific publication. Though this situation has improved significantly through internet-based model repositories, some authors still prefer to keep the source code of their model confidential. Though this excludes the model in a strict sense from the scientific discourse, it is possible to specify the employed relations in a way that an independent re-programming (eventually on a different software and hardware base) should be possible and can arrive at qualitatively (or even quantitatively) comparable results. Describing a model in such a way that the reader knows which relations have been used for model specification is crucial for the scientific discourse. How models should be documented is in itself a topic of a discourse. The ecological modelling practice has arrived at practical solutions which can be found in the respective literature (see e.g. the Journal "Ecological Modelling").

It is largely a matter of consensus which issues are required for model documentation in the scientific literature. The same general requirements hold as in other fields of science, which we summarize here with respect to the usual organization of a scientific paper. Good examples where the organization of scientific communication can be studied are Day and Gastel (2006); McMillan (2006); Alley (1996), and Goben and Swan (1990). Therefore, we concentrate on the particular aspects which relate to the necessary documentation for ecological model issues. In general, the "Authors Instructions" of the particular journals provide additional technical information regarding the requirements and formal standards of a journal.

The Abstract

Is the first part and should be written when the paper is finalized, outlining what the article offers to the reader specifying the topics dealt with and for which aspects results are provided and discussed and why they are relevant.

The Introductory Part

This part should contain a clear specification of the problem that was investigated. What makes the problem scientifically relevant? For this, a consideration of previous work and the state-of-the-art on which the developed model builds is necessary. With respect to the model development a description (or reference) of the performed systems analysis is important. This section should contain the identified system components, the relevant processes and their interrelations. Depending on the focus of the publication this part might also be placed in the following section.

The Material and Methods Section

Here a description is expected as to how and where the data were acquired and a detailed documentation of model design and processes should be given. This can be done either through an extensive model description (underlying assumptions, equations, relations, semantic structures) or through the quotation of a technical paper published (or available) elsewhere. If applicable, an explanation of the acquisition of the database used for model calibration and validation should be provided. The description should mention also the technical means, the specification of the hardware and software that was used.

Model Description Section

If a model is not only applied but scientifically described for the first time, the material and methods section can be shorter and a separate section for the model description should be inserted. The equations or the formal specifications have to be described in a way that (in an ideal case) the reader can re-programme the model on the basis of the description. Thus model documentation should contain relevant parts of the model and the performed simulations. Eventually these are:

- The used software versions, programming language, libraries, compiler, operating system and hardware requirements.
- A detailed model description comprising:
  1. An overview on the model structure with sub-modules (if applicable). A structural diagram and flow charts (see Sect. 2.1), eventually with sub-diagrams which illustrate crucial details.
  2. Information on the mathematical equations and applied rules systems; this should also include the integration methods (in case of differential equations) or scheduling information (for discrete models).
- Parameters for the basic simulations. Eventually a distinction of parameters into those referring directly to model processes and those constituting external or environmental influences should be pursued. This part should also contain a justification of parameters and relation to empirical studies and knowledge.
- Initial conditions for standard simulations.
- Results from sensitivity analysis and validation procedures.
- Initial conditions and parameters for further calculations and definition of scenario configurations.

Due to the wide range of different model types and intentions it is unreasonable to always follow the same standardized scheme of model documentation, as it needs to be meaningfully adjusted to specific needs. In some cases standardization (e.g. Grimm et al. 2006; Schmolke et al. 2010) can provide an orientation as to which parts should be included in a comprehensive model description. We regard it as sufficient to fulfil the overall requirement of reproducibility and established standards of scientific communication. In our opinion, modelling does not pose conditions other than those relevant in science in general. The development of easily accessible repositories improves the conditions to allow scientific peers to exchange models and to independently verify the correctness of given results.

Frequently, some extended parts which are not directly necessary for the direct understanding of model organization and functioning can be placed in an annex or as accompanying material on the internet. Specialized scientific journals have standards on how to document the model code itself and how to allow in a crucial case an investigation that could prove that the results are actually obtained through the model application (and not fictitious). If the code is treated as proprietary, it must be held available in the case of the requirement for formal inspection. In case that there should be reason to investigate the integrity of the scientific work, the authors must be able to prove to have followed the scientific code of conduct.[1]

Results Section

Model output can be represented within figures, tables or statistics, frequently in comparison with empirical data. The specific algorithms used for their generation should be explained in the text. If the text has a more methodological focus, the results of sensitivity analysis and the model validation are also expected here.

Discussion Section

The discussion should encompass the model development (if applicable) and the results to the research question as posed in the introduction and compared with findings of other authors in the same or related fields. Validity considerations and application ranges are of special interest here, as well as the relevance of the results for further application and development.

23.6 Concluding Remarks

It is important to be aware that the model development does not end with code writing. Following the availability of a fully functional model, an extended phase of model evaluation is necessary which comprises the model calibration and determination of parameters, the identification of model sensitivity to changes of single parameters or a combination of parameters and the comparison of model performance and structure with empirically determined values and causal

[1]e.g. US National Science Foundation (http://www.nsf.gov/bfa/dias/policy/rcr.jsp), German Science Foundation (http://www.dfg.de/foerderung/rechtliche_rahmenbedingungen/gwp/index.html).

Chapter 24
Perspectives in Ecological Modelling

Fred Jopp, Broder Breckling, Hauke Reuter, and Donald L. DeAngelis

24.1 Ecological Modelling: A Matured Discipline

In the chapters of this textbook we have presented a broad panorama of the network of discourse from which Ecological Modelling emerged and grew. Starting from the very early days, we have proceeded to give an overview of a wide spectrum of currently available approaches. Then, after looking at a selection of prominent model applications, we discussed how to assess model validity.

In the beginning, ecological modelling was largely influenced by approaches outside of biology. Ecology was one of the disciplines that started relatively late to use quantitative methods and theory. One relevant impetus for considering quantitative relations came from economics (Malthus 1798). Quantification of human interference with natural systems has always been relevant in agriculture. With the development of agricultural chemistry (Liebig 1831), the targeted adaptation of quantitative methods to production-oriented ecosystems became important. The quest to understand density-dependent regulation (Verhulst 1838; Pearl 1927), predator–prey interactions, and species competition sparked the borrowing of differential equations from classical mechanics (Lotka-Volterra equations; Lotka 1925; Volterra 1926) to analyze ecological dynamics. While the differential equation approach was successfully applied to areas of population ecology, it was structurally not as adequate for addressing the heterogeneity within populations (e.g., age, size and spatial structure) as it was for addressing dynamic and equilibrium behaviour at the level of whole populations and simple food webs. While analysis of food webs tended to be dominated by a few standardized approaches, such as energy or biomass budget models (Chap. 5, Ecopath), a wider variety of approaches emerged to deal with heterogeneity within populations. The challenge that drove methodological development for a number of decades was how to cope with the difficulties of representing temporal and spatial heterogeneities. With the development of object-oriented approaches (Dahl et al. 1968; Kaiser 1976, 1979;

F. Jopp (✉)
Department of Biology, University of Miami, P.O. Box 249118 Coral Gables, FL 33124, USA
e-mail: fredjopp@bio.miami.edu

F. Jopp et al. (eds.), *Modelling Complex Ecological Dynamics*,
DOI 10.1007/978-3-642-05029-9_24,

Hill 1996) and its import into ecological modelling (Hogeweg and Hesper 1983; Huston and DeAngelis 1988; DeAngelis and Gross 1992; Wolff 1994). It is reasonable to state that there now exists an adequate spectrum of approaches to deal with the broad spectrum of problems in ecology, and these approaches can be applied individually or in combination with other approaches, whatever is required by the problem.

This successful development of a highly flexible repertoire of quantitative approaches is why we consider ecological modelling to be a mature discipline. It consists of a large body of methodologies appropriate for the full spectrum of ecological research, from the assessment of small-scale and short-term individual behavioural pattern (autecology) through various scales and levels of organization (population, ecosystems, landscapes and biomes) up to processes on the biosphere level. Because of the wide range of approaches available, both the student who is interested getting his or her bearings in this field, and the researcher already working in ecology, need some help and orientation to identify the most reasonable approaches for a given problem.

24.2 Model Categorization Provides an Orientation but Hybridization and Mixing of Approaches Remain Constitutive

The model categorization we have developed in this text for such orientation is not an ontological one and the structure we used is not mandatory, as it does not constitute the only reasonable approach. Our design is somewhat subjective and only one of a number of possible ways of presenting an overview. All models have the common property that they can represent an existing or an imaginary ecological situation in a formal setting that uses particular forms of abstraction. The process of abstraction entails selecting and conceptually isolating particular relations from their natural interaction context, while leaving other relations out of consideration. Since ecology deals with such a wide variety of different situations, the modeller is required to have not only a profound expertise in ecology but also needs special creativity in how to handle abstractions. A given modelling application does not necessarily need to fall fully into one of the prototypical categories of model types outlined above. "Model hybridization" is a frequent feature employed in modelling practice. Thus, modelling applied to ecological problems is not a cookbook discipline of routine applications and standard procedures. It does require a background of knowledge of what the scientific standards for models are and how to apply them in a specific ecological situation, but major contributions to science usually involve innovative modelling. Thus, model categorization has a largely didactical and educational purpose. In the actual practice of science, such categorizations play more a role for orientation rather than for imposing strict rules.

We presented our categorization system of ecological models without ignoring that a strict categorization might result not only in logical problems (see Chaps. 2 and 4) but could also lead to inadequate practical results. As far as we see this issue, sharp boundaries between model types are not reasonable. For intellectual inspiration such categories are only important insofar as they help to identify and understand real world processes.

24.3 Structurally and Functionally Realistic Model Construction Tends to Generate Stable Model Behaviour

From the experience of many years of applying models, using different techniques and approaches, we arrived at an interesting conclusion. The more abstract a model is, i.e. the more it ignores the structural heterogeneities of the system being modelled, the more effort is required to parameterize it to meet a specific situation. On the other hand, if the model adequately represents the structural characteristics of the system being modelled, parameterization frequently tends to be not such a difficult matter. That is, model behaviour is not too sensitive to parameter choice, and it is relatively easy to decide on the range of parameters that are biologically reasonable. With the appropriate caution, this observation can be turned around: if we observe crucial sensitivities in a model, it may be a hint that it might be structurally inappropriate for the way the ecological system actually operates. Why this can often be the case is explained below.

Organisms that cannot complete their whole life cycle within the environment that is available to them will quickly vanish from the community of a particular ecosystem. Only organisms that are compatible with a particular environmental setting, and the uncertainty and variability it involves, can persist in the particular system. The considerable losses of biodiversity that accompany environmental alterations indicate that the persistence of organisms is not a trivial issue. It also implies that existence requires an environment that meets stringent conditions. In particular, environmental properties are needed that can maintain sufficient population densities as well as prevent mass proliferations and the resulting overall system instabilities that might lead to collapses. This can be considered as a standard situation, from which of course, exceptions and deviations exist.

Now, modelling such a situation may be difficult. Maintaining a model system in reasonably stable conditions often requires very careful fine-tuning of the model to meet those observed conditions. It is quite likely that in order to make the model mimic those observed conditions, one needs to apply mechanisms in the model that differ from the way the biotic system actually functions. With differential equation based models it is a frequent experience that parameter fine-tuning is required to an extent that is biologically unreasonable (for example, if making the model system behaviour properly depends on the fifth decimal of a parameter).

With object-oriented modelling, however, it is usually possible to approximate the actual structure to the ecological system quite well (see Chap. 12). Consequently, the model results tend to be surprisingly robust when parameterization is chosen according to biologically realistic values. The robustness of model results can be related to the appropriateness of the model assumptions with regard to the problem under investigation.

When one begins to work on an ecological model, one is seldom completely free to choose its logical structure. The core task of the modelling effort is to concentrate on the main driving forces of the complex ecological dynamics that one wants to describe, and so one is compelled to include only the necessary structural components. In other words: Modelling usually should follow Ockham's razor and one should prefer the simplest explanation that conforms to the observations. This implies that it may be necessary to change and adapt the structure of a model when this is *ecologically required* (see Chap. 12). By a change which is *ecologically required*, we mean that either relevant changes in the modelled object structure of biological entities (e.g. status or motivation of the individual) or the modelled environmental structure (e.g. temporal and/or spatial) of the simulated model universe have to be made. With equation-based models, which are mainly parameter driven (see Chaps. 6 and 7), it is frequently more difficult to adapt the model interaction structure, than with the object-oriented approach.

24.4 Across-Scale Modelling and Multi-level Modelling

How simple and straightforward the first ecological models of Verhulst in the nineteenth century now seem – the increase of a population in time is proportional to the population itself minus a constant times the square of the population size and – voilá! – logistic growth emerges. Current progress in modelling is not just a matter of adding building blocks to this ancestral model. The focus of scientific progress in ecological modelling is on developing modelling systems that successfully can deal with the emergence of qualitatively new and unexpected properties that arise from the interaction of particular components. This is a frequent result in object-oriented modelling; e.g., model predators are allowed to randomly consume prey, both populations proliferate – and wow! – the whole spatial area/community enters into coherent oscillations. Though the emerging spatio-temporal pattern is not a property of any single individual interaction, it emerges as a result of the overall ongoing activity.

Starting from the interactions of individuals and their behaviours, through population dynamics, community structure development, ecosystem and landscape processes, and on upwards to the biosphere scale; each step is characterized by self-organization effects that result in emergent properties (Jantsch 1980). The modelling of self-organization processes, is, in our opinion, the supreme objective of ecological modelling and allows to achieve a deeper understanding of how natural processes interrelate (e.g. Kauffman 1993; Gell-Mann 1994). Approximating these interdependencies that occur in the form of emergent phenomena makes ecological modelling the major approach for reconstructing ecological entities in terms of structural and functional relations.

The understanding of emergent properties is basically linked with across-scale modelling and the combination of model descriptions on different organizational levels. Within the modelling workflow, the natural processes are abstracted and reconstructed in computable program code, which allows for arbitrary numbers of repetitions under varying boundary and initial conditions. Aside from the pure coding process, which can in principle be automated using our current understanding of artificial intelligence, we believe that the moment of inspiration, which enables the researcher to formulate the ecological hypotheses, will remain the genuine and authentic domain of the modeller. The better science becomes in simulating the dynamics of complex ecological phenomena across scales and levels as theoretically formulated processes, the more relevant the next task will become. This is to expand our understanding of how processes and succession of natural systems respond and interact with human interferences, and vice versa: how ecosystem dynamics set boundary conditions and thus influence self-structuring processes of social activities and systems. However, this is by no means a trivial task.

24.5 Coupling Models of Natural and Human Systems

It is notoriously difficult to model the interaction of human subjects with each other. If our current understanding were better here, we would not stumble blindly into economic surprises and crises. The ongoing current crisis teaches how far away we are from such a crucial system understanding. Nevertheless, the feedback between social processes and ecological dynamics is one of the most demanding current fields of interdisciplinary research and development (see Costanza 1991). The relation of natural and social dynamics is a new interdisciplinary approach, called social-ecological research, which tries to connect ecological transformation of the society with social justice and economic demands (e.g. PT DLR 2008).

Due to the increase of the economic and ecological pressures on natural and human systems, the question arises whether the modelling of "symbiotic" entities, like agro-ecological systems and the associated social structures (of the farmers as a starting point) would not gain from being depicted together? Imagine, as an example, the situation of the introduction of genetically modified oilseed rape in an agronomic system of Southern Europe, where all the hybridization partners have their centres of origin and, hence, might give rise to a new variety of, e.g. herbicide resistant weeds. In this context, also other new evolutionary implications and self-organized dispersal events may be imagined. Would not the outcome of such a field trial with, for instance, genetically-modified oilseed rape have impacts on the societal structure of the participating farmers (Knispel et al. 2008; Beckie et al. 2003)? Surely this would be the case, and hence, models that work on these phenomena should reasonably be coupled (see Reuter et al. 2010).

24.6 Modelling Complex Ecological Dynamics: Predictions About Its Future!

Ecological Modelling will improve and expand. Models will further increase in their ability to represent the complexity of nature. Model generation, at least for the repetitive parts of models, will be conducted to a greater extent by employing artificial intelligence in model development. Iteration processes will be to a larger extent enabled to iterate iterations, i.e., to proceed to a meta-iteration level. Model-based adaptive management strategies will further improve. On-site or remote sensor driven system interfacing will improve, as well. But will this overall development allow more precise predictions of nature? In some limited fields this could be the case. But concerning strategic issues, nature will remain the source of surprise that it always has been. One of the surprises about modelling: Model-based argumentation can help to explain why this is the case.

Certain small-scale molecular level interactions in non-linear, energy dissipating systems can give rise to altered behaviour, which is then successively amplified to macroscopic levels. Looking at the detailed processes, we can well understand WHY evolutionary processes as a basis of ecological dynamics will proceed in setting alternating goal functions towards which adaptation then proceeds in continuously changing directions. We may also understand the HOW of the processes.

Fig. 24.1 Different views of the Zugspitze (Bavarian Alps), highest peak in Germany. The site is located in the Wetterstein Mts. range in the Northern Limestone Alps

But due to the complex nature of the participating processes, an anticipation of the exact WHERETO will remain limited in the future.

For those who consider models as a crucial means to get the entire range of relevant natural dynamics under anticipatory control, this is certainly a disappointing perspective. Natural self-organization will continue to create qualitatively new relations, system behaviours and organismic properties. Society is not operating on a passive substrate but is required to interact with its living and changing environment. For science this remains a real challenge: Ecological modelling will not run out of demanding new questions that urgently require investigation, in order to come up with new solutions in managing and developing natural resources in a more sustainable way. Ecological modellers are not likely to run out of job opportunities and will continue to contribute new options to enable reasonable interaction pattern of social requirements and natural dynamics. Modelling will require successive updating of its repertoire of methods and applications.

Often, the hallmark of natural science is assessed in terms of its ability to make predictions, which, in turn, can be studied and confirmed by experiments. Ecological models are often applied to cases where concluding experiments are not possible, such as the effects of global climate change, but where appropriate predictions are vital. In the light of the famous quote "predictions are difficult, specifically about the future!", how much confidence do we have in using ecological models for predictions about real world systems?

As in (Fig. 24.1), showing different perspectives of the mountain peak Zugspitze, there are different possible views, attitudes and opinions regarding the role, reliability and relevance of models in ecology.

Your own approach and practise in modelling will help to determine which views and perspectives are to be favoured!

Glossary

Additional terms and definitions are available from **www.mced-ecology.org**

Agent-based Modelling Modelling concept which emerged predominantly in the context of the social sciences to depict human interactions. It uses the same basic concept as *individual-based modelling* (which emerged in ecology). Both terms can be used synonymously.

Alternative Stable States Concept that describes ecological systems with diverse stable states that depend on key drivers. The system can change from one state to another (e.g. algae- vs. coral-dominated reef systems), when a critical threshold is exceeded; this phenomenon is addressed as phase-shift or regime shift (→ *hysteresis*).

Artificial Intelligence Visionary branch of informatics that deals with the automatization of intelligent behaviour, important in research and development; originally based on the idea of mimicking human-like intelligence.

Autonomously Acting Agents Autonomously Acting Agents in → *agent-based modelling*, system behaviour emerges on the basis of the interaction of "agents" which change their state as a response to external influences and their internal conditions. The concept is comparable to individual-based modelling, in which the low-level component is referred to as individual.

Bayesian Inference A method of modern statistics, based on the Bayses' Theorem, which describes the relation of two reverse conditional probabilities.

Beer-Lambert Function Basic absorptiometric law that describes the absorption of light dependent on the concentration of an absorbing component in the material through which the light is travelling.

Black-Box Approach In general, this is any kind of machinery or construct in which only input/output relations are focused on, without requiring considerations of internal operation.

F. Jopp et al. (eds.), *Modelling Complex Ecological Dynamics*,
DOI 10.1007/978-3-642-05029-9, © Springer-Verlag Berlin Heidelberg 2011

Bootstrapping A resampling with replacement method in statistics for drawing conclusions from data that are subject to random variation.

Bray-Curtis Index An ecological metric that is used to describe the dissimilarity of different sites using the species number at each site in relation to the total species number.

Carrying Capacity A population size which does not allow a further increase because of resource or environmental limitations.

Cause-Effect Chain The philosophical assumption that any effect has an antecedent cause, allows a chain- or net-like arrangement of observed interactions that are considered for modelling or other forms of description.

Chaotic Dynamics Special behaviour of dynamic systems with strong dependency on initial conditions which can make long-term predictions impossible.

Classical Logic Logic is a subfield of mathematics concentrated on aspects of reasoning. Classical logic is based on a number of axioms: law of identity (if a statement is true, then it is true), law of the excluded middle (a statement is either true or false), law of non-contradiction (a statement cannot be both true and false). Classical logic is bivalent: the truth value of a statement may only be either true or false. Statements can be connected by logical operators (and, or, if..then, ...), whose definition allows to identify the truth value of the resulting statement. There are also types of nonclassical, multivalued logic, e.g. fuzzy logic, which rejects the laws of excluded middle and non-contradiction, allowing a statement to be both true and false.

Computer Scenarios In computer experiments, different sets of conditions are grouped into alternative cases and the reaction of the model components to the variation of these characteristics is assessed through simulation. Scenarios are widely used to evaluate the outcomes of possible future situations (e.g. global climate change situation at increased temperatures) on the target variables.

Cybernetics This is an interdisciplinary approach of control and regulation of complex systems, like machinery or living organisms, founded by Norbert Wiener. A classic example is the regulation of temperature by a thermostat.

Ecological Niche Describes the position of a species in a formalized multidimensional space of environmental variables and variables describing resource requirements – e.g. the tolerated range of temperature, salinity, or minimum quantities of soil nutrients.

Ecosystem Indicators Measurable features in management and conservation ecology which are used to document certain ecosystem states (like ecosystem health). An indicator has to be measurable more easily than the target that it indicates. To be usable as an indicator, it needs to have a defined relationship to the ecological condition that is indicated.

Ecosystem Resilience The ability of an ecological system to resist external disturbances and to maintain its primary state with state variables, driving variables, and parameters. To trigger a → *phase-shift* the disturbance has to be stronger in higher resilient systems.

Eigenvalue It is a number λ satisfying the equation Av = λv, the non-zero vector v being the **eigenvector** of the → *square matrix* A. In general, a matrix acts on a vector by changing both its magnitude and its direction. A matrix acts on an eigenvector by multiplying its magnitude by a factor, which is positive if its direction is unchanged and negative if its direction is reversed. This factor is the eigenvalue associated with that eigenvector. They both give important information about the matrix, and can be used e.g. in matrix factorization.

Eigenvector See *eigenvalue.*

Emergent Properties On the basis of interactions between lower-level components, new qualities can appear on higher hierarchical integration levels which represent more than the sum of the constituting elements.

Exergy A thermodynamical concept, that indicates the sum of the energy in a system that can realize work, when the system is brought into a thermodynamic equilibrium.

Feedback Process It describes the mechanism in a → *cause-effect chain,* in which the output signal loops back and influences the input conditions.

Fractal A geometric term to describe objects which show self-similarity on different levels of magnification. Any part of the object thus has the same type of basic pattern/structure as the whole. The geometry of many natural phenomena like coastlines, snowflakes, clouds, dispersal pathways, can be described using fractal geometry. Plotting the measured quantities versus scale on a log-log graph gives a straight line with the slope indicating the fractal dimension.

Functional Groups Collection of organisms that respond to environmental stimuli in a similar way or have similar properties with respect to the investigated questions.

Gaussian Distribution A very common continuous probability distribution, based on the central limit theorem stating that a collection of independent random variables with $n \to \infty$, is normally distributed around a single mean value (normal distribution).

General Systems Theory An interdisciplinary approach for explaining complex phenomena by their system's character, founded by the biologist Ludwig von Bertalanffy, closely related to → *cybernetics.*

Generic Model A model that describes general properties of an ecological system and can be easily adapted to different specific situations.

Goal Functions In the ecological context adapted principles from thermodynamics to describe systems being far from equilibrium and which have the tendency to develop towards a particular state.

Hierarchy Theory This was derived from → *general systems theory* to deal with complexities in a system that is spanned over a range of hierarchical interacting levels in space and time. For ecology, these principles were successfully adapted since the works of Allen and Starr during the 1980s.

Hysteresis For systems with a memory effect, the strong correlation between input signal and output reaction can be impaired. Then, multiple systems' states can shift rapidly from one state to another. As a consequence, predictions for future time intervals might be impossible.

Inference This (in logic) is the transition from premises to conclusion, in such a way that they are logically connected. The typical form of the logical connection for inference is the so-called modus ponens: "if p then q": premises are taken and a conclusion is returned. If the premises are true, they guarantee the truth of the conclusion. An inference system is composed by a set of "if...then" rules, and it provides the support for mapping from a given input to an output.

Information Criteria Measure of the goodness of fit of a statistical model which formally describes the relation between model complexity and accuracy. Some examples are the Akaike's information criteria (AIC), the area under the curve (AUC), the coefficient of determination (R2), or the Nash-Sutcliffe model efficiency coefficient.

Intermediate Disturbance Hypothesis Concept that hypothesizes that local species diversity is highest for intermediate levels of disturbance; because very low disturbance levels favour competitively superior species, whereas exceptionally high disturbance levels increase the risk of extinction for most of the species.

Lacunarity Analysis Statistical analysis to describe the scale-dependent distribution of gap sizes.

Landscape Fragmentation A major threat to local biodiversity relates to the fragmentation of former coherent habitats as population size may fall below a critical threshold and dispersal and exchange between populations may be reduced or prevented.

Landscape Metrics General term that is used for the available collection of metrics for analyzing and interpreting landscape composition and configuration.

Metapopulation An ecological concept describing the overall dynamics of a number of sub-populations in space and time. It assumes partially independent dynamics of the sub-populations, a risk of local extinction and recolonization events. Thus the overall population may survive under conditions where the sub-populations would go extinct. In nature conservation this concept has also been discussed in the context of increasing → *landscape fragmentation.*

Michaelis-Menten Equation Mathematical model for enzyme kinetics which quantitatively relates the reaction speed of the enzyme-substrate complex to the substrate concentration.

Monod Kinetics Mathematical model that describes enzymatically mediated chemical reactions depending on the concentration of the substrates. See also → *Michaelis-Menten kinetics.*

Monte Carlo Approach Stochastic procedure, based on the law of large numbers; uses repeated random sampling techniques to solve complex problems numerically.

Nonlinear Systems This do not react proportionally to input signals in every case and usually cannot be solved by first order equations. Many ecological relationships are nonlinear, e.g. predator-prey population interactions tend to be proportional to the product of the population sizes.

Numerical Approximation When no explicit analytical solution exists for a mathematical problem or the solution cannot be obtained without reasonable costs, numerical computing can be introduced for calculating close estimates.

Percolation Theory Percolation describes the movement of entities or fluids in porous materials. In landscape ecology it has been used to describe the movement of organisms in fragmented landscapes and to detect critical levels of connectivity between habitat patches.

Phase Shifts See *alternative stable states* and *hysteresis.*

Phenotypic Plasticity The ability of an organism to change its phenotype (physical shape) in response to changes in the environmental conditions.

Poisson Distribution A discrete probability distribution that can be obtained by performing repeated random experiments (e.g. Bernoulli experiments) where occurrence is independent of the former events.

Population Viability Analysis A mathematical tool often applied in conservation biology, which calculates the probability that a population becomes extinct under a set of environmental conditions within a given time span.

Random Walk Mathematical description of a trajectory which is constructed from successive random elements. In ecology it is often used to describe the movement of organisms.

Self-Organization A process in which an overall systems state emerges from parallel and distributed interaction of its constituent elements without any central steering instance. Usually, nonlinear and nonequilibrium processes are involved. Many ecosystem states can be described as a result of self-organized processes.

Social-Ecological Systems A combined system of social and ecological components and drivers which interact and give rise to results which could not be understood on the basis of sociological or ecological considerations alone.

Soft Computing An emerging computer science area, inspired by biological systems and human mind, used to model complex systems arising in management science, medicine, biology and ecology. It consists of a variety of techniques, including fuzzy logic, neural networks, genetic algorithms, Bayesian networks. They all differ from conventional (hard) computing since they tolerate uncertainty, partial truth, approximation, lack of categoricity, and imprecision to achieve tractability, robustness and low solution cost. Soft computing approaches emulate characteristics of human reasoning such as the learning, training, and other types of high-order cognitive power.

Square Matrix This is a matrix which has the same number of rows and columns. An $n \times n$-matrix is a square matrix of order n. The Identity Matrix $\mathbf{I_n}$ is a square matrix of order n in which all elements on the main diagonal are equal to 1 and all other elements are equal to 0. The inverse $\mathbf{A^{-1}}$ of a square matrix is defined as: $\mathbf{A} \times \mathbf{A^{-1}} = \mathbf{I_n}$. If an inverse exists, **A** is called invertible or non-singular.

State Variable In dynamical systems, it is understood that the state of the system is completely characterized by the states of the variables within the system.

Trophic Efficiency The ratio of production which relates one trophic level in terms of energy and biomass with the adjacent level.

Trophic Level The hierarchical position which an organism occupies in the trophic web structure.

Weibull Distribution A continuous probability distribution in statistics that is used to describe, e.g. lifespans and failure rates of compound aggregates in analytical quality assurance procedures.

References

Aars J, Ims R (2002) Intrinsic and climatic determinants of population demography: the winter dynamics of tundra voles. Ecology 83:3449–3456

Abel DJ, Kilby PJ, Davis JR (1994) The systems integration problem. Int J Geogr Inf Syst 8 (1):1–12

Abelson H, diSessa AA (1982) Turtle geometry. MIT Press, Cambridge

Ackerly DD, Schwilk DW, Webb CO (2006) Niche evolution and adaptive radiation: testing the order of trait divergence. Ecology 87:S50–S61

Adriaenssens V, De Baets B, Goethals PLM, De Pauw N (2004) Fuzzy rule-based models for decision support in ecosystem management. Sci Total Environ 319:1–12

Allee WC (1931) Animal aggregations: a study in general sociology. University of Chicago Press, Chicago

Allen KR (1971) Relation between production and biomass. J Fish Res Board Can 28:1573–1581

Allen LJS (2003) An introduction to stochastic processes with applications to biology. Pearson, Upper Saddle River, NJ

Allen TFH, Starr TB (1982) Hierarchy: perspectives for ecological complexity. University of Chicago Press, Chicago

Allen TFH, Starr TB (1992) Towards a unified ecology. Columbia University Press, New York

Alley M (1996) The craft of scientific writing, 3rd edn. Prentice Hall, New Jersey [and accompanying web site: http://filebox.vt.edu/eng/mech/writing/]

Alpaydin E (2010) Introduction to machine learning, 2nd edn. MIT Press, Cambridge, MA

Alstad DN (2007) Populus. Simulations of population biology; at http://www.cbs.umn.edu/populus/, University of Minnesota

Altunkaynak A, Ozger M, Calkmakci M (2005) Fuzzy logic modelling of the dissolved oxygen fluctuations in Golden Horn. Ecol Modell 189:436–446

Andersen KP, Ursin E (1977) A multispecies extension to the Beverton and Holt theory of fishing, with accounts of phosphorus circulation and primary production. Medd Fra Dan Fisk Hav NS 7:319–435

Anderson TR (2005) Plankton functional type modelling: running before we can walk? J Plankt Res 27(11):1073–1081

Andrieu B (guest ed) (1999) Architectural modelling of plants. Agronomie (Special issue) 19:161–328

Ang JPO, De Wreede RE (1990) Matrix models for algal life history stages. Mar Ecol Progr Ser 59:171–181

Anscombe FJ (1973) Graphs in statistical analysis. Am Stat 27:17–21

Appice A, Dzeroski S (2007) Stepwise induction of multi-target model trees. In: Kok JN, Koronacki J, Lopez de Mantaras R, Matwin S, Mladenic D, Skowron A (eds) ECML 2007. LNCS 4701, pp 502–509. Springer, Heidelberg

Araújo MB, Guisan A (2006) Five (or so) challenges for species distribution modelling. J Biogeogr 33:1677–1688

Araújo MB, Thuiller W, Pearson RG (2006) Climate warming and the decline of amphibians and reptiles in Europe. J Biogeogr 33:1712–1728

Argent RM (2004) An overview of model integration for environmental applications-components, frameworks and semantics. Environ Modell Softw 19(3):219–234

Argent RM, Voinov A, Maxwell T, Cuddy SM, Rahman JM, Seaton S, Vertessy RA, Braddock RD (2006) Comparing modelling frameworks – A workshop approach. Environ Modell Softw 21 (7):895–910

Aronson RB, Macintyre IG, Wapnick CM, O'Neill MW (2004) Phase shifts, alternative states, and the unprecedented convergence of two reef systems. Ecology 85:1876–1891

Arreguín-Sánchez F, Hernández-Herrera A, Ramírez-Rodríguez M, Pérez-España H (2004a) Optimal management scenarios for the artisanal fisheries in the ecosystem of La Paz Bay, Baja California Sur, Mexico. Ecol Modell 172:373–382

Arreguín-Sánchez F, Zetina-Rejón M, Manickchand-Heileman S, Ramírez-Rodríguez M, Vidal L (2004b) Simulated response to harvesting strategies in an exploited ecosystem in the southwestern Gulf of Mexico. Ecol Modell 172:421–432

Aspinall R, Pearson D (2000) Integrated geographical assessment of environmental condition in water catchments: linking landscape ecology, environmental modelling and GIS. J Environ Manage 59:20

Atanasova N, Todorovski L, Dzeroski S, Kompare B (2006) Constructing a library of domain knowledge for automated modelling of aquatic ecosystems. Ecol Modell 194:14–36

Atanasova N, Todorovski L, Dzeroski S, Kompare B (2008) Application of automated model discovery from data and expert knowledge to a real-world domain: Lake Glumsø. Ecol Modell 212:92–98

Austin MP (2002) Spatial prediction of species distribution: an interface between ecological theory and statistical modelling. Ecol Modell 157:101–118

Backhaus JO, Hainbucher D (1987) A finite difference general circulation model for shelf seas and its application to low frequency variability on the North European Shelf. In: Nihoul JC, Jamart BM (eds) Three dimensional models of marine and estuarine dynamics, vol 45, Elsevier oceanography series. Elsevier, Amsterdam, pp 221–244

Bak RPM, Meesters EH (1998) Coral population structure: the hidden information of colony size-frequency distributions. Mar Ecol Progr Ser 162:301–306

Barnsley MF (1988) Fractals everywhere. Academic Press, Boston

Bascompte J, Solé RV, Martinez N (1997) Population cycles and spatial patterns in snowshoe hares: an individual-oriented simulation. J Theor Biol 187:213–222

Basnyat P, Teeter LD, Lockaby BG, Flynn KM (2000) The use of remote sensing and GIS in watershed level analyses of non-point source pollution problems. For Ecol Manage 128 (1–2):65–73

Battisti A, Stastny M, Buffo E, Larsson S (2006) A rapid altitudinal range expansion in the pine processionary moth produced by the 2003 climatic anomaly. Glob Change Biol 12: 662–671

Batzli GO (1996) Population cycles revisited (Grand Forks, North Dakota, USA, June 1996). Trends Ecol Evol 11:488–489

Bauer S, Berger U, Hildenbrandt H, Grimm V (2002) Cyclic dynamics in simulated plant populations. Proc R Soc Lond B Biol Sci 269:2443–2450

Beale CM, Lennon JJ, Yearsley JM, Brewer MJ, Elston DA (2010) Regression analysis of spatial data. Ecol Lett 13:246–264

Beckie HJ, Warwick SI, Nair H, Séguin-Swartz G (2003) Gene flow in commercial fields of herbicide-resistant canola (*Brassica napus*). Ecol Appl 13:1276–1294

Becks L, Hilker FM, Malchow H, Jürgens K, Arndt H (2005) Experimental demonstration of chaos in a microbial food web. Nature 435:1226–1229

Beddington JR (1975) Mutual interference between parasites or predators and its effect on searching efficiency. J Anim Ecol 44:331–340

Begon M, Mortimer M, Thompson DJ (1996) Population ecology, 3rd edn. Blackwell, Oxford

Behrend H, Dannowski R (2005) Nutrients and heavy metals in the Odra river system emissions from point and diffuse sources, their loads, and scenario calculations on possible changes. Weissensee, Berlin

Behrendt H, Kornmilch M, Korol R, Stronska M, Pagenkopf WG (1999) Point and diffuse nutrient emissions and transports in the Odra basin and its main tributaries. Acta Hydroch Hydrob 27 (5):274–281

Behrendt H, Huber P, Kornmilch M, Opitz D, Schmoll O, Scholz G, Uebe R (2000) Nutrient emissions into river basins of Germany. UBA

Bell JF (1999) Tree based methods. In: Fielding AH (ed) Machine learning methods for ecological applications. Kluwer, Dordrecht

Bellman RE (1957) Dynamic programming. Princeton University Press, Princeton, NJ

Bellwood DR, Hughes TP, Folke C, Nystrom M (2004) Confronting the coral reef crisis. Nature 429(6994):827–833

Belousov BP (1959) Periodicheskaya reaktsiya i ego mekhanizm. Sbornik Referatov po Radiatsionnoi Meditsine za 1958 god, pp 145–149

Bénard H (1900) Les tourbillons cellulaires dans une nappe liquide. Rev Gén Sci Pures Appl Bull Assoc Fr Av Sci 11:1261–1271, 1309–1328

Benton T, Grant A (1999) Elasticity analysis as an important tool in evolutionary and population ecology. Trends Ecol Evol 14:467–471

Betts MG, Ganio LM, Huso MMP, Som NA, Huettmann F, Bowman J, Wintle BA (2009) Comment on "Methods to account for spatial autocorrelation in the analysis of species distributional data: a review". Ecography 32:374–378

Bieber C, Ruf T (2005) Population dynamics in wild boar *Sus scrofa*: ecology, elasticity of growth rate and implications for the management of pulsed resource consumers. J Appl Ecol 42:1203–1213

Bini LM, Diniz-Filho JAF, Rangel TFLVB, Akre TSB, Albaladejo RG, Albuquerque FS, Aparicio A, Araújo MB, Baselga A, Beck J, Bellocq MI, Böhning-Gaese K, Borges PAV, Castro-Parga I, Chey VK, Chown SL, de Marco P, Dobkin DS, Jr Ferrer-Castán D, Field R, Filloy J, Fleishman E, Gómez JF, Hortal J, Iverson JB, Kerr JT, Kissling WD, Kitching IJ, León-Cortés JL, Lobo JM, Montoya D, Morales-Castilla I, Moreno JC, Oberdorff T, Olalla-Tárraga MA, Pausas JG, Qian H, Rahbek C, Rodríguez MA, Rueda M, Ruggiero A, Sackmann P, Sanders NJ, Terribile LC, Vetaas OR, Hawkins BA (2009) Coefficient shifts in geographical ecology: an empirical evaluation of spatial and non-spatial regression. Ecography 32:193–204

Bjorholm S, Svenning JC, Skov F, Balslev H (2005) Environmental and spatial controls of palm (Arecaceae) species richness across the Americas. Glob Ecol Biogeogr 14:423–429

Bjornstad ON, Peltonen M, Liebhold AM, Baltensweiler W (2002) Waves of larch budmoth outbreaks in the European Alps. Science 298:1020–1023

Blockeel H, Struyf J (2002) Efficient algorithms for decision tree cross-validation. J Mach Learn Res 3:621–650

Blockeel H, De Raedt L, Ramon J (1998) Top-down induction of clustering trees. In: Proceedings of 15th international conference on machine learning, Morgan Kaufmann, San Mateo, CA, pp 55–63

Bolker BM (2008) Ecological models and data in R. Princeton University Press, Princeton, NJ

Boonstra R (1994) Population cycles in microtines: the senescence hypothesis. Evol Ecol 8:196–219

Borcard D, Legendre P (2002) All-scale spatial analysis of ecological data by means of principal coordinates of neighbour matrices. Ecol Modell 153:51–68

Borja Á, Dauer DM (2008) Assessing the environmental quality status in estuarine and coastal systems: comparing methodologies and indices. Ecol Indic 8:331–337

Borja Á, Franco J, Valencia V, Bald J, Muxica I, Belzunce MJ, Solaun O (2004) Implementation of the European water framework directive from the Basque country (northern Spain): a methodological approach. Mar Poll Bull 48:209–218

Borja A, Galparsoro I, Solaun O, Muxika I, Tello EM, Uriarte A, Valencia V (2006) The European Water Framework Directive and the DPSIR, a methodological approach to assess the risk of failing to achieve good ecological status. Estuar Coast Shelf S 66(1–2):84–96

Bossel H (1986) Dynamics of forest dieback: systems analysis and simulation. Ecol Modell 34:259–288

Bossel H (1992) Modellbildung und simulation. Vieweg, Braunschweig

Bossel H (1996) Treedyn3 forest simulation model. Ecol Modell 90:187–227

Botkin DB, Janak JF, Wallis JR (1972) Some ecological consequences of a computer model of forest growth. J Ecol 60:849–872

Bouchon J, de Reffye P, Barthélémy D (eds) (1997) Modélisation et simulation de l'architecture des végétaux. Science Update, INRA, Paris

Boulant N, Garnier A, Curt T, Lepart J (2009) Disentangling the effects of land use, shrub cover and climate on the invasion speed of native and introduced pines in grasslands. Divers Distrib 15:1047–1059

Brandmeyer JE, Karimi HA (2000) Coupling methodologies for environmental models. Environ Modell Softw 15(5):479–488

Brauer F, van den Driessche P, Wu J (eds) (2008) Mathematical epidemiology. Lecture notes in mathematics, Vol 1945. Springer, Berlin

Braune E, Richter O, Söndgerath D, Suhling F (2008) Voltinism flexibility of a riverine dragonfly along thermal gradients. Glob Change Biol 14:470–482

Breckling B (1990) Singularität und Reproduzierbarkeit in der Modellierung Ökologischer Systeme, Diss Thesis, Universität Bremen

Breckling B (1996) An individual based model for the study of pattern and process in plant ecology: an application of object oriented programming. Ecosystems 4:241–254

Breckling B, Müller F (1997) Der Ökosystembegriff aus heutiger Sicht – Grundstrukturen und Funktionen von Ökosystemen. In: Fränzle O, Müller F, Schröder W (eds) Handbuch der Umweltwissenschaften. Ecomed, Landsberg

Breckling B, Reuter H, Middelhoff U (1997) An object oriented modelling strategy to depict activity pattern of organisms in heterogeneous environments. Environ Model Assess 2:95–104

Breckling B, Reuter H, Middelhoff U (2000) Self-organisation simplified: simulating spatial structures which emerge from non-directed ecological interactions. Proceedings of the 26th conference of the association of SIMULA users, pp 75–85

Breckling B, Müller F, Reuter H, Hölker F, Fränzle O (2005) Emergent properties in individual-based ecological models – introducing case studies in an ecosystem research context. Ecol Modell 186:376–388

Breckling B, Middelhoff U, Reuter H (2006) Individual-based models as tools for ecological theory and application: understanding the emergence of organisational properties in ecological systems. Ecol Modell 194:102–113

Breiman L, Friedman JH, Olshen RA, Stone CJ (1984) Classification and regression trees. Wadsworth, Belmont

Breine JJ, Maes J, Quataert P, Van den Bergh E, Simoens I, Van Thuyne G, Belpaire C (2007) A fish-based assessment tool for the ecological quality of the brackish Schelde estuary in Flanders (Belgium). Hydrobiologia 575:141–159

Brickman D, Ådlandsvik B, Thygesen UH, Parada C, Rose K, Hermann AJ, Edwards K (2009) Particle tracking. In: North EW, Gallego A, Petitgas P (eds) Manual of recommended practices for modelling physical–biological interactions during Fish Early Life, ICES Cooperative Research Report No. 295, pp 9–19

Broekhuizen N, Stahl JC, Sagar PM (2003) Simulating the distribution of Southern Buller's Albatross using an individual-based population model. J Appl Ecol 40:678–691

Broennimann O, Treier UA, Müller-Schärer H, Thuiller W, Peterson AT, Guisan A (2007) Evidence of climatic niche shift during biological invasion. Ecol Lett 10:701–709

Brooks JL, Dodson SI (1965) Predation, body size, and composition of plankton. Science 150:28–35

Brotons L, Thuiller W, Araújo MB, Hirzel AH (2004) Presence-absence versus presence-only modelling methods for predicting bird habitat suitability. Ecography 27:437–448

Buckley YM, Brockerhoff E, Langer L, Ledgard N, North H, Rees M (2005) Slowing down a pine invasion despite uncertainty in demography and dispersal. J Appl Ecol 42:1020–1030

Buck-Sorlin G, Kniemeyer O, Kurth W (2007) A grammar-based model of barley including virtual breeding, genetic control and a hormonal metabolic network. In: Vos J, Marcelis LFM, de Visser PHB, Struik PC, Evers JB (eds) Functional-structural plant modelling in crop production. Springer, Berlin, pp 243–252

Buddemeier RW, Smith SV (1999) Coral adaptation and acclimatization: a most ingenious paradox. Am Zool 39(1):1–9

Bullock JM, Kenward RE, Hails RS (2002) Dispersal ecology. Blackwell, Oxford

Bullock JM, Pywell RF, Coulson-Phillips SJ (2008) Managing plant population spread: prediction and analysis using a simple model. Ecol Appl 18:945–953

Buuren JV, Engelen G, Ven Kvd (2002) The DSS WadBOS and EU policies implementation, The Changing Coast. EUROCOAST/EUCC: Porto, Portugal, pp. 533–540

Bythell JC, Sheppard C (1993) Mass mortality of Caribbean shallow corals. Mar Pollut Bull 26:296–297

Bythell JC, Gladfelter EH, Bythell M (1993) Chronic and catastrophic natural mortality of three common Caribbean reef corals. Coral Reefs 12:143–152

Carl G, Dormann CF, Kühn I (2008) A wavelet-based method to remove spatial autocorrelation in the analysis of species distributional data. Web Ecol 8:22–29

Carpenter SR, Kitchell JF, Hodgson JR (1986) Cascading Trophic Interactions and Lake Productivity. Bioscience 35(10):634–639

Caswell H (1976) Community structure: a neutral model analysis. Ecol Monogr 46:327–354

Caswell H (1978) A general formula for the sensitivity of population growth rate to changes in life history parameters. Theor Popul Biol 14:215–230

Caswell H (1986) Life cycle models for plants. Lect Math Life Sci 18:171–223

Caswell H (2001) Matrix population models: construction, analysis and interpretation, 2nd edn. Sinauer, Sunderland, MA

Caswell H, Nisbet RM, de Roos AM, Tuljapurkar S (1996) Structured population models: many methods, a few basic concepts. In: Tuljapurkar S, Caswell H (eds) Structured-population models in marine, terrestrial, and freshwater systems. Chapman & Hall, New York, pp 3–17

Cavanagh RD, Lambin X, Ergon T, Bennett M, Graham IM, Van Soolingen D, Begon M (2004) Disease dynamics in cyclic populations of field voles (*Microtus agrestis*): cowpox virus and vole tuberculosis (*Mycobacterium microti*). Proc R Soc Lond B Biol Sci 271:859–867

Champagnat N, Ferrière R, Méléard S (2006) Unifying evolutionary dynamics: from individual stochastic processes to macroscopic models. Theor Popul Biol 69:297–321

Charnel MA (2008) A individual-based model of a tritrophic ecology. Ecol Modell 218:195–206

Chick JH, Geddes P, Trexler JC (2008) Periphyton mat structure mediates trophic interactions in a subtropical wetland. Wetlands 28:378–389

Chitty D (1960) Population increase in the vole and their relevance to general theory. Can J Zool 38:99–113

Chitty D (1967) The natural selection of self-regulatory behavior in animal populations. Proc Ecol Soc Aust 2:51–78

Christensen V (1995) A model of trophic interactions in the North Sea in 1981, the Year of the Stomach. Dana 11(1):1–19

Christensen V, Pauly D (1992) ECOPATH II – a software for balancing steady-state ecosystem models and calculating network characteristics. Ecol Modell 61:169–185

Christensen V, Pauly D (1993) Flow characteristics of aquatic ecoystems. In: Christensen V, Pauly D (eds) Trophic models of aquatic ecosystems. ICLARM Conference Proceedings, Vol 26. pp 338–352

Christensen V, Pauly D (1998) Changes in models of aquatic ecosystems approaching carrying capacity. Ecol Appl 8(1):S104–S109

Christensen V, Walters CJ (2004) Ecopath with Ecosim: methods, capabilities and limitations. Ecol Modell 172:109–139

Claeys G (2000) The survival of the fittest and the origins of social Darwinism. J Hist Ideas 61:223–229

Clark JS (1998) Why trees migrate so fast: confronting theory with dispersal biology and the paleorecord. Am Nat 152:204–224

Clark P, Boswell R (1991) Rule induction with CN2: some recent improvements. In: Kodratoff Y (ed) EWSL 1991, LNCS 482. Springer, Heidelberg, pp 151–163

Clark JS, Lewis M, Horvath L (2001) Invasion by extremes: population spread with variation in dispersal and reproduction. Am Nat 157:537–554

Clobert J, Danchin E, Dhondt AA, Nichols JD (2001) Dispersal. Oxford University Press, Oxford

Colasanti RL, Hunt R (1997) Resource dynamics and plant growth: a self-assembling model for individuals, populations and communities. Funct Ecol 11:133–145

Colbach N, Clermont Dauphin C, Meynard JM (2001) GeneSys: a model of the influence of cropping system on gene escape from herbicide tolerant rapeseed crops to rape volunteers: II. Genetic exchanges among volunteer and cropped populations in a small region. Agric Ecosyst Environ 83:255–270

Comins HN, Hassell MP, May RM (1992) The spatial dynamics of host-parasitoid systems. J Anim Ecol 61:735–748

Connell JH (1997) Disturbance and recovery of coral communities. Coral Reefs 16(Suppl): S101–S113

Conover WJ (1971) Practical nonparametric statistics. Wiley, New York

Cook RM, Sinclair A, Stefansson G (1997) Potential collapse of North Sea cod stocks. Nature 385:521–522

Corbet PS (1999) Dragonflies: behaviour and ecology of Odonata. Harley Books, Colchester

Corbet PS, Suhling F, Söndgerath D (2006) Voltinism of Odonata: a review. Int J Odonatol 9:1–44

Costanza R (ed) (1991) Ecological economics: the science and management of sustainability. Columbia University Press, New York, 525 pp

Costanza R, Ruth M (1998) Using dynamic modeling to scope environmental problems and build consensus. Environ Manage 22(2):183–195

Costanza R, Sklar FH (1985) Articulation, accuracy and effectiveness of mathematical models: a review of freshwater wetland applications. Ecol Modell 27:45–68

Council NR (2001) Grand challenges in environmental sciences, committee on grand challenges in environmental sciences, Oversight Commission for the Committee on Grand Challenges in Environmental Sciences. National Academy Press, Washington, DC, 106 pp

Courchamp F, Berec L, Gascoigne J (2008) Allee effects in ecology and conservation. Oxford University Press, Oxford

Cournède PH (2009) Système dynamique de la croissance des plantes. Université Montpellier II, HDR thesis

Cox DR, Oakes D (1984) Analysis of survival data. Chapman/Hall, London

Cucherousset J, Carpentier A, Paillisson JM (2007) How do fish exploit temporary waters throughout a flooding episode? Fish Manag Ecol 14:269–276

Cury PM, Shin YJ, Planque B, Durant JM, Fromentin JM, Kramer-Schadt S, Stenseth NC, Travers M, Grimm V (2008) Ecosystem oceanography for global change in fisheries. Trends Ecol Evol 23:338–346

Cushing JM, Costantino R, Dennis B, Desharnais RA, Henson S (2003) Chaos in ecology. Experimental nonlinear dynamics, Theoretical ecology series. Academic, Amsterdam

Dahl OJ, Myhrhaug B, Nygaard K (1968) SIMULA67: common base language (2. Rev. 1970) Publ 22. Norsk regnesentral/Norwegian Computing Center, Oslo

Dale VH (ed) (2003) Ecological modeling for resource management, vol XVII. Springer, Berlin

Dale MRT, Dixon P, Fortin MJ, Legendre P, Myers DE, Rosenberg MS (2002) Conceptual and mathematical relationships among methods for spatial analysis. Ecography 25:558–577

Darwin C (1859) On the origin of species. John Murray, London

Day RA, Gastel B (2006) How to write and publish a scientific paper, 6th edn. Cambridge University Press, Cambridge

De Kroon H, Plaisier A, van Groenendael J, Caswell H (1986) Elasticity: the relative contribution of demographic parameters to population growth rate. Ecology 67:1427–1431

De Kroon H, van Groenendael J, Ehrlen J (2000) Elasticities: a review of methods and model limitations. Ecology 81(3):607–618

de Laplace PS (1814) A philosophical essay on probabilities. Wiley/Chapman & Hall, New York/ London

DeAngelis DL, Gross LJ (eds) (1992) Individual-based models and approaches in ecology. Chapman & Hall, New York

DeAngelis DL, Mooij WM (2005) Individual-based modeling of ecological and evolutionary processes. Ann Rev Ecol Evol Syst 36:147–168

DeAngelis DL, Goldstein RA, O'Neill RV (1975) A model for trophic interaction. Ecology 56:881–892

DeAngelis DL, Cox D, Coutant CC (1979) Cannibalism and size dispersal in young-of-the-year largemouth bass: experiment and model. Ecol Modell 8:133–148

DeAngelis DL, Persson L, Rosemond AD (1996) Interaction of productivity and consumption. In: Polis GA, Winemiller KO (eds) Food webs. Integration of patterns and dynamics. Chapman & Hall, New York, pp 109–112

DeAngelis DL, Loftus WF, Trexler JC, Ulanowicz RE (1997) Modeling fish dynamics and effects of stress in a hydrologically pulsed ecosystem. J Aquat Ecosys Stress Recov 6(1):1–13

DeAngelis DL, Gross LJ, Huston MA, Wolff WF, Fleming DM, Comiskey EJ, Sylvester SM (1998) Landscape modeling for Everglades ecosystem restoration. Ecosystems 1:64–75

DeAngelis DL, Trexler JC, Loftus WF (2005) Life history trade-offs and community dynamics of small fishes in a seasonally pulsed wetland. Can J Fish Aquat Sci 62:781–790

DeAngelis DL, Trexler JC, Cosner C, Obaza A, Jopp F (2010) Fish population dynamics in a seasonally varying wetland. Ecol Modell 221(8):1131–1137

Debeljak M, Džeroski S, Adamič M (1999) Interactions among the red deer (*Cerus elaphus* L.) population, meteorological parameters and new growth of the natural regenerated forest in Sneznik, Slovenia. Ecol Modell 121:51–61

Debeljak M, Džeroski S, Jerina K, Kobler A, Adamič M (2001) Habitat suitability modelling of red deer (*Cervus elaphus* L.) in South-Central Slovenia. Ecol Modell 138:321–330

Debeljak M, Cortet J, Demšar D, Krogh PH, Džeroski S (2007) Hierarchical classification of environmental factors and agricultural practices affecting soil fauna under cropping systems using Bt-maize. Pediobiologia 51:229–238

Debeljak M, Squire G, Demšar D, Young M, Džeroski S (2008) Relations between the oilseed rape volunteer seedbank, and soil factors, weed functional groups and geographical location in the UK. Ecol Modell 212:138–146

Debeljak M, Kocev D, Towers W, Jones M, Griffiths B, Hallett P (2009) Potential of multi-objective models for risk-based mapping of the resilience characteristics of soils: demonstration at a national level. Soil Use Manage 25:66–77

Demšar D, Džeroski S, Larsen T, Struyf J, Axelsen J, Bruns-Pedersen M, Krogh PH (2006) Using multi-objective classification to model communities of soil microarthropods. Ecol Modell 191:131–143

Den Boer PJ (1981) On the survival of populations in a heterogenous and variable environment. Oecologia 50:39–53

Denzer R (2005) Generic integration of environmental decision support systems – state-of-the-art. Environ Modell Softw 20(10):1217–1223

Derby FW, Kumar M (2005) Marine information system: suggested improvements, OCEANS. pp 2435–2438

DeVantier BA, Feldman AD (1993) Review of GIS applications in hydrologic modeling. J Water Resour Plan Manage 119(2):246–261

Devaux C, Lavigne C, Austerlitz F, Klein EK (2007) Modelling and estimating pollen movement in oilseed rape (Brassica napus) at the landscape scale using genetic markers. Mol Ecol 16:487–499

Dolk DR, Kottemann JE (1993) Model integration and a theory of models. Decis Support Syst 9 (1):51–63

Dong Q, McCormick PV, Sklar FH, DeAngelis DL (2002) Structural instability, multiple stable states, and hysteresis in periphyton driven by phosphorus enrichment in the Everglades. Theor Popul Biol 61(1):1–13

Dormann CF (2007a) Effects of incorporating spatial autocorrelation into the analysis of species distribution data. Glob Ecol Biogeogr 16:129–138

Dormann CF (2007b) Promising the future? Global change predictions of species distributions. Bas Appl Ecol 8:387–397

Dormann CF (2009) Response to Comment on "Methods to account for spatial autocorrelation in the analysis of species distributional data: a review". Ecography 32:379–381

Dormann CF, McPherson J, Araújo MB, Bivand R, Bolliger J, Carl G, Davies RG, Hirzel A, Jetz W, Kissling WD, Kühn I, Ohlemüller R, Peres-Neto PR, Reineking B, Schröder B, Schurr FM, Wilson R (2007) Methods to account for spatial autocorrelation in the analysis of species distributional data: a review. Ecography 30:609–628

Dormann CF, Gruber B, Winter M, Herrmann D (2010) Evolution of the climate niche in European mammals? Biol Lett 6:229–232

Dorn NJ, Trexler WF, Gaiser EE (2006) Exploring the role of large predators in marsh food webs: evidence for a behaviorally-mediated trophic cascade. Hydrobiologia 569:375–386

Dovciak M, Frelich LE, Reich PB (2005) Pathways in old-field succession to white pine: seed rain, shade, and climate effects. Ecol Monogr 75:363–378

Duncan RP, Cassey P, Blackburn TM (2009) Do climate envelope models transfer? A manipulative test using dung beetle introductions. Proc R Soc B Biol Sci 276:1449–1457

Džeroski S, Drumm D (2003) Using regression trees to identify the habitat preference of the sea cucumber (*Holothuria leucospilota*) on Rarotonga, Cook Islands. Ecol Modell 170:219–226

Džeroski S, Grbović J (1995) Knowledge discovery in a water quality database. In: Proceedings of 1st international conference on knowledge discovery and data mining. AAAI Press, Menlo Park, CA, pp 81–86

Džeroski S, Todorovski L, Bratko I, Kompare B, Križman V (1999) Equation discovery with ecological applications. In: Fielding AH (ed) Machine learning methods for ecological application. Kluwer, Dordrecht

Džeroski S, Gjorgjioski V, Slavkov I, Struyf J (2007) Analysis of time series data with predictive clustering trees. In: Džeroski S, Struyf J (eds) KDID 2006. LNCS 4747. Springer, Heidelberg, pp 63–80

Edgar GA (1990) Measure, topology, and fractal geometry. Springer, New York

Elith J, Graham CH (2009) Do they? How do they? WHY do they differ? On finding reasons for differing performances of species distribution models. Ecography 32:66–77

Elith J, Leathwick J (2009a) Conservation prioritisation using species distribution models. In: Moilanen A, Wilson KA, Possingham HP (eds) Spatial conservation prioritization: quantitative methods and computational tools. Oxford University Press, Oxford

Elith J, Leathwick JR (2009b) Species distribution models: ecological explanation and prediction across space and time. Ann Rev Ecol Evol Syst 40:677–697

Elton C (1927) Animal ecology. Sidgwick & Jackson, London (reprint 1966, Methuen, London)

Eschenbach C (2005) Emergent properties modelled with the functional structural tree growth model ALMIS: computer experiments on resource gain and use. Ecol Modell 186:470–488

Evangelista PH, Kumar S, Stohlgren TJ, Jarnevich CS, Crall AW, Norman JB III, Barnett DT (2008) Modelling invasion for a habitat generalist and a specialist plant species. Divers Distrib 14(5):808–817

Fath B, Patten BC (2000) Ecosystem theory: network environ analysis. In: Jørgensen SE, Müller F (eds) Handbook of ecosystem theories and management. CRC, Boca Raton, pp 345–360

Fauth PT, Gustafson EJ, Rabenold KN (2000) Using landscape metrics to model source habitat for Neotropical migrants in the midwestern US. Landscape Ecol 15:621–631

Fegeas RG, Claire RW, Guptill SC, Anderson KG, Hallam CH (1983) Land use and land cover digital data Geological Survey circular 895-E. US Geological Survey, Reston, VA

Fennema RJ, Neidrauer CJ, Johnson RA, MacVicar TK, Perkins WA (1994) A computer model to simulate natural Everglades hydrology. In: Davis SM, Ogden JC (eds) Everglades – the ecosystem and its restoration. St Lucie, Delray Beach, FL, pp 249–289

Ferber J (1999) Multi-agent systems – an introduction to distributed artificial intelligence. Addison-Wesley, Boston

Ferrari JR, Lookingbill TR, McCormick B, Townsend PA, Eshleman KN (2009) Surface mining and reclamation effects on flood response of watersheds in the central Appalachian Plateau region. Water Resour Res 45:W04407. doi:10.1029/2008WR007109

Field JC, Francis RC, Aydin K (2006) Top-down modeling and bottom-up dynamics: linking a fisheries-based ecosystem model with climate hypotheses in the Northern California Current. Prog Oceanogr 68:238–270

Fielding AH (1999) An introduction to machine learning methods. In: Fielding AH (ed) Machine learning methods for ecological applications. Kluwer, Dordrecht

Fielding AH (2002) What are the appropriate characteristics of an accuracy measure? In: Scott JM, Heglund PJ, Morrison ML, Haufler JB, Raphael MG, Wall WA, Samson FB (eds) Predicting species occurrences: issues of accuracy and scale. Island, Washington, DC, pp 271–280

Filippi JB, Bisgambiglia P (2004) JDEVS: an implementation of a DEVS based formal framework for environmental modelling. Environ Modell Softw 19(3):261–274

Fisher RA (1937) The wave of advance of advantageous genes. Ann Eugen 7:355–369

Fitz HC, Sklar FH (1999) Ecosystem analysis of phosphorus impacts and altered hydrology in the Everglades: a landscape modeling approach. In: Reddy KR, O'Connor GA, Schelske CL (eds) Phosphorus biogeochemistry in subtropical ecosystems. Lewis, Boca Raton, FL, pp 585–620

Fleury V (1999) A possible connection between dendritic growth in physics and plant morphogenesis. CR Acad Sci III Vie 322:725–734

Forrester JW (1961) Industrial dynamics. MIT Press, Cambridge, MA

Forrester JW (1968) Principles of systems: text and workbook chapters 1 through 10, 2nd edn. Wright-Allen, Cambridge, MA

Forrester DM, Kürten KE, Kusmartsev FV (2007) Magnetic cellular automata and the formation of glassy and magnetic structures from a chain of magnetic particles. Phys Rev B 75:014416

Fortin MJ, Dale MRT (2009) Spatial analysis: a guide for ecologists. Cambridge University Press, Cambridge, p 380

Fourcaud T, Zhang X, Stokes A, Lambers H, Körner C (guest eds) (2008) Plant growth modelling. Ann Bot (Special Issue) 101:1053–1293

Fox CJ, Geffen AJ, Blyth R, Nash RDM (2003) Temperature-dependent development rates of plaice (*Pleuronectes platessa* L.) eggs from the Irish Sea. J Plankton Res 25(11):1319–1329

Fränzle O (2006) Complex bioindication and environmental stress assessment. Ecol Indic 6:114–136

Fränzle O, Kappen L, Blume HP, Dierßen K (eds) (2008) Ecosystem Organization of a complex landscape. Longterm Research in the Bornhöved Lake District, Germany, vol 202, Ecological studies. Springer, Heidelberg, pp 297–317

Freda K (1995) Decision support for natural resources management: models, GIS, and expert systems. AI Appl 9(3):3–19

Fréon P, Cury P, Shannon L, Roy C (2005) Sustainable exploitation of small pelagic fish stocks challenged by environmental and ecosystem changes: a review. Bull Mar Sci 76(2):385–462

Frijters D, Lindenmayer A (1974) A model for the growth and flowering of Aster novae-angliae on the basis of table <1, 0> L-systems. In: Rozenberg G, Salomaa A (eds) L-systems. Springer, Berlin, pp 24–52

Fulton EA, Smith ADM, Johnson CR (2003) Effect of complexity on marine ecosystem models. Mar Ecol Prog Ser 253:1–16

Gallego A, Heath MR (2003) The potential role of gadoid settlement on the stock- recruitment relationship: numerical experiments using bio-physical modeling simulations. ICES CM P:11

Gallego A, Heath MR, Wright P, Marteinsdottir G (1999) An empirical model of growth in the pelagic early life history stages of North Sea haddock. ICES CM Y:13

Gallego A, North EW, Petitgas P (2007) Introduction: status and future directions in modelling physical-biological interactions in the early life of fish. Mar Ecol Prog Ser 347:121–126

Gallienne CP, Robins DB, Pilgrim DA (1996) Measuring abundance and size distribution of zooplankton using the optical plankton counter in underway mode. Underwater Technol 21 (4):15–21

Gardner M (1970) Mathematical games: the fantastic combinations of John Conway's new solitaire game "life". Sci Am 223:120–123

Gardner RH (1999) RULE: a program for the generation of random maps and the analysis of spatial patterns. In: Klopatek JM, Gardner RH (eds) Landscape ecological analysis: issues and applications. Springer, New York, pp 280–303

Gardner RH, O'Neill RV (1990) Pattern, process and predictability: the use of neutral models for landscape analysis. In: Turner MG, Gardner RH (eds) Quantitative methods in landscape ecology. The analysis and interpretation of landscape heterogeneity, Ecological Studies Series. Springer, New York, pp 289–307

Gardner RH, Urban DL (2003) Model validation and testing: past lessons, present concerns, future prospects. In: Canham CD, Cole JC, Lauenroth WK (eds) Models in ecosystem science. Princeton University Press, Princeton, NJ, pp 184–203

Gardner RH, Urban DL (2007) Neutral models for testing landscape hypotheses. Landsc Ecol 22:15–29

Gardner RH, Walters S (2001) Identifying patches and connectivity of landscapes. In: Gergel SE, Turner MG (eds) Learning landscape ecology: a practical guide to concepts and techniques. Springer, New York, pp 112–128

Gardner RH, Milne BT, Turner MG, O'Neill RV (1987) Neutral models for the analysis of broad-scale landscape pattern. Landsc Ecol 1:19–28

Gardner RH, Jopp F, Cary GJ, Verburg PH (2008a) World congress highlights need for action. Landsc Ecol 23:1–2

Gardner RH, Lookingbill TR, Townsend PA, Ferrari J (2008b) A new approach for rescaling land cover data. Landsc Ecol 15:513–526

Garnier A, Lecomte J (2006) Using a spatial and stage-structured invasion model to assess the spread of feral populations of transgenic oilseed rape. Ecol Modell 194:141–149

Garnier A, Pivard S, Lecomte J (2008) Measuring and modelling anthropogenic secondary seed dispersal along roadverges for feral oilseed rape. Basic Appl Ecol 9:533–541

Gaston KJ (2009) Geographic range limits of species. Proc R Soc B, Biol Sci 276:1391–1393

Gell-Mann M (1994) The Quark and the Jaguar: adventures in the simple and the complex. WH Freeman, New York

Gerisch G (1968) Cell aggregation and differentiation in Dictyostelium. In: Moscona AA, Monroy A (eds) Curr Top Dev Biol 3:157–197

Gierer A, Meinhardt H (1972) A theory of biological pattern formation. Kybernetik 12:30–39

Gilpin ME (1979) Spiral chaos in a predator-prey model. Am Nat 113:306–308

Glansdorff P, Prigogine I (1971) Thermodynamic theory of structure stability and fluctuations. Wiley, New York

Gnauck A (2000) Fundamentals of ecosystem theories from general systems analysis. In: Joergensen SE, Müller F (eds) Handbook of ecosystem theories and management. CRC Press, Boca Raton, pp 75–88

Goben G, Swan J (1990) The science of scientific writing. Am Sci 78:550–558

Godin C, Sinoquet H (guest eds) (2005) Functional-structural plant modelling. New Phytol 166:705–708, 771–894

Goel NS, Rozehnal I (1991) Some non-biological applications of L-systems. Int J Gen Syst 18:321–405

Goodchild MF (1999) Principles of geographical information systems, 2nd edn. Environ Plan B Plann Des 26(2):315–316

Goodchild MF, Parks BO, Steyaert LT (1993) Environmental modeling with GIS. Oxford University Press, New York

Goodchild MF, Steyaert LT, Parks BO, Johnston C, Maidment D, Crane M, Glendinning S (1996) GIS and environmental modeling: progress and research issues. GIS World books, Fort Collins

Goodchild MF, Egenhofer MJ, Fegeas R (1997) Interoperating GISs: report of a specialist meeting held under the auspices of the Varenius Project. NCGIA, Santa Barbara, p 63

Graf RF, Bollmann K, Suter W, Bugmann H (2005) The importance of spatial scale in habitat models: capercaillie in the Swiss Alps. Landscape Ecol 20:703–717

Grant WE, Swannack TM (2007) Ecological modeling: a common-sense approach to theory and practice. Wiley, New York

Grant J, Curran KJ, Guyondet TL, Tita G, Bacher C, Koutitonsky V, Dowd M (2007) A box model of carrying capacity for suspended mussel aquaculture in Lagune de la Grande-Entrée, Îles-de-la-Madeleine, Québec. Ecol Modell 200:193–206

Grime JP (1973) Competitive exclusion in herbaceous vegetation. Nature 242:344–347

Grimm V, Railsback S (2005) Individual-based modelling & ecology. Princeton University Press, Princeton

Grimm V, Berger U, Bastiansen F, Eliassen S, Ginot V, Giske J, Goss-Custard J, Grand T, Heinz S, Huse G, Huth A, Jepsen JU, Jørgensen C, Mooij WM, Müller B, Pe'er G, Piou C, Railsback SF, Robbins AM, Robbins MM, Rossmanith E, Rüger N, Strand E, Souissi S, Stillman RA, Vabø R, Visser U, DeAngelis DL (2006) A standard protocol for describing individual-based and agent-based models. Ecol Modell 198:115–126

Gross LJ, DeAngelis DL (2001) Multimodeling: new approaches for linking ecological models. In: Scott JM, Heglund PJ, Morrison M, Raphael M, Haufler J, Wall B (eds) Predicting species occurrences: issues of scale and accuracy. Island, Covello, CA

Groves LR (1962) Now it can be told: the story of the Manhattan Project. Da Capo, New York

Guénette S, Christensen V, Pauly D (2008) Trophic modelling of the Peruvian upwelling ecosystem: towards reconciliation of multiple datasets. Prog Oceanogr 79:326–335

Guisan A, Thuiller W (2005) Predicting species distributions: offering more than simple habitat models. Ecol Lett 8:993–1009

Guisan A, Zimmermann NE (2000) Predictive habitat distribution models in ecology. Ecol Modell 135:147–186

Guisan A, Broennimann O, Engler R, Vust M, Yoccoz NG, Lehmann A, Zimmermann NE (2006) Using niche-based models to improve the sampling of rare species. Conserv Biol 20:501–511

Guisan A, Graham CH, Elith J, Huettmann F, NCEAS Species Distribution Modelling Group (2007) Sensitivity of predictive species distribution models to change in grain size. Divers Distrib 13:332–340

Guo Y, Ma Y, Zhan Z, Li B, Dingkuhn M, Luquet D, de Reffye P (2006) Parameter optimization and field validation of the functional-structural model GREENLAB for maize. Ann Bot 97:217–230

Gustafson EJ (1998) Quantifying landscape spatial pattern: what is the state of the art? Ecosystems 1:143–156

Gustafson EJ, Gardner RH (1996) The effect of landscape heterogeneity on the probability of patch colonization. Ecology 77:94–107

Haan CT, Storm DE (1996) Nonpoint source pollution modeling (with GIS). In: Singh VP, Fiorentino M (eds) Geographical information systems in hydrology: water science and technology library. Kluwer, Dordrecht, pp 323–338

Haeckel E (1866) Generelle Morphologie der Organismen. Verlag Georg Reimer, Berlin

Haeckel E (1868) Natürliche Schöpfungsgeschichte. Gemeinverständliche wissenschaftliche Vorträge über die Entwickelungslehre im Allgemeinen und diejenige von Darwin, Goethe und Lamarck im Besonderen, über die Anwendung derselben auf den Ursprung des Menschen und andere damit zusammenhängende Grundfragen der Naturwissenschaft. Verlag Georg Reimer, Berlin

Hagen-Zanker A, Lajoie G (2008) Neutral models of landscape change as benchmarks in the assessment of model performance. Landsc Urban Plan 86:284–296

Hairston NG, Smith FE, Slobodkin LE (1960) Community structure, population control and competition. Am Nat 879:421–425

Hajkova P, Hajek M, Apostolova I, Zeleny D, Dite D (2008) Shifts in the ecological behaviour of plant species between two distant regions: evidence from the base richness gradient in mires. J Biogeogr 35:282–294

Haken H (1977) Synergetics – an introduction: nonequilibrium phase-transitions and self-organization in physics, Chemistry and Biology. Springer, Berlin

Hampe A, Petit RJ (2005) Conserving biodiversity under climate change: the rear edge matters. Ecol Lett 8:461–467

Hanan J, Prusinkiewicz P (guest eds) (2008) Functional-structural plant modelling. Funct Plant Biol (Special Issue) 35(9/10):i–iii/739–1090

Hancock JF, Grumet R, Hokanson SC (1996) The opportunity for escape of engineered genes from transgenic crops. HortScience 31:1080–1085

Hannon B, Ruth M (2001) Dynamic modeling, 2nd edn. Springer, Berlin

Hanski I, Gilpin ME (1997) Metapopulation biology: ecology, genetics, and evolution. Academic, San Diego, CA

Hanski I, Korpimäki E (1995) Microtine rodent dynamics in northern Europe: parameterized models for the predator-prey interaction. Ecology 76(3):840–850

Hargis CD, Bissonette JA, David JL (1998) The behavior of landscape metrics commonly used in the study of habitat fragmentation. Landsc Ecol 13:167–186

Harper JL (1977) Population biology of plants. Academic Press, London

Harrell FE Jr (2001) Regression modeling strategies – with applications to linear models, logistic regression, and survival analysis. Springer, New York/Heidelberg

Harrison TD, Whitfield AK (2004) A multi-metric fish index to assess the environmental condition of estuaries. J Fish Biol 65:683–710

Harvey B (1997) Computer science logo style, 2nd edn., 3 vols. MIT Press, Cambridge, MA. Also available at http://www.cs.berkeley.edu/~bh/

Hassall C, Thompson DJ (2008) The effects of environmental warming on Odonata: a review. Int J Odonat 11:131–151

Hassall C, Thompson DJ, French GC, Harvey IF (2007) Historical changes in the phenology of British Odonata are related to climate. Glob Change Biol 13:933–941

Hastie T, Tibshirani R, Friedman JH (2009) The elements of statistical learning: data mining, inference, and prediction, 2nd edn. Springer, Berlin

Hastings A, Powell T (1991) Chaos in a three-species food chain. Ecology 72(3):896–903

Hawking SW (1988) A brief history of time: from the big bang to black holes. Bantam, New York

Haydon DT, Shaw DJ, Cattadori IM, Hudson PJ, Thirgood SJ (2002) Analysing noisy time-series: describing regional variation in the cyclic dynamics of red grouse. Proc R Soc Lond B Biol Sci 269:1609–1617

Heath MR, Gallego A (1998) Biophysical modelling of the early life stages of haddock in the North Sea. Fish Oceanogr 7:110–125

Heath MR, Gallego A (2000) Modelling the spatial and temporal structure of survivorship to settlement in North Sea and west of Scotland haddock. ICES CM N:11

Heath MR, MacKenzie BR, Ådlandsvik B, Backhaus JO, Begg GA, Drysdale A, Gallego A, Gibb F, Gibb I, Harms IH, Hedger R, Kjesbu OS, Logemann K, Marteinsdottir G, McKenzie E, Michalsen K, Nielsen E, Scott BE, Strugnell G, Thorsen A, Visser A, Wehde H, Wright PJ (2003) An operational model of the effects of stock structure and spatio-temporal factors on recruitment. Final report of the EU-STEREO project. FAIR-CT98-4122. Dec 1998–Feb 2002. Fisheries Res Serv Contract Rep No 10/03

Heath MR, Kunzlik PA, Gallego A, Holmes SJ, Wright PJ (2008) A model of meta-population dynamics for North Sea and West of Scotland cod – The dynamic consequences of natal fidelity. Fish Res 93:92–116

Heckman CW (1998) The pantanal of poconé. Kluwer, Dordrecht

Hector A, Schmid B, Beierkuhnlein C, Caldeira MC, Diemer M, Dimitrakopoulos PG, Finn J, Freitas H, Giller PS, Good J, Harris R, Högberg P, Huss-Danell K, Joshi J, Jumpponen A, Körner C, Leadley PW, Loreau M, Minns A, Mulder CPH, O'Donovan G, Otway SJ, Pereira JS, Prinz A, Read DJ, Scherer-Lorenzen M, Schulze ED, Siamantziouras ASD, Spehn EM, Terry AC, Troumbis AY, Woodward FI, Yachi S, Lawton JH (1999) Plant diversity and productivity experiments in European grasslands. Science 286:1123–1127

Hemmerling R, Smoleňová K, Kurth W (2010) A programming language tailored to the specification and solution of differential equations describing processes on networks. Proc of LATA 2010, Trier, Germany, May 24–28, 2010

Heppell SS, Caswell H, Crowder LB (2000) Life histories and elasticity patterns: perturbation analysis for species with minimal demographic data. Ecology 81:654–665

Herborg LM, Rudnick DA, Siliang Y, Lodge DM, MacIsaac HJ (2007) Predicting the range of Chinese mitten crabs in Europe. Conserv Biol 21:1316–1323

Herman GT, Schiff GL (1975) Simulation of multi-gradient models of organisms in the context of L-systems. J Theor Biol 54:35–46

Hickling R, Roy DB, Hill JK, Thomas CD (2005) A northward shift of range margins in British Odonata. Glob Change Biol 11:502–506

Higgins SI, Richardson DM, Cowling RM (2001) Validation of spatial simulation model of a spreading alien plant population. J Appl Ecol 38:571–584

Hijmans RJ, Graham CH (2006) The ability of climate envelope models to predict the effect of climate change on species distributions. Glob Change Biol 12:2272–2281

Hilker FM (2005) Spatiotemporal patterns in models of biological invasion and epidemic spread. Logos Verlag, Berlin

Hilker FM, Malchow H, Langlais M, Petrovskii SV (2006) Oscillations and waves in a virally infected plankton system: Part II: Transition from lysogeny to lysis. Ecol Complex 3:200–208

Hill DRC (1996) Object oriented analysis and simulation. Addison Wesley, Harlow

Hill NA, Pedley TJ (2005) Bioconvection. Fluid Dyn Res 37:1–20

Hinkel J (2009) The PIAM approach to modular integrated assessment modelling. Environ Modell Softw 24(6):739–748

Hobbs RJ, Hobbs VJ (1987) Gophers and grassland: a model of vegetation response to patchy soil disturbance. Vegetatio 69:141–146

Hoch R, Gabele T, Benz J (1998) Towards a standard for documentation of mathematical models in ecology. Ecol Modell 113(1–3):3–12

Hogeweg P, Hesper B (1974) A model study on biomorphological description. Patt Recogn 6:165–179

Hogeweg P, Hesper B (1979) Heterarchical selfstructuring simulation systems: concepts and applications in biology. In: Zeigler BP, Elzas MS, Klir GJ, Ören TI (eds) Methodolgy in systems modelling and simulation. North Holland, Amsterdam, at http://www-binf.bio.uu.nl/pdf/Hogeweg79.pdf

Hogeweg P, Hesper B (1983) The ontogeny of the interaction structure in bumble bee colonies: a MIRROR model. Behav Ecol Sociobiol 12:271–283

Hölker F, Breckling B (2002) Influence of activity in a heterogeneous environment on the dynamics of fish growth: an individual-based approach of roach. J Fish Biol 60:1170–1189

Hölker F, Breckling B (2005) A spatiotemporal individual-based fish model to investigate emergent properties at the organismal and the population level. Ecol Modell 186:406–426

Holmes EE, Lewis MA, Banks JE, Veit RR (1994) Partial differential equations in ecology: spatial interactions and population dynamics. Ecology 75:17–29

Holt RD (2003) On the evolutionary ecology of species' ranges. Evol Ecol Res 5:159–178

Holt RD, Barfield M (2009) Trophic interactions and range limits: the diverse roles of predation. Proc R Soc B, Biol Sci 276:1435–1442

Holt RD, Gaines MS (1992) Analysis of adaptation in heterogenous landscapes: implications for the evolution of fundamental niches. Evol Ecol 6:433–447

Holyoak M, Leibold M, Holt RD (2005) Metacommunities: spatial dynamics and ecological communities. University of Chicago Press, Chicago

Homer C, Huang CQ, Yang LM, Wylie B, Coan M (2004) Development of a 2001 National Land-Cover Database for the United States. Photogr Eng Rem Sens 70:829–840

Homer C, Dewitz J, Fry J, Coan M, Hossain N, Larson C, Herold N, McKerrow A, VanDriel J, Wickham J (2007) Completion of the 2001 National Land Cover Database for the conterminous United States. Photogr Eng Rem Sens 73:337–341

Hooper DU, Chapin FS, Ewel JJ, Hector A, Inchausti P, Lavorel S, Lawton JH, Lodge DM, Loreau M, Naeem S, Schmid B, Setälä H, Symstad AJ, Vandermeer J, Wardle DA (2005) Effects of biodiversity on ecosystem functioning: a consensus of current knowledge. Ecol Monogr 75:3–35

Hörnfeldt B (1978) Synchronous population fluctuations in voles, small game, owls, and tularemia in Northern Sweden. Oecologia 32:141–152

Howeth JG, Leibold M (2008) Planktonic dispersal dampens temporal trophic cascades in pond metacommunities. Ecol Lett 11(3):245–257

Hu BG, Jaeger M (eds) (2003) Plant growth modeling and applications. Proceedings – PMA03. Tsinghua University Press, Beijing

Hughes TP (1994) Catastrophes, phase shifts and large-scale degradation of a Caribbean coral reef. Science 265:1547–1551

Hughes TP, Connell JH (1999) Multiple stressors on coral reefs: a long term perspective. Limnol Oceanogr 44:932–940

Hughes TP, Rodrigues MJ, Bellwood DR, Ceccarelli D, Hoegh-Guldberg O, McCook L, Moltschaniwskyj N, Pratchett MS, Steneck RS, Willis B (2007) Phase shifts, herbivory, and the resilience of coral reefs to climate change. Curr Biol 17:360–365

Huitu O, Koivula M, Korpimäki E, Klemola T, Norrdahl K (2003) Winter food supply limits growth of northern vole populations in the absence of predation. Ecology 84:2108–2118

Huston M, DeAngelis DL, Post W (1988) New computer models unify ecological theory. BioScience 38:682–691

Hutchinson WF, Carvalho GR, Rogers SI (2001) Marked genetic structuring in localised spawning populations of cod *Gadus morhua* in the North Sea and adjoining waters, as revealed by microsatellites. Mar Ecol Prog Ser 223:251–260

Huth A, Wissel C (1994) The simulation of fish schools in comparison with experimental data. Ecol Modell 75:135–145

Icaga Y (2007) Fuzzy evaluation of water quality classification. Ecol Indic 7:710–718

ICES (2003) Report of the ICES advisory committee on fishery management, 2002. ICES Co-op. Res Rep 255

Inform Software Corporation (2001) FuzzyTECH professional edition. Inform Software Corp, Chicago

Iwasa Y, Kazunori S, Nakashima S (1991) Dynamic modeling of wave regeneration (Shimagare) in subalpine *Abies* forests. J Theor Biol 152:143–158

Jacquemyn H, Brys R, Neubert MG (2005) Fire increases invasive spread of *Molinia caerulea* mainly through changes in demographic parameters. Ecol Appl 15:2097–2108

Jager HI, King AW (2004) Spatial uncertainty and ecological models. Ecosytems 7:841–847

Jakeman AJ, Letcher RA, Norton JP (2006) Ten iterative steps in development and evaluation of environmental models. Environ Modell Softw 21:602–614

Jang RJ-S, Gulley N (1995) The fuzzy logic toolbox for use with MATLAB. The MathWorks Inc, Natick

Janssen PHM, Heuberger PSC (1995) Calibration of process-oriented models. Ecol Modell 83:55–66

Jantsch E (1980) The self-organizing universe: scientific and human implications of the emerging paradigm of evolution. Pergamon Press, New York

Jedrzejewski W, Jedrzejewska B (1996) Rodent cycles in relation to biomass and productivity of ground vegetation and predation in the Palearctic. Acta Theriol 41:1–34

Jeffers JNR (1999) Genetic algorithms I. In: Fielding AH (ed) Machine learning methods for ecological applications. Kluwer, Dordrecht

Jepsen JU, Madsen AB, Karlsson M, Groth D (2005) Predicting distribution and density of European badger (*Meles meles*) setts in Denmark. Biodivers Conserv 14:3235–3253

Jerina K, Debeljak M, Džeroski S, Kobler A, Adamič M (2003) Modelling the brown bear population in Slovenia: a tool in the conservation management of a threatened species. Ecol Modell 170:453–469

Jetz W, Rahbek C (2002) Geographic range size and determinants of avian species richness. Science 297:1548–1551

Jiménez-Valverde A, Lobo JM, Hortal J (2008) Not as good as they seem: the importance of concepts in species distribution modelling. Divers Distrib 14:885–890

Johnson NL, Kotz S (1969) Discrete distributions. Houghton Mifflin, New York

Johnson NL, Kotz S (1970) Continuous univariate distributions-1. Houghton Mifflin, New York

Johnson SN, Zhang X, Crawford JW, Gregory PJ, Young IM (2007) Egg hatching and survival time of soil-dwelling insect larvae: a partial differential equation model and experimental validation. Ecol Modell 202:493–502

Johst K, Brandl R, Eber S (2002) Metapopulation persistence in dynamic landscapes: the role of dispersal distance. Oikos 98:263–270

Jopp F (2003) Empirical analyses and modelling of dispersal events of invertebrates from a fen-wetland system. Dissertation Thesis. Free University Berlin

Jopp F (2006) Comparative studies of the dispersal of the Great Ramshorn (*Planorbarius corneus* L.) in heterogeneous environments – A modelling approach. Limnologica 36:17–25

Jopp F, Lange C (2007) Improving data interpretation of fragmentary data-sets on invertebrate dispersal with permutation-tests. Acta Oecol 31:102–108

Jopp F, Reuter H (2005) Dispersal of carabid beetles – emergence of distribution patterns. Ecol Modell 18:389–405

Jopp F, DeAngelis DL, Trexler JC (2010) Modeling seasonal dynamics of small fish cohorts in fluctuating freshwater marsh landscapes. Landsc Ecol 25:1041–1054

Jørgensen SE (1976) A eutrophication model for a lake. Ecol Modell 2:147–165

Jørgensen SE (ed) (1979) Handbook of environmental data and ecological parameters. Pergamon, Oxford

Jørgensen SE (1982) Exergy and buffering capacity in ecological systems. In: Mitsch WJ, Ragade RK, Bosserman RW, Dillon JA Jr (eds) Energetics and systems. Ann Arbor Science, Ann Arbor, pp 61–72

Jørgensen SE (1986a) Fundamentals of ecological modeling. Elsevier, Amsterdam

Jørgensen SE (1986b) Structural dynamic model. Ecol Modell 31:1–9

Jørgensen SE (1992) Development of models able to account for changes in species composition. Ecol Modell 62:195–208

Jørgensen SE (1995) The growth rate of zooplankton at the edge of chaos: ecological models. J Theor Biol 175:13–21

Jørgensen SE (2002) Integration of ecosystem theories – a pattern, 3rd edn. Kluwer, Dordrecht

Jørgensen SE (2008) Overview of the model types available for development of ecological models. Ecol Modell 215:3–9

Jørgensen SE, Bendoricchio G (2001) Fundamentals of ecological model, vol 21, 3rd edn, Developments in environmental modelling. Elsevier, Amsterdam

Jørgensen SE, Mejer HF (1979) A holistic approach to Ecol Model. Ecol Modell 7:169–189

Jørgensen SE, Mejer HF (1981) Exergy as a key function in ecological models. In: Mitsch WJ, Bosserman RW, Klopatek JM (eds) Energy and ecological modelling. Elsevier, Amsterdam, pp 587–590

Jørgensen SE, Mejer HF (1983) Trends in ecological modelling. In: Lauenroth WK, Skogerboe GV, Flug M (eds) Analysis of ecological systems: state-of-the-art in ecological modelling. Elsevier, Amsterdam, pp 21–26

Jørgensen SE, Müller F (2000) Ecosystems as Complex Systems. In: Jørgensen SE, Müller F (eds) Handbook of ecosystem theories and management. CRC, Boca Raton, pp 5–20

Jørgensen SE, Jacobsen OS, Høi I (1973) A prognosis for a lake. Vatten 29:382–404

Jørgensen SE, Mejer HF, Friis M (1978) Examination of a lake model. Ecol Modell 4:253–278

Jørgensen SE, Kamp-Nielsen L, Mejer HF (1982) Comparison of a simple and a complex sediment phosphorus model. Ecol Modell 16:99–124

Jørgensen SE, Jørgensen LA, Mejer HF, Kamp-Nielsen L (1983) A water quality model for the Upper Nile system. In: Lauenroth WK, Skogerboe GV, Flug M (eds) Analysis of ecological systems: state-of-the-art in ecol model. Elsevier, Amsterdam, pp 631–639

Jørgensen SE, Kamp-Nielsen L, Christensen T, Windolf-Nielsen JR, Westergaard B (1986a) Validation of a prognosis based upon a eutrophication model. Ecol Modell 32:165–182

Jørgensen SE, Kamp-Nielsen L, Jørgensen LA (1986b) Examination of the generality of eutrophication models. Ecol Modell 32:251–266

Jorquera H, Pérez R, Cirpiano A, Espejo A, Letelier MV, Acuña G (1998) Forecasting ozone daily maximum levels at Santiago, Chile. Atmos Environ 32:3415–3424

Judson OP (1994) The rise of the individual-based models in ecology. Trends Ecol Evol 9:9–14

Jung SI, Choi I, Jin H, Lee D, Cha H, Kim Y, Lee J (2009) Size-dependent mortality formulation for isochronal fish species based on their fecundity: an example of Pacific cod (*Gadus macrocephalus*) in the eastern coastal areas of Korea. Fish Res 97:77–85

Kaiser H (1976) Quantitative description and simulation of stochastic behaviour in dragonflies (*Aeschna cyanea*). Acta Biotheor 25:163–210

Kaiser H (1979) The dynamics of populations as the result of the properties of individual animals. Fortschr Zool 25:109–136

Kaitala V, Alaja S, Ranta E (2001) Temporal self-similarity created by spatial individual-based population dynamics. Oikos 94:273–278

Kampichler C, Džeroski S, Wieland R (2000) The application of machine learning techniques to the analysis of soil ecological data bases: relationships between habitat features and Collembola community characteristics. Soil Biol Biochem 32:197–209

Karwowski R, Prusinkiewicz P (2003) Design and implementation of the L+C modeling language. Electron Notes Theor Comput Sci 86(2):19, http://algorithmicbotany.org/papers/l+c.tcs2003.pdf

Kastner-Maresch A, Kurth W, Sonntag M, Breckling B (eds) (1998) Individual-based structural and functional models in ecology. Bayreuther Forum Ökologie 52, pp 243

Kauffman S (1993) The origins of order: self-organization and selection in evolution. Oxford University Press, Oxford

Kaw A (2008) Introduction to Matrix Algebra. available at: www.autarkaw.com

Kawasaki K, Takasu F, Caswell H, Shigesada N (2006) How does stochasticity in colonization accelerate the speed of invasion in a cellular automaton model? Ecol Res 21:334–345

Kay D, McDonald AT, Stapleton CM, Wyer MD, Fewtrell L (2006) Europe: a challenging new framework for water quality. Proceedings of the Institution of Civil Engineers, Water Management 159(5):58–64

Kearney M (2006) Habitat, environment and niche: what are we modelling? Oikos 115:186–191

Keen PGW (1978) Decision support systems: an organizational perspective. Addison-Wesley, Reading, MA

Kendal WS (1995) A probabilistic model for the variance to mean power-low in ecology. Ecol Modell 80:293–297

King AW, With KA (2002) Dispersal success on spatially structured landscapes: when do spatial pattern and dispersal behavior rally matter? Ecol Modell 147:23–39

Kingsland SE (1995) Modeling nature. Episodes in the history of population ecology, 2nd edn. University of Chicago Press, Chicago

Kirk RE (1968) Experimental design: procedures for the behavioral sciences. Brooks/Cole, Belmont, CA

Klemola T, Pettersen T, Stenseth N (2003) Trophic interactions in population cycles of voles and lemmings: a model-based synthesis. Adv Ecol Res 33:75–160

Klepper O (1997) Multivariate aspects of model uncertainty analysis: tools für sensitvity analyis and calibration. Ecol Modell 83:55–66

Kniemeyer O (2008) Design and implementation of a graph-grammar based language for functional-structural plant modelling. Doctoral dissertation, BTU Cottbus, 432 pp. at http://nbn-resolving.de/urn/resolver.pl?urn=urn:nbn:de: kobv:co1-opus-5937

Kniemeyer O, Buck-Sorlin G, Kurth W (2004) A graph-grammar approach to Artificial Life. Artif Life 10:413–431

Knispel AL, McLachlan SM (2009) Landscape-scale distribution and persistence of genetically modified oilseed rape (*Brassica napus*) in Manitoba, Canada. Environ Sci Pollut Res Int. doi:10.1007/s11356-009-0219-0

Knispel AL, McLachlan SM, Van Acker RC, Friesen LF (2008) Gene flow and multiple herbicide resistance in escaped canola populations. Weed Sci 56(1):72–80

Kobler A, Adamič M (1999) Brown bears in Slovenia: identifying locations for construction of wildlife bridges across highways. In: Proceeding of the 1999 international conference on wildlife ecology and transportation. Rep No FL-ER-73-99. Florida Dept Transportation, Tallahassee, FL, pp 29–38

Köhler P, Chave J, Riera B, Huth A (2003) Simulating long-term response of tropical wet forests to fragmentation. Ecosystems 6:129–143

Kolmogorov A, Petrovsky I, Piskunov N (1937) Étude de l'equation de la diffusion avec croissance de la quantité de matière et son application à un problème biologique. Bull Univ Moscou Sér Int A 1:1–25

Kompare B, Džeroski S (1995) Getting more out of data: automated modelling of algal growth with machine learning. In: Proceedings of international conference on coastal ocean space utilization, University of Hawaii, pp 209–220

Konikow LF, Bredehoeft JD (1992) Ground-water models cannot be validated. Adv Water Resour 15:75–83

Korpimäki E, Oksanen L, Oksanen T, Klemola T, Norrdahl K, Banks PB (2005) Vole cycles and predation in temperate and boreal zones of Europe. J Anim Ecol 74:1150–1159

Kosko B (1994) Fuzzy systems as universal approximators. IEEE Trans Comput 43:1329–1333

Kramer PA (2003) Synthesis of coral reef health indicators for the western Atlantic: results of the AGRRA program (1997–2000). Atoll Res Bull 496:1–58

Krebs CJ (2002) Ecology: the experimental analysis of distribution and abundance, 5th edn. Benjamin-Cummings, Zug, Switzerland

Krummel JR, Gardner RH, Sugihara G, O'Neill RV, Coleman PR (1987) Landscape patterns in a disturbed environment. Oikos 48:321–324

Kuguru BL, Mgaya YD, Ohman MC, Wagner GM (2004) The reef environment and competitive success in the *Corallimorpharia*. Mar Biol 145:875–884

Kühn I, Bierman SM, Durka W, Klotz S (2006) Relating geographical variation in pollination types to environmental and spatial factors using novel statistical methods. New Phytol 72:127–139

Kurth W (2007) Specification of morphological models with L-systems and relational growth grammars. Image – J Interdisc Image Sci 5, Special Issue 50–79

Kurth W, Sloboda B (2001) Sensitive growth grammars specifying models of forest structure, competition and plant-herbivore interaction. Proceedings of the IUFRO 4.11 Congress, Greenwich, UK, June 25–29, 2001; at http://cms1.gre.ac.uk/conferences/iufro/Proc/kurthsloboda.pdf (accessed Feb. 7, 2010)

Lacroix G, McCloghrie P, Huret M and North EW (2009) Hydrodynamic models. In: North EW, Gallego A, Petitgas P (eds) Manual of recommended practices for modelling physical–biological interactions during Fish Early Life, ICES Cooperative Research Reports 295:3–8

Lam D, Leon L, Hamilton S, Crookshank N, Bonin D, Swayne D (2004) Multi-model integration in a decision support system: a technical user interface approach for watershed and lake management scenarios. Environ Modell Softw 19(3):317–324

Lamarck JB (1815–1822) Histoire naturelle des animaux sans vertèbres présentant les caractères généraux et particuliers de ces animaux. Déterville, Paris

Langmead O, Sheppard C (2004) Coral reef community dynamics and disturbance: a simulation model. Ecol Modell 175:271–291

Langton C (ed) (1994) Artificial life. SFI Studies in the Sciences of Complexity. Addison-Wesley, New York

Lanwert D (2007) Funktions-/Strukturorientierte Pflanzenmodellierung in E-Learning-Szenarien. Doctoral Dissertation, University of Göttingen, 209 pp. at http://resolver.sub.uni-goettingen.de/purl/?webdoc-1692

Laperriére V, Badarotti D, Banos A, Müller JP (2009) Structural validation of an individual based model plague epidemic simulation. Ecol Compl 6:102–112

Latimer AM, Wu S, Gelfand AE, Silander JA (2006) Building statistical models to analyze species distributions. Ecol Appl 16:33–50

Lavorel S, Davies I, Noble I (2000) LAMOS: a landscape modelling shell. In: Hawkes BC, Flannigan MD (eds) Landscape Fire Modeling Workshop, pp 25–28

Le Corff J, Horvitz CC (2005) Population growth versus population spread of an ant-dispersed neotropical herb with a mixed reproductive strategy. Ecol Modell 188:41–51

Legendre P (1993) Spatial autocorrelation: trouble or new paradigm? Ecology 74:1659–1673

Leibold MA, Holyoak M, Mouquet N, Amarasekare P, Chase JM, Hoopes MF et al (2004) The metacommunity concept: a framework for multi-scale community ecology. Ecol Lett 7:601–613

Leis JM (2007) Behaviour as input for modelling dispersal of fish larvae: behaviour, biogeography, hydrodynamics, ontogeny, physiology and phylogeny meet hydrography. Mar Ecol Progr Ser 347:185–193

Lek S, Guegan JF (1999) Application of artificial neural networks in ecological modelling. Ecol Modell 120:2–3

Lek-Ang S, Deharveng L, Lek S (1999) Predictive models of collembolan diversity and abundance in a riparian habitat. Ecol Modell 120:247–260

Lennon JJ, Greenwood JJD, Turner JRG (2000) Bird diversity and environmental gradients in Britain: a test of the species-energy hypothesis. J Anim Ecol 69:581–598

Leontief WW (1951) The structure of the U.S. economy, 2nd edn. Oxford University Press, New York

LePage C, Cury P (1997) Population viability and spatial fish reproductive strategies in constant and changing environments: an individual-based modelling approach. Can J Fish Aquat Sci 54:2235–2246

LeRoux X, Sinoquet H (guest eds) (2000) 2nd International Workshop on functional-structural tree models. Ann Forest Sci (Special Issue) 57:393–621

Leslie PH (1945) On the use of matrices in certain population mathematics. Biometrika 33:183–212

Leslie PH (1948) Some further notes on the use of matrices in population mathematics. Biometrika 35:213–245

Lessios HA (1988) Mass mortality of *Diadema antillarum* in the Caribbean: what have we learned? Ann Rev Ecol Syst 19:371–393

Levin SA (1992) The problem of pattern and scale in ecology. Ecology 73(6):1943–1967

Lewis MA, Pacala S (2000) Modelling and analysis of stochastic invasion processes. J Math Biol 41:387–429

Li XZ, He HS, Bu RC, Wen QC, Chang Y, Hu YM, Li YH (2005) The adequacy of different landscape metrics for various landscape patterns. Patt Recogn 38:2626–2638

Li SC, Chang Q, Peng FA, Wang YL (2009) Indicating landscape fragmentation using L-Z complexity. Ecol Indic 9:780–790

Liebig J (1831) Ueber einen neuen Apparat zur Analyse organischer Körper, und über die Zusammensetzung einiger organischen Substanzen. Ann Phys 21:1–47

Lima SL, Zollner PA (1996) Towards a behavioral ecology of ecological landscapes. Trends Ecol Evol 11:131–136

Lin Y, Cobourn WG (2007) Fuzzy system models combined with nonlinear regression for daily ground-level ozone predictions. Atmos Environ 41:3502–3513

Lindborg R, Eriksson O (2004) Historical landscape connectivity affects present plant species diversity. Ecology 85:1840–1845

Lindeman RL (1941) Seasonal food-cycle dynamics in a Senescent Lake. Am Midl Nat 26:636–673

Lindeman R (1942) The trophic dynamic aspect of ecology. Ecology 23:399–418

Lindenmayer A (1968) Mathematical models for cellular interactions in development. J Theor Biol 18:280–315

Lischke H, Zimmermann NE, Bolliger J, Rickebusch S, Löffler TJ (2006) TreeMig: a forest-landscape model for simulating spatio-temporal patterns from stand to landscape scale. Ecol Modell 199:409–420

Liston SE (2006) Interactions between nutrient availability and hydroperiod shape macroinvertebrate communities in Florida Everglades marshes. Hydrobiologia 569:343–357

Littler MM, Littler DS, Lapointe BE (1993) Modification of tropical reef community structure due to cultural eutrophication: the southwest coast of Martinique. Proceedings of the 7th international coral reef symposium, vol 1, pp 335–343

Livingstobe D, Raper J (1994) Modelling environmental systems with GIS: theoretical barriers to progress. Innov GIS 1:229–240

Lobo JM, Jimenez-Valverde A, Real R (2008) AUC: a misleading measure of the performance of predictive distribution models. Glob Ecol Biogeogr 17:145–151

Logemann K, Backhaus JO, Harms IH (2004) SNAC: a statistical emulator of the north-east Atlantic circulation. Ocean Model 7(1–2):97–110

Łomnicki A (1988) Population ecology of individuals. Princeton University Press, Princeton, NJ

Lookingbill TR, Kaushal SS, Gardner RH, Elmore AJ, Morgan RP, Hilderbrand RH, Eshleman KN, Boynton WR, Palmer MA, Dennison WC (2009) Altered ecological flows blur boundaries in urbanizing watersheds. Ecol Soc 14(2):10

Lorenz EN (1963) Deterministic nonperiodic flow. J Atmos Sci 20(2):130–141

Losos JB (2008) Phylogenetic niche conservatism, phylogenetic signal and the relationship between phylogenetic relatedness and ecological similarity among species. Ecol Lett 11:995–1003

Lotka AJ (1925) Elements of physical biology. Williams & Wilkins, Baltimore, MD, Reprint 1956: Elements of mathematical biology. Dover, New York

Loucks DP, Kindler J, Fedra K (1984) Interactive water resources modeling and model use: an overview. Water Resour Res 21(2):95–102

Lough RG, Broughton EA (2007) Development of micro-scale frequency distributions of plankton for inclusion in foraging models of larval fish, results from a video plankton recorder. J Plankton Res 29(1):7–17

Lowe VPW (1969) Population dynamics of the Red Deer (*Cervus elaphus* L.) on Rhum. J Anim Ecol 38:425–427

Luther R (1906) Räumliche Fortpflanzung chemischer Reaktionen. Z Elektrochem 12:596–600

Maasen S, Prinz W, Roth G (eds) (2003) Voluntary action. Brains, minds, and sociality. Oxford University Press, Oxford

Mackay DS, Robinson VB (2003) A multiple criteria decision support system for testing integrated environmental models. Fuzzy Set Syst 113(1):53–67

Mackey BG, Lindenmayer DB (2001) Towards a hierarchical framework for modelling the spatial distribution of animals. J Biogeogr 28:1147–1166

Maggini R, Lehmann A, Zimmermann NE, Guisan A (2006) Improving generalized regression analysis for the spatial prediction of forest communities. J Biogeogr 33:1729–1749

Mahecha MD, Schmidtlein S (2008) Revealing biogeographical patterns by nonlinear ordinations and derived anisotropic spatial filters. Glob Ecol Biogeogr 17:284–296

Makler-Pick V, Gal G, Gorfine M, Hipsey MR, Yohay Carmel Y (2010) Sensitivity analysis for complex ecological models – a new approach. Environ Modell Softw. doi:10.1016/j.envsoft.2010.06.010

Malchow H, Schimansky-Geier L (1985) Noise and diffusion in bistable nonequilibrium systems, vol 5, Teubner-Texte zur Physik. Teubner-Verlag, Leipzig

Malchow H, Petrovskii SV, Medvinsky AB (2002) Numerical study of plankton–fish dynamics in a spatially structured and noisy environment. Ecol Modell 149:247–255

Malchow H, Hilker FM, Petrovskii SV (2004a) Noise and productivity dependence of spatiotemporal pattern formation in a prey-predator system. Discrete Cont Dyn S B 4(3):707–713

Malchow H, Hilker FM, Petrovskii SV, Brauer K (2004b) Oscillations and waves in a virally infected plankton system. Part I: The lysogenic stage. Ecol Compl 1(3):211–223

Malchow H, Hilker FM, Sarkar RR, Brauer K (2005) Spatiotemporal patterns in an excitable plankton system with lysogenic viral infection. Math Comput Model 42:1035–1048

Malchow H, Petrovskii SV, Venturino E (2008) Spatiotemporal patterns in ecology and epidemiology: theory, models, and simulation, CRC mathematical and computational biology series. CRC Press, Boca Raton

Malthus TR (1798) An Essay on the Principle of Population (1798 1st edn, plus excerpts 1803 2nd edn), Introduction by Philip Appleman, and assorted commentary on Malthus edited by Appleman. Norton Critical Editions

Mandelbrot BB (1967) How long is the coast of Britain? Statistical self-similarity and fractional dimension. Science 156:636–638

Mandelbrot BB (1983) The Fractal Geometry of Nature, Revisedth edn. Freeman, New York

Manson SM (2003) Validation and verification of multi-agent models for ecosystem management. In: Janssen M (ed) Complexity and ecosystem management: the theory and practice of multi-agent approaches. Edward Elgar, Northampton, MA, pp 63–74

Marchini A (2010) Fuzzy indices of ecological conditions: review of techniques and applications. In: Vargas RE (ed) Fuzzy logic: theory, programming and applications. Novascience, New York, pp 115–172

Marchini A, Facchinetti T, Mistri M (2009) F-IND: a framework to design fuzzy indices of environmental conditions. Ecol Indic 9:485–496

Marsili-Libelli S, Checchi N (2005) Identification of dynamic models for horizontal subsurface constructed wetlands. Ecol Modell 187:201–218

Martis MS (2006) Validation of simulation based models: a theoretical outlook. Electron J Bus Res Meth 4(1):39–46

Matsinos YG, Troumbis AY (2002) Modeling competition, dispersal and effects of disturbance in the dynamics of a grassland community using a cellular automaton model. Ecol Modell 149:71–83

Mayr E (1982) The growth of biological thought: diversity, evolution, and inheritance. Belknap Press (Harvard University Press, Cambridge, MA

Mazaris AD, Fiksen O, Matsinos YG (2005) Using an individual-based model for assessment of sea turtle population viability. Popul Ecol 47:179–191

McCarthy J (1963) Situations and actions and causal laws. Stanford Artificial Intelligence Project Memo 2, Stanford

McClanahan TR (1995) A coral reef ecosystem-fisheries model: impacts of fishing intensity and catch selection on reef structure and processes. Ecol Modell 80:1–19

McGarigal K, Cushman SA, Neel MC, Ene E (2002) FRAGSTATS: spatial pattern analysis program for categorical maps. University of Massachusetts, Amherst, MA

McManus JW, Polsenberg JF (2004) Coral-algal phase shifts on coral reefs: ecological and environmental aspects. Prog Oceanogr 60:263–279

McMillan V (2006) Writing papers in the biological sciences, 4th edn. Bedford Books, New York

Meadows DH (1972) The limits to growth – a report for the club of Rome's project on the predicament of mankind. Universe Books, New York

Meadows DH, Meadows DL, Randers J, Behrens WW III (1972) The limits to growth. Universe Books, New York

Medlock J, Kot M (2003) Spreading disease: integro-differential equations old and new. Math Biosci 184:201–222

Mejer HF, Jørgensen SE (1979) Exergy and ecological buffer capacity. Ecol Modell 7:829–846
Meyer C (2000) Matrix analysis and applied linear algebra. Society for Industrial and Applied Mathematics, Philadelphia, PA
Meyer WB, Turner BL (1992) Human-population growth and global land-use cover change. Ann Rev Ecol Syst 23:39–61
Meynard CN, Quinn JF (2007) Predicting species distributions: a critical comparison of the most common statistical models using artificial species. J Biogeogr 34:1455–1469
Middelhoff U, Reuter H, Breckling B (2009) onward) GeneTraMP, a spatio-temporal model of the dispersal and persistence of transgenes in feral, volunteer and crop plants of oilseed rape and related species. Ecol Indic. doi:10.1016/j.ecolind.2009.03.006 (available online)
Midgley GF, Hughes GO, Thuiller W, Rebelo AG (2006) Migration rate limitations on climate change-induced range shifts in Cape Proteaceae. Divers Distrib 12:555–562
Miller TJ (2007) Contribution of individual-based coupled physical–biological models to understanding recruitment in marine fish populations. Mar Ecol Progr Ser 347:127–138
Mitchell T (1997) Machine learning. McGraw-Hill, New York
Mitro MG (2001) Ecological model testing: verification, validation, or neither? Bull Ecol Soc Am 82:235–236
Moloney CL, Jarre A, Arancibia H, Bozec YM, Neira S, Jean-Paul Roux JP, Shannon LJ (2005) Comparing the Benguela and Humboldt marine upwelling ecosystems with indicators derived from inter-calibrated models. ICES J Mar Sci 62:493–502
Monsi M, Saeki T (1953) Über den Lichtfaktor in den Pflanzengesellschaften und seine Bedeutung für die Stoffproduktion. Jap J Bot 14:205–234
Mora C (2008) A clear human footprint in the coral reefs of the Caribbean. Proc R Soc London B 275:767–773
Mosteller F (1948) A k-sample slippage test for an extreme population. Ann Math Stat 19:58–65
Moustakas A, Silvert W, Dimitromanolakis A (2006) A spatially explicit learning model of migratory fish and fishers for evaluating closed areas. Ecol Modell 192:245–258
Müller F (1992) Hierarchical approaches to ecosystem theory. Ecol Modell 63:215–242
Müller F (1999) Ökosystemare Modellvorstellungen und Ökosystemmodelle in der angewandten Landschaftsökologie. In: Schneider-Sliwa R, Schaub D, Gerold G (eds) Angewandte Landschaftsökologie – Grundlagen und Methoden. Springer, Berlin Heidelberg New York, pp 25–46
Müller F, Leupelt M (eds) (1997) Eco targets, goals, functions and orientors. Springer, New York, 619 pp
Mumby PJ (2006) The impact of exploiting grazers (Scaridae) on the dynamics of Caribbean coral reefs. Ecol Appl 16:747–769
Mumby PJ (2009) Phase shifts and the stability of macroalgal communities on Caribbean coral reefs. Coral Reefs 28:761–773
Mumby PJ, Hastings A (2008) The impact of ecosystem connectivity on coral reef resilience. J Appl Ecol 45:854–862
Murray JD (2003) Mathematical biology. II. Spatial models and biomedical applications, vol 18, 3rd edn, Interdisciplinary Applied Mathematics. Springer, Berlin
Muscatine L (1990) The role of symbiotic algae in carbon and energy flux in reef corals. In: Dubinsky Z (ed) Ecosystems of the world: coral reefs. Elsevier, Amsterdam, pp 75–87
Myers JH (1988) Can a general hypothesis explain population cycles of forest lepidoptera. Adv Ecol Res 18:179–242
Myers RA, Hutchings JA, Barrowman NJ (1997) Why do fish stocks collapse? The example of cod in Atlantic Canada. Ecol Appl 7(1):91–106
Nagelkerken I, Roberts CM, van der Velde G, Dorenbosch M, van Riel MC, Cocheret de la Morinière E, Nienhuis PH (2002) How important are mangroves and seagrass beds for coral-reef fish? The nursery hypothesis tested on an island scale. Mar Ecol Progr Ser 244:299–305
Naiman RJ (1992) Watershed management: balancing sustainability and environmental change. Springer, New York

Nathan R, Getz WM, Revilla E, Holyoak M, Kadmon R, Saltz D, Smouse PE (2008) A movement ecology paradigm for unifying organismal movement research. Proc Natl Acad Sci USA 105:19052–19059
Neel MC, McGarigal K, Cushman SA (2004) Behavior of class-level landscape metrics across gradients of class aggregation and area. Landsc Ecol 19:435–455
Nehrbass N, Winkler E, Müllerová J, Pergl J, Pyšek P, Perglová I (2007) A simulation model of plant invasion: long-distance dispersal determines the pattern of spread. Biol Invasions 9:383–395
Nekola JC (1999) Paleorefugia and neorefugia: the influence of colonization history on community pattern and process. Ecology 80:2459–2473
Neubert MG, Caswell H (2000) Demography and dispersal: calculation and sensitivity analysis of invasion speed for structured populations. Ecology 81:1613–1628
Nielsen SN (1995) Optimization of exergy in a structural dynamic model. Ecol Modell 77:111–122
Nielsen SN (1997) Examination and Optimization of different exergy forms in Macrophyte Societies. Ecol Modell 102:115–127
Nielsen SN (2009) Modelling in biological, ecological & environmental science. University of Coimbra, Scriptum
Nielsen SN, Müller F (2000) Emergent properties of ecosystems. In: Jørgensen SE, Müller F (eds) Handbook of ecosystem theories and management. CRC Press, Boca Raton, pp 177–194
Nijhout HF, Wray GA, Kremen C, Teragawa CK (1986) Ontogeny, phylogeny and evolution of form: an algorithmic approach. Syst Zool 35:445–457
Nogués-Bravo D (2009) Predicting the past distribution of species climatic niche. Glob Ecol Biogeogr 18:521–531
Norrdahl K (1995) Population cycles in northern small mammals. Biol Rev 70(621):637
Norstroem AV, Nystroem M, Lokrantz J, Folke C (2009) Alternative states on coral reefs: beyond coral macroalgal phase shifts. Mar Ecol Progr Ser 376:295–306
NRC (2003) NEON: addressing the nation's environmental challenges, National Research Council. National Academies Press, Washington, DC
NVIDIA Corp. (2006) DirectX 10: the next-generation graphics API. Technical brief; at http://forum.greycomputer.de/download/handbuecher/nvidia/nvidia_microsoft_directx-10_technical_brief.pdf. Accessed Feb 6, 2010
Nystroem M, Folke C, Moberg F (2000) Coral reef disturbance and resilience in a human-dominated environment. Trends Ecol Evol 15:413–417
Nyström P, Brönmark C, Graneli W (1996) Patterns in benthic food webs: a role for omnivorous crayfish? Freshw Biol 36:631–646
O'Neill RV, DeAngelis DL, Waide JB, Allen THF (1986) A hierarchical concept of ecosystems, vol 23, Monographs in population ecology. Princeton University Press, Princeton
O'Neill RV, Krummel J, Gardner RH, Sugihara G, Jackson B, DeAngelis DL, Milne B, Turner MG, Zygmunt B, Christensen S, Dale VH, Graham RL (1987) Indices of landscape pattern. Landsc Ecol 1:153–162
Ocampo-Duque W, Ferré-Huguet N, Domingo JL, Schuhmacher M (2006) Assessing water quality in rivers with fuzzy inference systems: a case study. Environ Int 32:733–742
Odum EP (1953/1971) Fundamentals of ecology. Saunders, Philadelphia, PA, p 574
Odum HT (1957) Trophic structure and productivity of Silver Springs, Florida. Ecol Monogr 27:55–112
Odum EP (1969) The strategy of ecosystem development. Science 104:262–270
Odum HT (1971) Environment, power and society. Wiley, New York
Odum EP (1977) Ecology – the common-sense approach. Ecologist 7(7):250–253
Odum HT (1983) Systems ecology. Wiley, Chichester
Odum EP (1997) Ecology – a bridge between science and society. Sinauer, Portland, OR
Oey LY, Chen P (1992) A model simulation of circulation in the northeast Atlantic shelves and sea. J Geophys Res 97:20087–20115

Ogden JC (1994) A comparison of wading bird nesting colony dynamics (1931–1946 and 1974–1989) as an indication of ecosystem conditions in the southern Everglades. In: Davis SM, Ogden JC (eds) Everglades: the system and its restoration. St. Lucie, Delray Beach, FL, pp 533–570

Ogris N, Jurc M (2007) Potential changes in the distribution of maple species (*Acer pseudoplatanus, A. campestre, A. platanoides, A. obtusatum*) due to climate change in Slovenia. In: Proceedings of the symposium on climate change influences on forests and forestry. University of Ljubljana, Slovenia

Ogris N, Jurc M (2010) Sanitary felling of Norway spruce due to spruce bark beetles in Slovenia: a model and projections for various climate change scenarios. Ecol Modell 221:290–302

Okey TA, Banks S, Born AF, Bustamante RH, Calvopiña M, Edgar GJ, Espinoza E, Fariña JM, Garske LE, Reck GK, Salazar S, Shepherd S, Toral-Granda V, Wallem P (2004) A trophic model of a Galápagos subtidal rocky reef for evaluating fisheries and conservation strategies. Ecol Modell 172:383–401

Okubo A (1980) Diffusion and ecological problems: mathematical models, vol 10, Biomathematics texts. Springer, Berlin

Okubo A, Levin S (2001) Diffusion and ecological problems: modern perspectives, vol 14, 2nd edn, Interdisciplinary applied mathematics. Springer, New York, 465 pp

Oli MK, Dobson FS (2001) Population cycles in small mammals: the alpha hypothesis. J Mammal 82:573–581

Oreskes N, Shrader-Frechette K, Belitz K (1994) Verification, validation, and confirmation of numerical models in the earth sciences. Science 263:641–646

Osher LJ, Matson PA, Amundson R (2003) Effect of land use change on soil carbon in Hawaii. Biogeochemistry 65:213–232

Overpeck JT, Cole JE (2006) Abrupt change in Earth's climate system. Ann Rev Environ Resour 31:1–31

Overpeck JT, Weiss JL (2009) Projections of future sea level becoming more dire. Proc Natl Acad Sci USA 106:21461–21462

Oxley T, ApSimon HM (2007) Space, time and nesting Integrated Assessment Models. Environ Modell Softw 22(12):1732–1749

Ozinga WA, Schaminée JHJ, Bekker RM, Bonn S, Poschlod P, Tackenberg O, Bakker J, van Groenendael JM (2005) Predictability of plant species composition from environmental conditions is constrained by dispersal limitation. Oikos 108(3):555–561

Paley W (1803) Natural theology: or, evidences of the existence and attributes of the deity collected from the appearances of nature. Whiting, Albany (re-issued by Cambridge University Press, 2009)

Paris CB, Irisson JO, Lacroix G, Fiksen Ø, Leis JM, Mullon C (2009) Connectivity. In: North EW, Gallego A, Petitgas P (eds) Manual of recommended practices for modelling physical–biological interactions during Fish Early Life. ICES Cooperative Research Report 295:63–76

Parmesan C (2006) Ecological and Evolutionary Responses to Recent Climate Change. Ann Rev Ecol Evol Syst 37:637–669

Parrot L, Kok R (2002) A generic, individual-based approach to modelling higher trophic levels in simulation of terrestrial ecosystems. Ecol Modell 154:151–178

Pascarella JB, Horvitz CC (1998) Hurricane disturbance and the population dynamics of a tropical understory shrub: megamatrix elasticity analysis. Ecology 79:547–563

Pascual M (1993) Diffusion-induced chaos in a spatial predator–prey system. Proc R Soc Lond B 251:1–7

Pascual M, Roy M, Guichard F, Flier G (2002) Cluster size distributions: signatures of self-organization in spatial ecologies. Philos Trans R Soc Lond B 357:657–666

Patten BC (1959) An introduction to the cybernetics of the ecosystem: the trophic-dynamic aspect. Ecology 40(2):221–231

Patten BC (1975) Systems Analysis and Simulation in Ecology. Academic, New York, 607 pp

Patten BC (1978) Systems approach to the concept of environment. Ohio J Sci 78:206–222

Patten BC, Odum EP (1981) The cybernetic nature of ecosystems. Am Nat 118:886–895

Pauly D, Christensen V, Walters C (2000) Ecopath, Ecosim, and Ecospace as tools for evaluating ecosystem impact of fisheries. ICES J Mar Sci 57:697–706

Pe'er G, Kramer-Schadt S (2008) Incorporating the perceptual range of animals into connectivity models. Ecol Modell 213:73–85

Peacor SD, Allesina S, Riolo RL, Hunter TS (2007) A new computational system, DOVE (Digital Organisms in a Virtual Ecosystem), to study phenotypic plasticity and its effects in food webs. Ecol Modell 205:13–28

Pearce JL, Boyce MS (2006) Modelling distribution and abundance with presence-only data. J Appl Ecol 43:405–412

Pearl R (1925) The biology of population growth. Knopf, New York

Pearl R (1927) The growth of populations. Q Rev Biol 2:532–548

Pearlstine LG, Smith SE, Brandt LA, Allen CR, Kitchens WM, Stenberg J (2002) Assessing state-wide biodiversity in the Florida Gap analysis project. J Environ Manage 66:127–144

Pearson RG (2007) Species' distribution modeling for conservation educators and practitioners. Synthesis. American Museum of Natural History. http://ncep.amnh.org

Pearson SM, Gardner RH (1997) Neutral models: useful tools for understanding landscape patterns. In: Bisonnette JA (ed) Wildlife and landscape ecol: effects of pattern and scale. Springer, New York, pp 215–230

Pearson RG, Dawson TP, Liu C (2004) Modelling species distributions in Britain: a hierarchical integration of climate and land-cover data. Ecography 27:285–298

Pearson RG, Thuiller W, Araújo MB, Brotons L, Martinez-Meyer E, McClean C, Miles L, Segurado P, Dawson TP, Lees D (2006) Model-based uncertainty in species range prediction. J Biogeogr 33:1704–1711

Peitgen HO (1992) Chaos and fractals – new frontiers of science. Springer, New York

Peitgen HO, Richter PH (1986) The beauty of fractals. Springer, Berlin

Penven P, Echevin V, Pasapera J, Cola F, Tam J (2005) Average circulation, seasonal cycle, and mesoscale dynamics of the Peru current system: a modeling approach. J Geophys Res 110 (C10021):1–21

Pereira A, Duarte P, Reis LP (2008) Agent-based ecological model calibration – on the edge of a new approach. In: Ramos C, Vale Z (eds) Proceedings of International Conference Knowledge Engineering and Decision Support, ISEP, Porto, Portugal, July 2004, pp 107–113

Pessel FD, Lecomte J, Emeriau V, Krouti M, Messean A, Gouyon PH (2001) Persistence of oilseed rape (*Brassica napus* L.) outside of cultivated fields. Theor Appl Genet 102:841–846

Pfenninger M, Nowak C, Magnin F (2007) Intraspecific range dynamics and niche evolution in *Candidula* land snail species. Biol J Linn Soc 90:303–317

Pfreundt J (1988) Modellierung der räumlichen Verteilung von Strahlung, Photosynthesekapazität und Produktion in einem Fichtenbestand und ihre. Beziehung zur Bestandesstruktur. Doctoral Dissertation, Universität Göttingen

Pfreundt J, Sloboda B (1996) The relation of local stand structure to photosynthetic capacity in a spruce stand: a model calculation. Lesnictví/Forestry 42:149–160

Phillips BL, Brown GP, Webb JK, Shine R (2006a) Runaway toads: an invasive species evolves speed and thus spreads more rapidly through Australia. Nature 439:803

Phillips SJ, Anderson RP, Schapire RE (2006b) Maximum entropy modeling of species geographic distributions. Ecol Modell 190:231–259

Phillips SJ, Elith J, Dudik M, Graham CH, Lehmann A, Leathwick J, Ferrier S (2009) Sample selection bias and presence-only distribution models: implications for background and pseudo-absence data. Ecol Appl 19:181–197

Pichancourt JB, Burel F, Auger P (2006) A hierarchical matrix model to assess the impact of habitat fragmentation on population dynamics: an elasticity analysis. CR Biol 329:31–39

Pivard S, Adamczyk K, Lecomte J, Lavigne C, Bouvier A, Deville A, Gouyon PH, Huet S (2008a) Where do the feral oilseed rape populations come from? A large-scale study of their possible origin in a farmland area. J Appl Ecol 45:476–485

Pivard S, Demšar D, Lecomte J, Debeljak M, Džeroski S (2008b) Characterizing the presence of oilseed rape feral populations on field margins using machine learning. Ecol Modell 221:147–154

Plotnick RE, Gardner RH, O'Neill RV (1993) Lacunarity indexes as measures of landscape texture. Landsc Ecol 8:201–211

Plotnick RE, Gardner RH, Hargrove WW, Prestegaard K, Perlmutter M (1996) Lacunarity analysis: a general technique for the analysis of spatial patterns. Phys Rev E 53:5461–5468

Polovina J (1984) Model of a coral reef ecosystem I. The ECOPATH Model and its application to French Frigate shoals. Coral Reefs 3:1–11

Power M (1993) The predictive validation of ecological and environmental models. Ecol Modell 68:33–50

Preston KL, Rotenberry JT, Redak RA, Allen MF (2008) Habitat shifts of endangered species under altered climate conditions: importance of biotic interactions. Glob Change Biol 14:2501–2515

Prigogine I, Stengers I (1984) Order out of chaos: man's new dialogue with nature. Bantam Books, New York

Prusinkiewicz P (1987) Applications of L-systems to computer imagery. Lect Notes Comput Sci 291:534–548

Prusinkiewicz P, Lindenmayer A (1990) The algorithmic beauty of plants. Springer, New York, http://algorithmicbotany.org/papers/abop/abop.pdf

PT DLR (2008) Social-ecological research framework concept 2007–2010, German Aerospace Center, Project Management Agency (PT DLR) Environment, Culture, Sustainability, 53227 Bonn, 38 pp

Pullar D (2004) SimuMap: a computational system for spatial modelling. Environ Modell Softw 19(3):235–243

Pyke CR (2004) Habitat loss confounds climate change impacts. Front Ecol Environ 2:178–182

Pykh YA, Efremova SS (2000) Equilibrium, stability and chaotic behavior in Leslie matrix models with different density-dependent birth and survival rates. Math Comput Simul 52:87–112

Quinlan JR (1986) Induction of decision trees. Mach Learn 1:81–106

Quinlan JR (1992) Learning with continuous classes. Proceedings of the 5th Australian joint conference on Artificial Intelligence. World Scientific, Singapore, pp 343–348

Quinn GP, Keough MJ (2002) Experimental design and data analysis for biologists. Cambridge University Press, Cambridge

R Development Core Team (2008) R: a language and environment for statistical computing. R Foundation for Statistical Computing, Vienna, Austria, http://www.R-project.org

Radtke E, Straškraba M (1980) Self-optimization in a phytoplankton model. Ecol Modell 9:247–268

Raick C, Soetaert K, Grégoire M (2006) Model complexity and performance: how far can we simplify? Progr Oceanogr 70:27–57

Randin CF, Dirnbock T, Dullinger S, Zimmermann NE, Zappa M, Guisan A (2006) Are niche-based species distribution models transferable in space? J Biogeogr 33:1689–1703

Rasmussen S, Karampurwala H, Vaiduangth F, Jensen KS, Hameroff S (1990) Computational connectionism within neurons: a model of cytoskeletal automata subserving neural networks. Physika 42D:428–449

Ratz A (1995) Long-term spatial pattern created by fire: a model oriented towards boreal forests. Int J Wildland Fire 5:25–34

Reaka-Kudla ML (1997) The global biodiversity of coral reefs: a comparison with rain forests. In: Reaka-Kudla ML, Wilson DE, Wilson EO (eds) Biodiversity II: understanding and protecting our biological resources. Joseph Henry/National Academy Press, Washington, DC, pp 83–108

Rebolj D, Sturm PJ (1999) A GIS based component-oriented integrated system for estimation, visualization and analysis of road traffic air pollution. Environ Modell Softw 14(6):531–539

Recknagel F, French M, Harkonen P, Yabunaka K (1997) Artificial neural network approach for modelling and prediction of algal blooms. Ecol Modell 96:11–28

Record NR, Pershing AJ, Runge JA, Mayo CA, Chen C (2010) Improving ecological forecasts of copepod community dynamics using genetic algorithms. J Mar Syst 82:96–110

Rehage JS, Trexler JC (2006) Assessing the net effect of anthropogenic disturbance on aquatic communities in wetlands: community structure relative to distance from canals. Hydrobiologia 569:359–373

Reiche EW (1996) WASMOD. Ein Modellsystem zur gebietsbezogenen Simulation von Wasser- und Stoffflüssen. ECOSYS 4:143–163

Reuter H (2005) Community processes as emergent properties: modelling multilevel interaction in small mammals communities. Ecol Modell 186:427–446

Reuter H, Breckling B (1994) Selforganization of fish schools: an object-oriented model. Ecol Modell 75:147–159

Reuter H, Breckling B (1999) Emerging properties on the individual level: modelling the reproduction phase of the European robin *Erithacus rubecula*. Ecol Modell 121:199–219

Reuter H, Hoelker F, Middelhoff U, Jopp F, Eschenbach C, Breckling B (2005) The concepts of emergent and collective properties in individual-based models – Summary and outlook of the Bornhoved case studies. Ecol Modell 186:489–501

Reuter H, Jopp F, Hölker F, Eschenbach C, Middelhoff U, Breckling B (2008) The ecological effect of phenotypic plasticity – analyzing complex interaction networks (COIN) with agent-based models. Ecol Inf 3:35–45

Reuter H, Jopp F, Calabrese J, Damgaard C, Matsinos Y, Blanco-Moreno JM, DeAngelis DL (2010) Ecological hierarchies and self-organisation – Pattern analysis, modelling and process integration across scales. Basic Appl Ecol (2010), doi:10.1016/j.baae.2010.08.002

Revilla E, Wiegand T (2008) Individual movement behavior, matrix heterogeneity, and the dynamics of spatially structured populations. Proc Natl Acad Sci USA 105:19120–19125

Richardson DM, Williams PA, Hobbs RJ (1994) Pine invasions in the Southern Hemisphere: determinants of spread and invadability. J Biogeogr 21:511–527

Rietkerk M, Dekker SC, de Ruiter PC, van de Koppel J (2004) Self-organized patchiness and catastrophic shifts in ecosystems. Science 305:1926–1929

Righton D, Quayle V, Hetherington S, Burt G (2007) Movements and distribution of cod in the southern North sea and English Channel: results from conventional and electronic tagging experiments. J Mar Biol Assoc UK 87:599–613

Riitters KH, O'Neill RV, Hunsaker CT, Wickham JD, Yankee DH, Timmins SP, Jones KB, Jackson BL (1995) A factor analysis of landscape pattern and structure metrics. Landsc Ecol 10:23–39

Rittenhouse CD, Thompson FR III, Dijak WD, Millspaugh JJ, Clawson RL (2010) Evaluation of habitat suitability models for forest passerines using demographic data. J Wildl Manag 74:411–422

Roberts DW (1987) A dynamical systems perspective on vegetation theory. Vegetatio 69:27–33

Rondinini C, Wilson KA, Boitani L, Grantham H, Possingham HP (2006) Tradeoffs of different types of species occurrence data for use in systematic conservation planning. Ecol Lett 9:1136–1145

Room PM, Hanan JS (1995) Virtual cotton: a new tool for research, management and training. In: Constable GA, Forrester NW (eds) Challenging the future. CSRIO, Australia, pp 40–44

Rosenzweig ML, Abramsky Z (1980) Microtine cycles: the role of habitat heterogeneity. Oikos 34:141–146

Rosenzweig ML, MacArthur RH (1963) Graphical representation and stability conditions of predator–prey interactions. Am Nat 97:209–223

Ross T, Gaboury I, Lueck R (2008) Simultaneous acoustic observations of turbulence and zooplankton. J Acoust Soc Am 123(5):3212

Rozenberg G (1973) T0L systems and languages. Inform Control 23:357–381

Rozenberg G (1997) Handbook of graph grammars and computing by graph transformations, vol 1, Foundations. World Scientific, Singapore

RuleQuest (2009) www.rulequest.com. Accessed 10 Sep 2010

Rumbaugh J, Blaha M, Premerlani W, Eddy F, Lorensen W (1991) Object oriented modelling and design. Prentice Hall, Englewood Cliffs

Rupp SP, Rupp P (2010) Development of an individual-based model to evaluate elk (*Cervus elaphus nelsoni*) movement and distribution patterns following the Cerro Grande Fire in north central New Mexico, USA. Ecol Modell 221:1605–1619

Rykiel EJ Jr (1996) Testing ecological models: the meaning of validation. Ecol Modell 90:229–244

Ryoke M, Nakamori Y, Heyes C, Makowski M, Schöpp W (2000) A simplified ozone model based on fuzzy rules generation. Eur J Oper Res 122:440–451

Salomonsen JR, Jensen JJ (1996) Use of a lake model to examine exergy response to changes in phytoplankton growth parameters and species composition. Ecol Modell 87:41–49

Salthe SN (1993) Development and evolution. Complexity and change in biology. Columbia University Press, New York

Samu F, Sunderland KD, Szinetar C (1999) Scale-dependent dispersal and distribution patterns of spiders in agricultural systems: a review. J Arachnol 27:325–332

Sanderson BL, Hrabik TR, Magnuson JJ, Post DM (1999) Cyclic dynamics of a yellow perch (*Perca avescens*) population in an oligotrophic lake: evidence for the role of intraspecic interactions. Can J Fish Aquat Sci 56:1534–1542

Sargent RG (1998) Verification and validation of simulation models. In: Medeiros DJ, Watson EF, Carson JS, Manivannan MS (eds) Proceedings of the 1998 winter simulation conference, pp 121–130

Scardi M, Cataudella S, Di Dato P, Fresi E, Tancioni L (2008) An expert system based on fish assemblages for evaluating the ecological quality of streams and rivers. Ecol Inform 3:55–63

Scheffer M, Hosper SH, Meijer ML, Moss B, Jeppesen E (1993) Alternative equilibria in shallow lakes. Trends Ecol Evol 8:275–279

Schellnhuber HJ, Tóth FL (1999) Earth system analysis and management. Environ Model Assess 4 (4):201–207

Schiegg K, Walters JA, Priddy JR (2005) Testing a spatially explicit, individual-based model of the population dynamics of red-cockaded woodpeckers. Ecol Appl 15:1495–1503

Schleiter IM, Borchardt D, Wagner R, Dapper T, Schmidt KD, Schmidt HH, Werner H (1999) Modelling water quality, bioindication and population dynamics in lotic ecosystems using neural networks. Ecol Modell 120:271–286

Schmolke A, Thorbek P, DeAngelis DL, Grimm V (2010) Ecological models supporting environmental decision making: a strategy for the future. Trends Ecol Evol 25:479–486

Schreiber H, Behrendt H, Constantinescu LT, Cvitanic I, Drumea D, Jabucar D, Juran S, Pataki B, Snishko S, Zessner M (2005) Point and diffuse nutrient emissions and loads in the transboundary Danube River Basin – I. A modelling approach. Arch Hydrobiol Suppl 158(1–2):197–220

Schreier H, Brown S (2002) Scaling issues in watersheds assessments. Water Policy 3(6): 475–489

Schröder B, Richter O (1999) Are habitat models transferable in space and time? Z Ökol Nat 8:195–205

Schröder U, Söndgerath D (1996) Computing the switching points of a growth model for winter wheat. Ecol Modell 88:1–8

Schurr FM, Midgley GF, Rebelo AG, Reeves G, Poschlod P, Higgins SI (2007) Colonization and persistence ability explain the extent to which plant species fill their potential range. Glob Ecol Biogeogr 16:449–459

Schweiger O, Settele J, Kudrna O, Klotz S, Kühn I (2008) Climate change can cause spatial mismatch of trophically interacting species. Ecology 89:3472–3479

Science Subgroup (1994) South Florida ecosystem restoration: scientific information needs. Management and Coordination Working Group, Interagency task-force on the South Florida Ecosystem, Miami, FL, USA

Searle SR (2006) Matrix algebra useful for statistics, Wiley series in probability and statistics. Wiley, New York

Segel LA, Jackson JL (1972) Dissipative structure: an explanation and an ecological example. J Theor Biol 37:545–559

Segurado P, Araújo MB (2004) An evaluation of methods for modelling species distributions. J Biogeogr 31:1555–1568

Seitz A (1982) (1984) Simulationsmodelle als Werkzeuge in der Populationsökologie (Bern. Verhandlungen der Gesellschaft für Ökologie 12:471–486

Selas V (1997) Cyclic population fluctuations of herbivores as an effect of cyclic seed cropping of plants: the mast depression hypothesis. Oikos 80:257–268

Selhorst T, Söndgerath D, Weigand S (1991) A Model Describing the Predator-prey-interaction between *Scolothrips longicornis* and *Tetranychus cinnabarinus* based upon the Leslie Theory. Ecol Modell 59:123–138

Serrano L, Serrano L (1996) Influence of groundwater exploitation for urban water supply on temporary ponds from the Donana National Park (SW Spain). J Environ Manage 46:229–238

SFWMD (1992) Surface water improvement and management plan for the Everglades, Supporting Information Document. SFWMD, West Palm Beach, Florida

Shannon LJ, Moloney CL, Jarre A, Field JG (2003) Trophic flows in the southern Benguela during the 1980s and 1990s. J Mar Syst 39:83–116

Shannon LJ, Field JG, Moloney CL (2004) Simulating anchovy-sardine regime shifts in the southern Benguela ecosystem. Ecol Modell 172:269–281

Shannon LJ, Neira S, Taylor MH (2008) Comparing internal and external drivers in the southern Benguela, and the southern and northern Humboldt upwelling ecosystems. Afr J Mar Sci 30 (1):63–84

Shao Q, Zhou C, Li H, Du Y, Su F, Ji M (2000) The framework of dynamic marine geographic information system and pilot study. Geoscience and Remote Sensing Symposium. IGARSS 2000. IEEE 5:2126–2128

Shepard RB (2005) Quantifying environmental impact assessments using fuzzy logic. Springer, New York

Sherman K, Ajayi T, Anang E, Cury P, Pierre Freon MCM, Hardman-Mountford NJ, AIbe C, Koranteng KA, McGlade J, Nauen C, Pauly D, Peter AG, Scheren M, Skjoldal HR, Tang Q, Guillaume Zabi S, Diaz-de-Leon A (2003) Suitability of the large marine ecosystem concept. Fish Res 64(2–3):197–204

Sherratt JA, Smith MJ (2008) Periodic travelling waves in cyclic populations: field studies and reaction-diffusion models. J R Soc Interface 5:483–505

Shigesada N, Kawasaki K (1997) Biological invasions: theory and practice. Oxford University Press, Oxford

Shin YJ, Cury P (2001) Exploring fish community dynamics through size-dependent trophic interactions using a spatialized individual-based model. Aquat Living Resour 14:65–80

Shugart WW, West DC (1977) Development of an Appalachian deciduous forest succession model and its application to assessment of the impact of the chestnut blight. J Environ Manage 5:161–179

Sieber M, Malchow H, Schimansky-Geier L (2007) Constructive effects of environmental noise in an excitable prey–predator plankton system with infected prey. Ecol Compl 4:223–233

Sieber M, Malchow H, Petrovskii SV (2010) Noise-induced suppression of periodic travelling waves in oscillatory reaction-diffusion systems. Proc R Soc A 466:1903–1917

Sievänen R, Mäkelä A, Nikinmaa E (guest eds) (1997) Functional-structural tree models. Silva Fennica (Special Issue) 31:237–380

Silvert W (1993) Object-oriented ecosystem modelling. Ecol Modell 68:91–118

Silvert W (1997) Ecological impact classification with fuzzy sets. Ecol Modell 96:1–10

Silvert W (2000) Fuzzy indices of environmental conditions. Ecol Modell 130:111–119

Silvertown J, Holtier S, Johnson J, Dale P (1992) Cellular automaton models of interspecific competition for space – the effect of pattern on process. J Ecol 80:527–534

Sirakoulis GCH, Karafyllidis I, Thanaikakis A (2000) A cellular automaton model the effects of population movement and vaccination on epidemic propagation. Ecol Modell 133:209–229

Skellam JG (1951) Random dispersal in theoretical populations. Biometrika 38:196–218

Skellam JG (1973) The formulation and interpretation of mathematical models of diffusionary processes in population biology. In: Bartlett MS, Hiorns R (eds) The mathematical theory of the dynamics of biological populations. Academic, New York, pp 63–85

Sloboda B, Pfreundt J (1989) Baum- und Bestandeswachstumsprozess. Ein systemanalytischer Ansatz mit Versuchsplanungskonsequenzen für die Durchforstung und Einzelbaumentwicklung. Bericht der Jahrestagung der Sektion Ertragskunde im Deutschen Verband Forstlicher Forschungsanstalten, Attendorn 1989, pp 17/1–17/25

Smith AR (1984) Plants, fractals, and formal languages. Comput Graph (ACM/SIGGRAPH) 18:1–10

Soberón J (2007) Grinnellian and Eltonian niches and geographic distributions of species. Ecol Lett 10:1115–1123

Soetaert K, Herman PMJ (2009) A practical guide to ecol model. Springer, Heidelberg

Solomon S et al (eds) (2007) IPCC, Climate Change (2007) – The physical science basis. Contribution of Working Group I to the Fourth Assessment Report of the Intergovernmental Panel on Climate Change. IPCC, Cambridge/New York, p 996

Söndgerath D, Müller-Pietralla W (1996) A model for the development of the cabbage root fly (*Delia radicum* L.) based on the extended Leslie model. Ecol Modell 91:67–76

Söndgerath D, Richter O (1990) An extension of the Leslie matrix model for describing population dynamics of species with several development stages. Biometrics 46:595–607

Söndgerath D, Schröder B (2002) Population dynamics and habitat connectivity affecting the spatial spread of populations: a simulation study. Landsc Ecol 17:57–70

Soons MB, Bullock JM (2008) Non-random seed abscission, long-distance wind dispersal and plant migration rates. J Ecol 96:581–590

Spain JD (1982) BASIC microcomputer models in biology. Addison-Wesley, London

Stankovski V, Debeljak M, Bratko I, Adamič M (1998) Modelling the population dynamics of Red deer (*Cervus elaphus* L.) with regard to forest development. Ecol Modell 108:145–153

Stauffer D (1985) Introduction to percolation theory. Taylor & Francis, London

Stenseth NC (1999) Population cycles in voles and lemmings: density dependence and phase dependence in a stochastic world. Oikos 87:427–461

Stenseth NC, Leirs H, Mercelis S, Mwanjabes P (2001) Comparing strategies for controlling an African pest rodent: an empirically based theoretical study. J Appl Ecol 38:1020–1103

Stoehr AM (1999) Are significance thresholds appropriate for the study of animal behaviour? Anim Behav 75(5):F22–F25

Stohlgren TJ, Chase TN, Pielke RA, Kittel TGF, Baron JS (1998) Evidence that local land use practices influence regional climate, vegetation, and stream flow patterns in adjacent natural areas. Glob Change Biol 4:495–504

Strakraba M (1979) Natural control mechanisms in models of aquatic ecosystems. Ecol Modell 6:305–321

Strong D (1999) Predator control in terrestrial ecosystems: the underground food chain of bush lupine. In: Olff H, Brown V, Drent R (eds) Herbivores: between plants and predators. Blackwell, Oxford, pp 577–602

Struss P (2009) Towards model integration and model-based decision support for environmental applications, 18th World IMACS/MODSIM Congress. Cairns, Australia

Struyf J, Džeroski S (2006) Constraint based induction of multi-objective regression trees. In: Bonchi F, Boulicaut JF (eds) KDID 2005, LNCS 3933. Springer, Heidelberg, pp 222–233

Struyf J, Zenko B, Blockeel H, Džeroski S (2010) Clus: a predictive clustering system. J Mach Learn Res (under review). Available for download from http://www.cs.kuleuven.be/~dtai/clus/. Accessed 10 Sep 2010

Su W, Mackey BG (1997) A spatially explicit and temporal dynamic simulation model of forested landscape ecosystems. In: McDonald AD, McAleer M (eds) MODSIM 97. International congress on modelling and simulation. The Modelling and Simulation Society of Australia, Hobart, pp 1635–1640

Sui DZ, Maggio RC (1999) Integrating GIS with hydrological modeling: practices, problems, and prospects. Comput Environ Urban Syst 23(1):33–51

Sun H, Benzie PW, Burns N, Hendry DC, Player MA, Watson J (1871) (2008) Underwater digital holography for studies of marine plankton. Philos Transact A Math Phys Eng Sci 366:1789–1806

Sundell J, Huitu O, Henttonen H, Kaikusalo A, Korpimäki E, Pietiänen H, Saurola P, Hanski I (2004) Large-scale spatial dynamics of vole populations in Finnland revealed by the breeding success of vole-eating avian predators. J Anim Ecol 73:167–178

Suzuki N, Murasawa K, Sakurai T, Nansai K, Matsuhashi K, Moriguchi Y, Tanabe K, Nakasugi O, Morita M (2004) Geo-referenced multimedia environmental fate model (G-CIEMS): model formulation and comparison to the generic model and monitoring approaches. Environ Sci Technol 38(21):5682–5693

Svenning JC, Skov F (2004) Limited filling of the potential range in European tree species. Ecol Lett 7:565–573

Tam J, Taylor MH, Blaskovic V, Espinoza P, Ballón RM, Díaz E, Wosnitza-Mendo C, Argüelles J, Purca S, Ayón P, Quipuzcoa L, Gutiérrez D, Goya E, Ochoa N, Wolff M (2008) Trophic modeling of the Northern Humboldt Current Ecosystem, Part I: comparing trophic linkages under La Niña and El Niño conditions. Progr Oceanogr 79:352–365

Tansley AB (1935) The use and abuse of vegetational terms and concepts. Ecology 16:284–307

Taylor KE (2001) Summarizing multiple aspects of model performance in a single diagram. J Geophys Res 106:7183–7192

Taylor MH, Tam J, Blaskovic V, Espinoza P, Ballón RM, Wosnitza-Mendo C, Argüelles J, Díaz E, Purca S, Ochoa N, Ayón P, Goya E, Gutiérrez D, Quipuzcoa L, Wolff M (2008a) Trophic modeling of the Northern Humboldt Current Ecosystem, Part II: elucidating ecosystem dynamics from 1995 to 2004 with a focus on the impact of ENSO. Progr Oceanogr 79:366–378

Taylor MH, Wolff M, Vadas F, Yamashiro C (2008b) Trophic and environmental drivers of the Sechura Bay Ecosystem (Peru) over an ENSO cycle. Helgol Mar Res 62(1):15–32. doi:10.1007/s10152-007-0093-4

Teh SY, DeAngelis DL, LdSL S, Miralles-Wilhelm FR, Smith TJ, Koh HL (2008) A simulation model for projecting changes in salinity concentrations and species dominance in the coastal margin habitats of the Everglades. Ecol Modell 213:245–256

Thiery JM, D'Herbes JM, Valentin C (1995) A model simulating the genesis of banded vegetation patterns in Niger. J Ecol 83:497–507

Thompson JN (1998) Rapid evolution as an ecological process. Trends Ecol Evol 13:329–332

Thuiller W (2004) Patterns and uncertainties of species' range shifts under climate change. Glob Change Biol 10:2020–2027

Thuiller W, Albert C, Araújo MB, Berry PM, Guisan A, Hickler T, Midgley GF, Paterson J, Schurr FM, Sykes MT, Zimmermann NE (2008) Predicting global change impacts on plant species' distributions: future challenges. Perspect Plant Ecol Evol Syst 9:137–152

Tingey DT (1989) Bioindicators in air pollution research – applications and constraints. In: Biologic markers of air pollution stress and damage in forests, Committee on biological markers of air pollution damage in trees. National Research Council, National Academy Press, Washington DC

Todorovski L, Dzeroski S (2006) Integrating knowledge-driven and data-driven approaches to modeling. Ecol Modell 194:3–13

Todorovski L, Džeroski S, Kompare B (1998) Modelling and prediction of phytoplankton growth with equation discovery. Ecol Modell 113:71–81

Toffoli T, Margolus N (1987) Cellular automata machines. MIT Press, Cambridge, MA

Travers M, Shin YJ, Jennings S, Cury P (2007) Towards end-to-end models for investigating the effects of climate and fishing in marine ecosystems. Progr Oceanogr 75:751–770

Trexler JC, Loftus WF, Jordan F, Chick JH, Kandl KL, McElroy TC, Bass OL Jr (2002) Ecological scale and its implications for freshwater fishes in the Florida Everglades. In: Porter JW, Porter

KG (eds) The Everglades, Florida Bay, and coral reefs of the Florida Keys: an ecosystem sourcebook. CRC Press, Boca Raton, pp 153–181

Trexler JC, Loftus WF, Perry S (2005) Disturbance frequency and community structure in a twenty-five year intervention study. Oecologia 145:140–152

Troitzsch KG (2004) Validating simulation models. In: Horton G (ed) 18th European simulation multiconference. Networked Simulations and Simulation Networks. SCS Publ House, Erlangen, pp 265–270

Tscherko D, Kandeler E, Bárdossy A (2007) Fuzzy classification of microbial biomass and enzyme activities in grassland soils. Soil Biol Biochem 39:1799–1808

Tsoar A, Allouche O, Steinitz O, Rotem D, Kadmon R (2007) A comparative evaluation of presence-only methods for modelling species distribution. Divers Distrib 13:397–405

Tunbridge A, Jones H (1995) An L-systems approach to the modelling of fungal growth. J Vis Comput Animat 6:91–107

Turchin P (1998a) Quantitative analysis of movement: measuring and modeling population redistribution in animals and plants. Sinauer, Sunderland MA

Turchin P (1998b) Quantitative analysis of movement: measuring and modeling population redistribution in plants and animals. Sinauer Associates, Sunderland, MA

Turchin P, Hanski I (2001) Contrasting alternative hypotheses about rodent cycles by translating them into parameterized models. Ecol Lett 4:267–276

Turing AM (1952) The chemical basis of morphogenesis. Philos Trans R Soc Lond B 237:37–72

Turk A (1992) Visualization in environmental management: beyond the buzz word. Landsc Urban Plan 21(4):253–255

Turner MG, Ruscher CL (1988) Changes in landscape patterns in Georgia, USA. Landsc Ecol 1:241–251

Ulanowicz RE (1986) Growth and development: ecosystems phenomenology. Springer, New York, pp 203

Ulanowicz RE (1997) Ecology, the ascendent perspective, Complexity in ecological systems series. Columbia University Press, New York, pp 201

van Griensven A, Meixner T, Grunwald S, Di Luzio A, Srinivasan R (2006) A global sensitivity analysis tool for the parameters of multi-variable watershed models. J Hydrol 324:10–23

Vaughan IP, Ormerod SJ (2005) The continuing challenges of testing species distribution models. J Appl Ecol 42:720–730

Vellend M, Verheyen K, Jacquemyn H, Kolb A, Van Calster H, Peterken G, Hermy M (2006) Extinction debt of forest plants persists for more than a century following habitat fragmentation. Ecology 87:542–548

Vens C, Struyf J, Schietgat L, Džeroski S, Blockeel H (2008) Decision trees for hierarchical multi-label classification. Mach Learn 73:185–214

Verhulst PF (1838) Notice sur la loi que la population poursuit dans son accroissement. Corr Math Phys 10:131–121

Veron JEN, Hoegh-Guldberg O, Lenton TM, Lough JM, Obura DO, Pearce-Kelly P, Sheppard CRC, Spalding M, Stafford-Smith MG, Rogers AD (2009) The coral reef crisis: the critical importance of <350 ppm CO_2. Mar Poll Bull 58:1428–1436

Vitousek PM, Mooney HA, Lubchenco J, Melillo JM (1997) Human domination of earth's ecosystems. Science 277:494–499

Vladusic D, Kompare B, Bratko I (2006) Modelling Lake Glumsø with Q2 learning. Ecol Modell 191:33–46

Vogelmann JE, Sohl TL, Campbell PV, Shaw DM (1998) Regional land cover characterization using Landsat thematic mapper data and ancillary data sources. Environ Mon Assess 51:415–428

Vogelmann JE, Howard SM, Yang L, Larson CR, Wylie BK, Van Driel N (2001) Completion of the 1990s National Land Cover Data Set for the Conterminous United States from Landsat Thematic Mapper data and ancillary data sources. Progr Eng Remote Sens 67:650–662

Voinov A, Fitz C, Boumans R, Costanza R (2004) Modular ecosystem modeling. Environ Modell Softw 19(3):285–304

Volterra V (1926) Variazioni e fluttuazioni del numero d'individui in specie animali conviventi. In: Mem R Accad Naz dei Lincei Ser VI 2(3):31–113 [Translation by Wells ME: Variations and Fluctuations of the number of individuals in animal species living together]; [Translated in Chapman RN 1931: Animal Ecology. McGraw Hill, New York]; also at http://icesjms.oxfordjournals.org/cgi/reprint/3/1/3.pdf

von Bertalanffy L (1949) Zu einer allgemeinen Systemlehre. Biol Generalis 19:114–129

von Bertalanffy L (1950) The theory of open systems in physics and biology. Science 111:23–29

von Bertalanffy L (1969/1976) General system theory: foundations, development, applications. Braziller, New York

von Linné C (1748) Systema naturae. Kiesewetter & Vandenhoeck, Stockholm

von Sperling DL, Behrendt H (2007) Application of the Nutrient Emission Model MONERIS to the Upper Velhas River Basin, Brazil. In: Gunkel G, Sobral M (eds) Reservoirs and river basins management: exchange of experience from Brazil, Portugal and Germany. Universitätsverlag TU Berlin, Berlin, pp 265–279

Vos J, Marcelis LFM, de Visser PHB, Struik PC, Evers JB (eds) (2007) Functional-structural plant modelling in crop production. Springer, Dordrecht

Wager H (1911) On the effect of gravity upon the movements and aggregation of *Euglena viridis* Ehrb., and other micro-organisms. Philos Trans R Soc Lond B 201:333–390

Wagner HH, Fortin MJ (2005) Spatial analysis of landscapes: concepts and statistics. Ecology 86:1975–1987

Wainwright J, Mulligan M (2004) Environmental modelling – finding simplicity in complexity. Wiley, Chichester

Walters C, Christensen V, Pauly D (1997) Structuring dynamic models of exploited ecosystems from trophic mass-balance assessments. Rev Fish Biol Fish 7:139–172

Walters C, Pauly D, Christensen V, Kitchell JF (2000) Representing density dependent consequences of life history strategies in aquatic ecosystems: EcoSim II. Ecosystems 3:70–83

Wang Q, Malanson GP (2007) Patterns of correlation among landscape metrics. Phys Geogr 28:170–182

Wang X, Homer M, Dyer SD, White-Hull C, Du C (2005) A river water quality model integrated with a web-based geographic information system. J Environ Manage 75(3):219–228

Ward G, Hastie T, Barry S, Elith J, Leathwick JR (2009) Presence-only data and the EM algorithm. Biometrics 65:554–563

Weinberger HF (1978) Asymptotic behavior of a model in population genetics. Lect Notes Math 648:47–96

WGBU (1997) World in transition: the research challenge. German Advisory Council on Global Change, Annual Report 1996, Springer Verlag, Berlin

White LP (1971) Vegetation stripes on sheet wash surfaces. J Ecol 59:615–622

White R, Engelen G (1993) Cellular automata and fractal urban form: a cellular modelling approach to the evolution of urban land-use patterns. Environ Plan A 25:1175–1199

White R, Engelen G, Uijee I (1997) Cell automata fractal urban form cellular modelling approach evolution urban land use patterns Environ Plan 24:323–343

Wickham JD, Riitters KH (1995) Sensitivity of landscape metrics to pixel size. Int J Remote Sens 16:3585–3594

Wickham JD, Stehman SV, Smith JH, Yang L (2004) Thematic accuracy of the 1992 National Land-Cover Data for the western United States. Remote Sens Environ 91:452–468

Williams MR, Filoso S, Lefebvre P (2004) Effects of land-use change on solute fluxes to floodplain lakes of the central Amazon. Biogeochem 68:259–275

Willmott CJ, Matsuura K (2005) Advantages of the mean absolute error (MAE) over the root mean square error (RMSE) in assessing average model performance. Clim Res 30:79–82

Winberg GG (1956) Rate of metabolism and food requirements of fishes. Translation Series of Fisheries Research Board of Canada, p 253

Witten IH, Frank E (2005) Data mining: practical machine learning tools and techniques. Morgan Kaufmann, San Francisco

Wolff WF (1994) An individual-oriented model of a wading bird nesting colony. Ecol Modell 72:75–114

Wolff M (2006) Biomass flow structure and resource potential of two mangrove estuaries: insights from comparative modelling in Costa Rica and Brazil. Rev Biol Trop 54(1):69–86

Wolfram S (1994) Cellular automata and complexity: collected papers. Addison-Wesley, New York

Worthington E (1975) The evolution of IBP, International biological programme synthesis series. Cambridge University Press, Cambridge

Wright PJ, Neat FC, Gibb FM, Gibb IM, Thoradarson H (2006) Evidence for metapopulation structuring in cod from the west of Scotland and North Sea. J Fish Biol Suppl C 69:181–199

Xu P (2004) Nutrient emissions into the Taihu Lake from the Southern Catchments. DAAD, p 28

Yang LM, Stehman SV, Smith JH, Wickham JD (2001) Thematic accuracy of MRLC land cover for the eastern United States. Remote Sens Environ 76:418–422

Yniguez A, DeAngelis DL, McManus J (2008) Allowing macroalgae growth forms to emerge: use of an agent-based model to understand the growth and spread of macroalgae in Florida coral reefs, with emphasis on *Halimeda tuna*. Ecol Modell 216:60–74

Yokozawa M, Hara T (1999) Global versus local coupling models and theoretical stability analysis of size-structure dynamics in plant populations. Ecol Modell 118(1):61–72

Young OR (2003) Environmental governance: the role of institutions in causing and confronting environmental problems. Int Environ Agreem P 3:377–393

Zabel CJ, Dunk JR, Stauffer HB, Roberts LM, Mulder BS, Wright A (2003) Northern spotted owl habitat models for research and management application in California (USA). Ecol Appl 13:1027–1040

Zalewski M (2002) Ecohydrology – the use of ecological and hydrological processes for sustainable management of water resources. Hydrolog Sci J 47(5):823–832

Zar JH (1996) Biostatistical analysis. Prentice Hall, Upper Saddle River, New Jersey

Zhang J, Gurkan Z, Jørgensen SE (2010) Application of eco-exergy for assessment of ecosystem health and development of structurally dynamic models. Ecol Modell 221:693–702

Zuo W, Lao N, Geng Y, Ma K (2008) GeoSVM: an efficient and effective tool to predict species' potential distribution. J Plant Ecol 1:143–145

Zurell D, Jeltsch F, Dormann CF, Schröder B (2009) Static species distribution models in dynamically changing systems: how good can predictions really be? Ecography 32:733–744

Index

A

Abax parallelepipedus, 22
Abstract representations, 5
Acari, 205
Across Trophic Level System Simulation (ATLSS), 295–297
Activator–inhibitor, 99
Age distribution, 20
Age-structured population dynamics model, 56, 120, 281
Agent-based models (ABMs), 164
Aggregated parameters, 327
Aggregation, diffusion-limited, 113
  simplification of data, 312
Agro-ecology, 204, 345
Akaike's information criterion (AIC), 189, 352
Algae, filamentous, 148
ALGOL (ALGOrithmic Language), 37
All purpose approach, 49
Allee effect, 96
Alnus glutinosa, 52
Alternative equilibrium, 77
Alternative stable states, 85
American alligator, 297
American crocodile, 297
Ammonium, 271
Apple snails, 299
Asio otus, 175
Attractor, 75, 87
AU model, 56
Autonomous system, 77
Auxiliaries, 48

B

Basin of attraction, 75, 88
Beetle dispersal, 20
Belousov-Zhabotinskii reaction, 93
Benthic layer, 280
Bifurcations, 84
Biocoenosis, 275
BIODEPTH, 115
Biomass, changes, time-series trends, 64
  decrease in food chain, 258
Biophysical models, 280
Bituminaria bituminosa, 116
Black box approach, 6
Boosted Regression Trees (BRT), 185
Bornhöved Lakes, nitrogen leaching, 23
Bornhöved Lakes Ecosystem Research, 23, 277
Bottom-up control, 258
Brassica napus, 206, 235
Bt-maize, 204

C

C4.5/C5.0, 202
Caeté estuary (Brazil), 62
Calibration, 26, 328
Cape Sable seaside sparrow, 297
Carabus hortensis, 22
Caribbean coral reefs, phase shifts, 249
Carrying capacity, 96
CART (Classification And Regression Trees), 188, 202
Cascading trophic interactions, 259
Catastrophe, mathematical term, 85
Cause–effect diagram, 45
Cellular automata, 39, 105ff, 170
  boundary conditions, 109
  cells, 107
  coral species, 245
  diffusion-limited aggregation, 113
  excitable media, 112
  Game of Life, 106, 110
  grid, 106, 107
  iteration, 109

Cellular automata (*cont.*)
neighbourhood, 108
spatial competition, 115, 245
Ulam, S., 106
von Neumann, J., 105
Wolfram S., 106
Ceteris paribus conditions, 89, 324
Changes, living systems, 20
Chaos, deterministic, 7, 82, 100
diffusion-induced, 101
wave of, 100
CLASS, 166
instance, 166
object, 50, 166
Class vs. attribute, 198
Classification and regression trees (CARTs), 188
Classification vs. regression, 198
Clethrionomys glareolus, 175
Club of Rome, 314
CLUS, 203
Cod, 280
Collapse, 77
Collembola, 205
Collinearity, 184
Compartments, 43, 47
Complexity vs. simplicity, 7, 8
Comprehensive Everglades Restoration Plan (CERP), 291, 293
Coniferous tree stand, eco-physiological model, 157
Connections, 47
Consistency check, 26
Consumption, 58
Context assessment, 43
Continuous Systems Modelling Program (CSMP), 271
Controls, 48
Conway's Game of Life, 106, 110
Coral reefs, food web, equation-based modelling, 242
parrotfish grazing, 249
phase shifts, 241
production, 243
Cross-diffusion, 102
Cross-validation (CV), 189
Cumulative frequency distributions (cfd), 220, 222
Curse of dimensionality, 185

D

D'Alembert, J.B., 30
Dahl, O.J., 37
Darwin, C., 31
de Laplace, P.S., 5
Decision support, 118, 134, 180, 314
Decision trees, 197ff
classification and regression trees, 203
habitat suitability, 205
induction, 199
machine learning, 198
multi-target trees, 199
predictive modelling, 198
pruning, 202
types, 199
Defuzzification, 136, 140
Delimitations, 16, 324
Demersal layer, 280
Descartes, R., 30, 31
Destabilizer–stabilizer, 99
Deterministic chaos, 7, 82, 100
Diadema antillarum, 249
Dictyota spp., 247
Diderot, D., 30
Diel vertical migration (DVM), 283
Differential equations, 33, 50, 67
ordinary (ODEs), 68ff, 94, 234
partial (PDEs), 93ff
Diffusion, 95, 266
Diffusion-limited aggregation, 113
Dimension of the system, 74
Dimensional reduction, 185, 187
Dipsacus sylvestris, 120
Direction field, 74
Discrete event scheduling, 171
Dispersal, 20, 95, 114, 231
beetle, 20
coral, 253
oil seed rape, 235
pines, 237
seed, 115
Dispersal kernels, 233
Dispersal vectors, 232
Diversity–productivity, 116
Domain of attraction, 75, 88
Dragonflies, voltinism, global change, 129
Dynamical systems, differential equations, 71
DynaMo, 47

E

Earthworms, 205
Ecological modelling, ancestors, 30
founders, 33
motivations, 3
Ecological theory, 44
Ecology, Odum, 35

ECOPATH, 55
Ecoranger, 59
ECOSIM, 56, 60
Ecospace module, 56
Ecosystem, Tansley A.B., 30
Ecosystem indicators, 60
Ecotrophic efficiency, 58
Eigenvalue, 124, 235
Eigenvector, 122
Eisenia fetida, 199
Elasticity analysis, 124
Emergent phenomena, 19
Emergent property, 19, 176, 247, 276, 344
Emigration–immigration, 57
Equilibria, multiple, 85
  periodic, 80
  single/multiple, 78
  stable, 71
  unstable, 72
Eulerian-Lagrangian models, 280
Eutrophication, 274
Everglades, 257, 291
Everglades Landscape Model (ELM), 295–297
Evolutionary theory, Darwin, 31
EwE, 55
Excitable media, 112
Explanatory variables, 183
Exploratory data plotting, 185
Explosion, 77

F
Feedback processes, 21
Feral-to-crop gene flow, 235
Florida panther, 297
Flow diagram (flow chart), 46
Flow equilibrium, von Bertalanffy, 34
Focal level, 20
Focus of investigation, 15
Food chains, 258 ff
Food dependent growth/reproduction, 53
Food webs, 56, 175, 242, 257
Foraging arena, 55, 60
  theory, 55
Forecasts, 5, 25, 129, 316
Forest decline, acidification, 36
Forest fires, 22
Forrester, J.W., 36
FORTRAN, 271
Fractals, 153
  grammar-based, 147
FRACTINT, fractal generator, 115
FRAGSTATS, 218
Framework constellation, 5
Functional-structural plant model (FSPM), 148, 157
Fuzzification, 135, 138
Fuzzy logic, 133ff
  defuzzication, 136. 140
  fuzzification, 135, 138
  inference, 135, 137
  membership function, 134, 138
  uncertain information, 134

G
Gadus morhua, 280
Galilei, G., 30, 31
GAM (generalised additive models), 183, 188, 189
Game of Life, 106, 110
G-CIEMS, 316
General systems theory, 49, 68
Generalised additive models (GAM) 183, 188, 189
Generalised linear models (GLM), 189
Genetically modified herbicide tolerant (GMHT) crops, 206, 235
Genetically modified plants, 4, 235, 345
  herbicide tolerant (GMHT), 206
  oil seed rape, 235, 345
Genotype/phenotype, 154
Geographic information systems (GIS), 205, 301, 306
  model integration, 306
Geo-Referenced Multimedia Environmental Fate Model (G-CIEMS), 316
Gilpin's spiral chaos attractor, 83
GLM (Generalised linear models), 183, 189
Global change, 129
  dragonflies, 129
  temperature trends, 2, 4
Gomphus vulgatissimus, 130
Grammar-based models, 147ff
  functional-structural plant model, 157
  L-systems, 150
  morphological development, 148
  parametric L-Systems, 152
  relational growth grammar, 154
  rewriting systems, 150
  self similarity, 153
  turtle geometry, 148
Grammars, 147
Graph grammars, 156
Grassland communities, competition/ dispersal, 114
Grasslands, invasions by pines, 237

Greenlab model, 149
Grid maps, 170
GroIMP (Growth-grammar related Interactive Modelling Platform), 147
Growth, exponential, 31, 70, 95
  logistic, 32, 70, 85, 96
Gulf of Nicoya (Costa Rica), 62

H
Habitat fragmentation, 219
Habitat loss, 219
Habitat modelling, decision trees, 205
Haeckel, E., 30, 31
Halimeda spp., 247
HAMSOM hydrodynamic model, 281
Hardwood hammocks, 17
Hierarchical structures, signal transfer, 19
Hierarchy theory, 20
Hirschfeldia incana, 116
Hogeweg, P., 39
Holothuria leucospilota, 206
Hopf bifurcation, 90
Hordeum geniculatum, 116
Horizontal diffusion, 285
Hypothesis testing, 211, 221
Hysteresis, 26, 85

I
Ichthyoplankton, 280
IDW (inverse distance weighted), 311
Individual-based models (IBM), 163ff, 242, 255
  activity control, 169
  behavioural repertoire, 53, 163
  event scheduling, 171
  life loop, 169
  object-orientation, 166
  macroalgae, 247
  physiological processes, 23, 51, 168, 175
  predator–prey interaction, 171
  predator–prey model (IPP), 171
  rodent communities, 174
  spatially-explicit, 165
Integrated environmental modelling (IEM), 301, 303, 318
Integration, 301, 308
Interactions, emergent phenomena, 19
  implications, 20
  relevance, 45
Inter disciplinary linkages, 36
Inter governmental Panel on Climate Change (IPCC), 129, 130
International Biological Programme (IBP), 36
International Council for the Exploration of the Sea (ICES), 280
Invasions, open habitats, 237
Inventory, biota, 5
Ips typographus, 204
Isocline, 75
Iteration, CA, 109
  Monte Carlo, 219
Iterative processes, 21

J
Joining, 310
Jørgensen, S.E., 36
Julia sets, 154

K
Kaiser, H., 39
Key factors, 45
Kriging (interpolate a random field), 311

L
Lagoecia cuminoides, 116
Lagrangian Individual-Centred Models, 280
Lake Glumsø, 269
Lake Okeechobee, 292
Land use, rainfall, 309
Land use and land cover (LUDA) database, 216
Landscape dynamics, 216
Landscape metrics, development/ interpretation, 219
Landscape models, neutral, 215
Landscapes, spatial patterns, 211
Laplace's demon, 5
Large marine ecosystem (LMEs), 57
Learning decision trees, 200
Leslie matrices, 119ff
  age class, 120
  age-structured, 120, 129
  biological time, 126
  eigenvalue, 122
  eigenvector, 122
  elasticity analysis, 124
  life cycle, 121
  stage-structured, 129
  survival rates, 119, 121, 128
Level-crossing phenomena, 20
Life loop, 169
Life-cycle, 121, 233
Limit cycle, 78, 80, 100
Lindeman, R., 35, 258
Lindenmayer, A., 148, 151

LOGO, 149
Lotka, A.J., 6, 33, 72
Lotka–Volterra (LV) model, 33, 49, 72, 341
Lotka–Volterra isoclines, 76
L-systems, 148, 150
  parametric, 152

M
M5, 203
Machine learning, 197, 276, 328
Macroalgae, growth patterns, 247
Malthus, R., 31, 32, 91, 95, 341
Mandelbrot, B.B., 39, 154, 217
Mandelbrot set, 154
Mangroves, 62
  vs. hardwood hammocks, 17
Marine protected areas (MPAs), 56
Maryland Piedmont maps, 221
Mass balance, automated, 59
  modelling, 55, 57
Mathematica, 106
Matrix models, 119
MaxEnt, 183
Meadows, D.H., 36
Membership functions, 134, 138
Microtus agrestis, 175
Minerals, 271
Missing data, 184
Modelling, across-scale, 344
  approaches, 53
  coherence/completeness, 18
  differential equation-based, 67
  individual-based, 37
  integrative capacity/power, 212, 296
  iterative processes, 21
  multi-level, 344
  population dynamics, 203
  potential, 16
  predictive, 197
Models, accuracy, 326
  age-structured, 56, 120, 281
  calibration, 26, 328
  categorization, 342
  communication, 336
  complexity, 14
  coupling, natural and human, 345
  decision making processes, 24
  description, 338
  development, 29, 43
  documentation, 336
  empirical/theoretical problems, 17
  formulation, 187
  general definition, 6, 13
  grammar-based, 147
  hybridization, 342
  hypotheses, 25
  individual-based (IBM), 163
  input, 20
  limitations, 24
  modular, 304
  monolithic approach, 304
  multi-scale problems, 23
  object structure, 51
  parameters, 327
  performance, assessment, 191
  predictive power, 25
  reevaluation, 26
  representations, simplicity, 6
  resilience, 241
  scenarios, 27
  stage-structured, 129, 231ff
  system analysis, 13
  validation, 330
Modular approach, 304
MONERIS (Modelling Nutrient Emissions in River Systems), 315
Monolithic approach, 304
Montastraea annularis, 249
Monte Carlo approach, 59, 219
Moore, E.F., 108
MT-SMOTI, 203
Multi-agent simulations (MAS), 164
Multi-objective (regression) trees, 199
Multi-target classification trees, 199
Multi-target trees, 199, 202
Multiple stability, 84
Multispecies Virtual Population Analysis (MSVPA), 56
Mustela nivalis, 175

N
Natural frequencies, 20
Natural systems model, 295
netCDF (network common data format), 313
Network structures, 34
Neural networks, 198
Neutral landscape models, 215
Newton, I., 30
Niche, 179, 194
Niche evolution, 195
Niche models, 194
Nitrate/nitrite, 271
Nitrogen, 271
Nodes/leaves, 200

Nutrient budgets, 275
Nygaard, K., 37, 38

O

O'Neill function, 126
Object, instantiation, 167
Object paradigm, 37
Object-oriented programming (OOP), 50, 154, 165, 166
Object-oriented systems analysis (OOSA), 43, 50
Ockham's razor, 344
ODE (ordinary differential equations), 94, 234
Odum, E.P., 35, 44
Oilseed rape (OSR), 206, 235
One-dimensional systems, 77
Optical plankton counters, 280
Ordinary differential equations (ODE), 68ff, 94, 234
  attractor, 75, 84, 87
  basin of attraction, 75, 88
  bifurcation, 85
  catastrophe, 85
  ceteris paribus conditions, 89
  direction field, 74
  domain of attraction, 75, 88
  dynamical systems, 71
  equilibrium, 71, 77, 80
  exponential growth, 70
  Gilpin attractor, 84
  Hopf bifurcation, 89
  hysteresis, 26, 85
  isocline, 75
  limit cycle, 78, 80
  Lotka-Volterra, 6, 33, 72
  Michaelis-Menten equation, 79
  oscillations, 78
  phase space, 73
  phase transition, 84, 85, 89
  pool metaphor, 69
  quasi-periodic oscillator, 82
  state space, 73
  trajectory, 75
Oscillations, damped amplitude, 78
  increasing amplitude, 79
  marginal stable, 78
  quasi-periodic, 82
Outliers, 184

P

Papert, S., 149
Parallel rewriting systems, 150
Parameter space, 187
Parameters, aggregated, 327
  calibration, 15
  identification, 328
  ranges, 336
Parametrization, 327
Partial differential equations, 93ff
  Allee effect, 96
  Belousov-Zhabotinskii reaction, 93
  Bénard convection cells, 93
  deterministic chaos, 100
  diffusion coefficient, 95
  Holling type, 99
  limit cycle oscillations, 100
  Malthusian assumption, 95
  propagation front, 93
  reaction-diffusion equation, 95
  Rosenzweig-MacArthur model, 99
  spatio-temporal dynamics, 93
  spiral waves, 100
  travelling waves, 93
  Turing patterns, 99
Particle release patterns, 286
Particle-tracking models, 280
Patten, B., 36
Pattern generating mechanisms, 112
PDE (partial differential equation), 93
Pearl, R., 32, 33, 341
Pedigree index, 59
Pelagic zone, 280
Phalaris coerulescens, 116
Phase shifts, 241
Phase space, 74
Phase transitions, 88, 336
Phenotype/genotype, 154
Phosphorus, 204, 259, 271, 297
Photosynthetic performance, 157
Physics, modelling, 31
Phytoplankton, 261, 271
Pinus nigra, 238
Pinus sylvestris, 238
Pityogenes chalcographus, 204
Plankton, 261, 271, 280
  ichthyoplankton, 280
  phytoplankton, 261, 271
  zooplankton, 261, 271
Plant development, 52
Pointer, 50
Pool metaphor, differential equations, 68, 69
Population growth, Malthus, 31
Population models, single, 94
  spatially explicit, 232
  stage-structured dynamics coupled with dispersal, 232

Population viability analysis (PVA), 166
Populations, age-structured, 120
 dynamics, 203
 red deer, 204
 long-term behaviour, 122
 oscillation, rodents, 177
POPULUS, 83
Predator–prey model, individual-based (IPP), 11, 171
 Lotka-Volterra, 6, 33, 72, 86, 171
 Rosenzweig-MacArthur, 99
Predictability, limits, 24
Predictions, complex ecological dynamics, 346
Predictive clustering trees (PCTs), 202
Predictive modelling, 197
 machine learning, 198
Preprocessing, 182
Primary production required (PPR), 57
Principal component analysis (PCA), 184
Procedural approach, 37
Production, 58
Programming, object-oriented, 37
 procedural, 37
Programming languages, ALGOL (ALGOrithmic Language), 37
 C++, 155
 CLASS, 166
 FORTRAN, 271
 JAVA, 155, 165
 L+C, 155
 LOGO, 149
 R, 218, 221
 SIMULA (SIMUlation LAnguage), 37, 165, 171
 SMALLTALK–80 37
 XL (eXtended L-system language), 155
Proisotoma minuta, 201
Projection matrix, 121
Protaphorura fimata, 201
Pruning, 202
Prusinkiewicz, P., 148

Q
Qrule, 217, 229
Quality index, fuzzy model, 141

R
R, 218, 221
Radio-tracking, 166
randomForest, 185
Random-with-constraints (RwC), 220
Rate level graph, 71
Reaction-diffusion systems, Turing patterns, 99
Reference variable, 50
Reflexive-transitive hull, 156
Regional oceanography model (ROM), 170
Regression tree, 199
Relational growth grammars (RGG), 154
Reliability, limits, 323
Resilience, 114, 242
Respiration, 58, 157
Response variables, 182
Result-probability distributions, 336
Rodent cycle model, 174, 175
RULE (Qrule), 218, 229
Rule induction, 198
Rule-based, 136, 148
Rutilus rutilus, 53

S
Schooling, fish, 18
Schwartz'/Bayesian Information Criterion (BIC), 189
Scolothrips longicornis, 120
Self-organization, 9
 paradigm, 39
 spatio-temporal structures, 51
Self-reproducing patterns, simulation, 105
Self-scaling random pattern, 112
SELF-similar structures, 9
Sensitivity analysis, 26, 329
Shoreline erosion, 242
Silver Springs ecosystem study, Odum, 35
Simplification, ecosystem models, 8
SIMULA (SIMUlation LAnguage), 37, 165, 171
Sinks, 48
SMALLTALK-80, 37
Snail kite, 297
Software, C4.5/C5.0, 202
 CLUS, 203
 Continuous Systems Modelling Program (CSMP), 271
 DynaMo, 47
 ECOPATH, 55
 ECOSIM, 56, 60
 EwE, 55
 FRACTINT, 115
 FRAGSTATS, 218
 G-CIEMS, 316
 GroIMP (Growth-grammar related Interactive Modelling Platform), 147
 M5, 203
 Mathematica, 106
 MONERIS (Modelling Nutrient Emissions in River Systems), 315

Software (*cont.*)
  MT-SMOTI, 203
  POPULUS, 83
  RULE (Qrule), 218
  Weka suite, 203
Soil organisms, 204
Sources, 48
South Florida Water Management Model (SFWMD), 295, 297
Spatial aggregation, 312
Spatial autocorrelation (SAC), 190
Spatial differences, temporal distinctions, 20
Spatial interpolation, 311
Spatial joining, 310
Spatially-explicit, 2, 55, 172, 232, 247, 262, 296, 308
Spatio-temporal continuum, 20
Species, range, 194
Species distribution models (SDMs), 179ff
  collinearity, 184
  cross-validation, 188
  machine learning, 189
  realized niche, 179, 194
  spatial autocorrelation, 190
Spiral waves, 100
Spline (approximate complex shapes), 311
SPREAD (Spatially-Explicit Reef Algae Dynamics), 247
Stage-structured integro-differential models, 231ff
State space, 74
Stationary states, 78
Statistical model of the North Atlantic circulation (SNAC), 281
Statistics, 181
Steady state, 78
Steady state models, 55ff
  ECOPATH, 56, 58
  ECOSIM, 56
  ecotrophic efficiency, 58
  mass balance modelling, 57
  Monte Carlo approach, 59
  Multispecies Virtual Population Analysis (MSVPA), 56
  pedigree routine, 59
Structural dynamic modelling (SDM), 50, 275
Structurally variable interaction networks, 50
System analysis, 14, 43ff
  cause-effect diagram, 43, 45
  flow diagram, 46
  general system theory, 43, 44ff
  object-oriented system analysis, 50
  von Bertalanffy, L., 44
Systems, dimensions, 74
  types, 53

T
Tansley, A.B. 30
Target patterns, 100
Temperature trends, global average surface temperature, 25
Tetranychus cinnabarinus, 120
Time dependent processes, 67
Time-dynamic simulation, 60
Top-down control, 258
Top-down induction of decision tress (TDIDT), 197, 200
Trajectory, 75
Transition matrix, 233
Transition probability matrices, 281
Transitional waters, 141
Tree species, invasive, 237
Trophic cascades, 257
Trophic modelling, 56
Tropical coral reefs, 241
Turing patterns, 99
Turing test, 334
Turtle geometry, 149

U
Ucides cordatus, 63
Ulam, S., 106
Ulva rigida, 204
Uncertainty, 59, 133
  fuzzy logic, 133

V
Validation, 26, 325, 330
  limits, 334
Validity range, 334
Validity window, 335
Variability/heterogeneity, simulating, 170
Verhulst, P.F., 32, 33
Video plankton recorders, 280
Virtual beech tree, 155
Virtual plants, 148
Visualization, 182
Volterra, V., 6, 33, 72, 91
von Bertalanffy, L., 34, 44, 49, 68
von Liebig, J., 30
von Neumann, J. 105
Vulnerability, 55

W
WadBOS, 317
Waddenzee (Netherlands), 317

Waddington, C.H., 36
WASMOD modelling, 23
Water Framework Directive (WFD), 315
Wave of chaos, 100
Wave pattern, spiral, 112
WEKA suite, 203
White-tailed deer, 297
Wolfram, S., 106

X

XL (eXtended L-system language), 147, 155

Z

Zadeh, L.A., 134
Zero growth isocline, 75
Zooplankton, 261, 271